# Renewable Polymers

# Renewable Polymers

## Processing and Chemical Modifications

Special Issue Editors

**Marc A. Dubé**
**Tizazu Mekonnen**

MDPI • Basel • Beijing • Wuhan • Barcelona • Belgrade • Manchester • Tokyo • Cluj • Tianjin

*Special Issue Editors*

Marc A. Dubé
University of Ottawa
Canada

Tizazu Mekonnen
University of Waterloo
Canada

*Editorial Office*
MDPI
St. Alban-Anlage 66
4052 Basel, Switzerland

This is a reprint of articles from the Special Issue published online in the open access journal *Processes* (ISSN 2227-9717) (available at: https://www.mdpi.com/journal/processes/special_issues/renewable_polymers).

For citation purposes, cite each article independently as indicated on the article page online and as indicated below:

LastName, A.A.; LastName, B.B.; LastName, C.C. Article Title. *Journal Name* **Year**, *Article Number*, Page Range.

ISBN 978-3-03928-766-6 (Pbk)
ISBN 978-3-03928-767-3 (PDF)

# Contents


**About the Special Issue Editors** . . . . . . . . . . . . . . . . . . . . . . . . . . . . . . . . . . . vii

**Tizazu Mekonnen and Marc A. Dubé**
Special Issue "Renewable Polymers: Processing and Chemical Modifications"
Reprinted from: *Processes* **2019**, 7, 398, doi:10.3390/pr7070398 . . . . . . . . . . . . . . . . . . . . . 1

**Joseph Khouri, Alexander Penlidis and Christine Moresoli**
Viscoelastic Properties of Crosslinked Chitosan Films
Reprinted from: *Processes* **2019**, 7, 157, doi:10.3390/pr7030157 . . . . . . . . . . . . . . . . . . . . . 5

**Shidan Cummings, Yujie Zhang, Niels Smeets, Michael Cunningham and Marc A. Dubé**
On the Use of Starch in Emulsion Polymerizations
Reprinted from: *Processes* **2019**, 7, 140, doi:10.3390/pr7030140 . . . . . . . . . . . . . . . . . . . . . 23

**Prachiben Panchal, Emmanuel Ogunsona and Tizazu Mekonnen**
Trends in Advanced Functional Material Applications of Nanocellulose
Reprinted from: *Processes* **2019**, 7, 10, doi:10.3390/pr7010010 . . . . . . . . . . . . . . . . . . . . . 63

**Hu Jin, Wen Nie, Yansong Zhang, Hongkun Wang, Haihan Zhang, Qiu Bao and Jiayi Yan**
Development of Environmental Friendly Dust Suppressant Based on the Modification of
Soybean Protein Isolate
Reprinted from: *Processes* **2019**, 7, 165, doi:10.3390/pr7030165 . . . . . . . . . . . . . . . . . . . . . 91

**Birendra B. Adhikari, Michael Chae, Chengyong Zhu, Ataullah Khan, Don Harfield,
Phillip Choi and David C. Bressler**
Pelletization of Torrefied Wood Using a Proteinaceous Binder Developed from Hydrolyzed
Specified Risk Materials
Reprinted from: *Processes* **2019**, 7, 229, doi:10.3390/pr7040229 . . . . . . . . . . . . . . . . . . . . . 109

**Nor Anizah Mohamad Aini, Nadras Othman, M. Hazwan Hussin, Kannika Sahakaro and
Nabil Hayeemasae**
Hydroxymethylation-Modified Lignin and Its Effectiveness as a Filler in Rubber Composites
Reprinted from: *Processes* **2019**, 7, 315, doi:10.3390/pr7050315 . . . . . . . . . . . . . . . . . . . . . 121

**Sepehr Kardan, Omar Garcia Valdez, Adrien Métafiot and Milan Maric**
Nitroxide-Mediated Copolymerization of Itaconate Esters with Styrene
Reprinted from: *Processes* **2019**, 7, 254, doi:10.3390/pr7050254 . . . . . . . . . . . . . . . . . . . . . 143

**Se-Ra Shin, Van Dung Mai and Dai-Soo Lee**
Chemical Recycling of Used Printed Circuit Board Scraps: Recovery and Utilization of Organic
Products
Reprinted from: *Processes* **2019**, 7, 22, doi:10.3390/pr7010022 . . . . . . . . . . . . . . . . . . . . . 161

**Ze Kan, Hao Shi, Erying Zhao and Hui Wang**
Preparation and Performance of Different Modified Ramie Fabrics Reinforced Anionic
Polyamide-6 Composites
Reprinted from: *Processes* **2019**, 7, 226, doi:10.3390/pr7040226 . . . . . . . . . . . . . . . . . . . . . 181

# About the Special Issue Editors

**Marc A. Dubé**, Ph.D., P.Eng., FCIC, FEIC is a Professor in the Department of Chemical and Biological Engineering at the University of Ottawa. He is a registered Professional Engineer in the Province of Ontario and is a Fellow of the Chemical Institute of Canada and a Fellow of the Engineering Institute of Canada. His research is focused on sustainable polymer reaction engineering, which includes the synthesis of new polymers from renewable materials, as well as the development of sustainable polymerization processes. His research group has expertise in the synthesis, characterization, and product performance measurement of multi-component polymers in bulk, solution, and emulsion polymerization.

**Tizazu H. Mekonnen**, Ph.D. is an Assistant Professor of Chemical Engineering at the University of Waterloo in Waterloo, ON. His research focuses on the design and modification of bio-resourced polymers, sustainable nanostructured polymers, carbohydrate polymers, multiphase functional materials, polymer blending and reactive extrusion, and rubber processing with applications in engineering materials, functional coatings, functional packaging, biodegradable and compostable materials, and personal care products. His research has produced 35 refereed journal publications and 9 patent applications. Prior to his academic position, he was a polymer scientist at E. I. DuPont.

*Editorial*

# Special Issue "Renewable Polymers: Processing and Chemical Modifications"

**Tizazu Mekonnen [1],* and Marc A. Dubé [2],***

[1]   Department of Chemical Engineering, Institute for Polymer Research (IPR), University of Waterloo, Waterloo, ON N2L 3G1, Canada
[2]   Department of Chemical and Biological Engineering, University of Ottawa, Ottawa, ON K1N 6N5, Canada
*   Correspondence: tizazu.mekonnen@uwaterloo.ca (T.M.); Marc.Dube@uOttawa.ca (M.A.D.)

Received: 21 June 2019; Accepted: 21 June 2019; Published: 26 June 2019

---

The use of renewable resources for polymer production is receiving substantial and ever-growing interest. While only about 8% of the petroleum produced globally is used for the manufacture of polymers, there are alarming environmental and safety concerns associated with both the feedstock used to produce polymers and their end-of-life disposal. One possible solution to mitigate these negative environmental consequences is to develop more sustainable polymers via the use of renewable raw materials, or to recycle synthetic polymers. Feedstock such as proteins, cellulose, starch, lignin, chitosan, gums, vegetable oils, terpenes, and polyphenols can be used for the manufacture of a variety of sustainable materials and products, including elastomers, plastics, hydrogels, flexible electronics, sensors, engineering polymers, and composites. Various novel processing technologies and chemical modification strategies are also being implemented to make this feedstock more suitable for polymeric materials.

This Special Issue of *Processes* brings together several papers from leading scientists and researchers active in the area of "Sustainable and Renewable Polymers, Processing, and Chemical Modifications". The collected papers include both original research and two review articles related to renewable feedstock for polymer applications, processes for the fabrication of renewable polymer-based nanomaterials, the design and modification of renewable polymers, and applications of renewable polymers. The Special Issue is available online at the following link: https://www.mdpi.com/journal/processes/special_issues/renewable_polymers. The contributions are summarized below.

Khouri et al. [1] investigated the viscoelastic properties of treated chitosan films to elucidate the potential crosslinking of chitosan with citric acid. The study experimentally demonstrates that the thermal treatment of chitosan films in the presence of citric acid causes ionic crosslinking. The viscoelastic studies confirm that citric acid can be a safe alternative crosslinking agent to other crosslinkers, such as glutaraldehyde and epichlorohydrin of chitosan, by employing a heterogeneous film preparation method.

Adhikari et al. [2] developed a renewable wood-pellet binder from a short chain proteinaceous material co-reacted with polyamidoamine epichlorohydrin resin. The main contribution of this work is an experimental demonstration on the use of an otherwise waste protein material for wood binder applications. This proof-of-concept work, based on bench-scale continuous pelletization trials, clearly indicates the feasibility of using waste protein-derived peptides as a binder to improve the durability of torrefied wood pellets.

In an excellent example of enhancing the sustainability of plastics via recycling, Shin et al. [3] demonstrate the recovery of a useful organic material from recycled printed circuit boards that used brominated epoxy as a matrix. The depolymerization of the thermosetting epoxy was conducted via glycolysis using polyethylene glycol (PEG) under basic conditions. The recovered organic product (pre-polyol) obtained after glycolysis was converted into a recycled polyol via the Mannich reaction and the addition of the polymerization of propylene oxide to prepare rigid polyurethane foams for thermal insulation and other applications.

The use of soybean protein isolate as a biodegradable and environmentally-friendly alternative dust suppressing agent was investigated by Jin et al. [4]. They propose the use of an anionic surfactant to de-structure the complex tertiary and secondary structures of the protein, resulting in a relaxed protein–surfactant complex. Such modifications expose the hydrophobic functional groups of the protein, which provide an enhanced adsorption between the protein-based formulated dust suppressant and the hydrophobic dust (coal dust).

The potential of renewable ramie fibers with a range of treatments for fiber reinforced composite material applications was investigated by Kan et al [5]. Vacuum assisted resin infusion molding is used to fabricate composite materials with a target of producing lightweight, but high specific strength and modulus structural materials for the automotive and aerospace industry. Pretreatment along with the use of a coupling agent modification of the ramie fabrics results in composites with enhanced flexural modulus, tensile strength, and dynamic mechanical properties.

Kardan et al. [6] produced itaconic diacids via the fermentation of sugars; itaconic diacids are one of the most promising sustainable feedstocks for renewable polymeric material development. This work entailed nitroxide-mediated polymerization (NMP) of di-n-butyl itaconate using the BlocBuilder family of unimolecular initiators to obtain statistical copolymers with enhanced elastomeric properties. The polymerizations as a function of temperature and their initial compositions were presented.

The work by Mohamad Aini et al. [7] uses modified lignin as a renewable complementary filler of rubber composites. The modification of lignin employs hydroxymethylation inorder to enhance the compatibility of lignin with the hydrophobic polybutadiene rubber matrix. The results of the study indicate that the inclusion of hydroxymethylated lignin in rubber composites weakens the filler–filler interaction and improves processability. Furthermore, the compatibility and high interfacial adhesion between the modified lignin and the rubber matrix increase the cure rate of the rubber compound, and hence, increase the crosslink density compared to the unmodified lignin composites.

Panchal et al. [8] show that nanocellulose, in the form of cellulose nanocrystals and cellulose nanofibers, obtained by the deconstruction of fibers from wood or other cellulosic plants, are very appealing materials for functional material applications. This is because of their sustainability attributes, excellent mechanical properties owing to their crystalline assembly, and their abundant surface hydroxyl groups, which make them amenable to chemical modification. This article critically reviews the recent progress in the surface modification chemistries and processes of nanocellulose as functional nanomaterials, as well as carriers/substrates of other functional materials. Modifications that provide superhydrophobicity, barrier, electrical, and antimicrobial properties for coating, packaging, and electronics applications are reviewed in detail.

Finally, Cummings et al. [9] survey the use of starch in polymerizations, with a focus on sustainable emulsion polymerization processes. Some of the starch forms included in this review article are granular, native, functionalized, and nano-sized starch materials. One of the remarkable merits of this work was that it clearly presented future strategies of characterization methods and applications of starch grafted latexes and films.

## References


1. Khouri, J.; Penlidis, A.; Moresoli, C. Viscoelastic Properties of Crosslinked Chitosan Films. *Processes* **2019**, *7*, 157. [CrossRef]
2. Adhikari, B.B.; Chae, M.; Zhu, C.Y.; Khan, A.; Harfield, D.; Choi, P.; Bressler, D.C. Pelletization of Torrefied Wood Using a Proteinaceous Binder Developed from Hydrolyzed Specified Risk Materials. *Processes* **2019**, *7*, 229. [CrossRef]
3. Shin, S.R.; Mai, V.D.; Lee, D.S. Chemical Recycling of Used Printed Circuit Board Scraps: Recovery and Utilization of Organic Products. *Processes* **2019**, *7*, 22. [CrossRef]

4. Jin, H.; Nie, W.; Zhang, Y.; Wang, H.; Zhang, H.; Bao, Q.; Yan, J. Development of Environmental Friendly Dust Suppressant Based on the Modification of Soybean Protein Isolate. *Processes* **2019**, *7*, 165. [CrossRef]
5. Kan, Z.; Shi, H.; Zhao, E.; Wang, H. Preparation and Performance of Different Modified Ramie Fabrics Reinforced Anionic Polyamide-6 Composites. *Processes* **2019**, *7*, 226. [CrossRef]
6. Kardan, S.; Garcia Valdez, O.; Métafiot, A.; Maric, M. Nitroxide-Mediated Copolymerization of Itaconate Esters with Styrene. *Processes* **2019**, *7*, 254. [CrossRef]
7. Mohamad Aini, N.A.; Othman, N.; Hussin, M.H.; Sahakaro, K.; Hayeemasae, N. Hydroxymethylation-Modified Lignin and Its Effectiveness as a Filler in Rubber Composites. *Processes* **2019**, *7*, 315. [CrossRef]
8. Panchal, P.; Ogunsona, E.; Mekonnen, T. Trends in Advanced Functional Material Applications of Nanocellulose. *Processes* **2019**, *7*, 10. [CrossRef]
9. Cummings, S.; Zhang, Y.; Smeets, N.; Cunningham, M.; Dubé, M.A. On the Use of Starch in Emulsion Polymerizations. *Processes* **2019**, *7*, 140. [CrossRef]

*Article*

# Viscoelastic Properties of Crosslinked Chitosan Films

**Joseph Khouri, Alexander Penlidis** and **Christine Moresoli** *

Department of Chemical Engineering, University of Waterloo, 200 University Avenue West, Waterloo,
ON N2L 3G1, Canada; j2khouri@uwaterloo.ca (J.K.); penlidis@uwaterloo.ca (A.P.)
* Correspondence: cmoresoli@uwaterloo.ca; Tel.: +1-519-888-4567 (ext. 35254)

Received: 17 February 2019; Accepted: 8 March 2019; Published: 14 March 2019

**Abstract:** Chitosan films containing citric acid were prepared using a multi-step process called heterogeneous crosslinking. These films were neutralized first, followed by citric acid addition, and then heat treated at 150 °C/0.5 h in order to potentially induce covalent crosslinking. The viscoelastic storage modulus, $E'$, and tan$\delta$ were studied using dynamic mechanical analysis, and compared with neat and neutralized films to elucidate possible crosslinking with citric acid. Films were also prepared with various concentrations of a model crosslinker, glutaraldehyde, both homogeneously and heterogeneously. Based on comparisons of neutralized films with films containing citric acid, and between citric acid films either heat treated or not heat treated, it appeared that the interaction between chitosan and citric acid remained ionic without covalent bond formation. No strong evidence of a glass transition from the tan$\delta$ plots was observable, with the possible exception of heterogeneously crosslinked glutaraldehyde films at temperatures above 200 °C.

**Keywords:** chitosan; crosslinking; viscoelasticity; citric acid; glutaraldehyde; heterogeneous crosslinking

---

## 1. Introduction

The search for biodegradable and non-toxic materials to offset the consumption and production of plastics for the food industry has led to research on polysaccharide edible films, including chitosan, the crustacean-derived aminopolysaccharide. Chitosan films have been studied either in stand-alone form as potential packaging or wrapping products or as coatings [1]. The physico-chemical properties of films prepared by the solvent casting method have been well studied, including in relation to the organic acids used in film preparation (e.g., acetic, lactic, citric acid, etc.) [2,3], the molecular weight of chitosan [3], and the degree of deacetylation [4,5]. The degree of deacetylation (DD) is the percentage of glucosamine residues with a primary amine side group at the C-2 position. The chemical structure of chitosan is given in Figure 1.

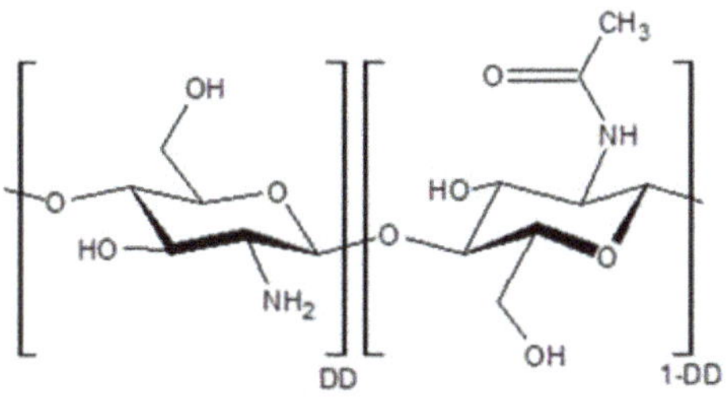

**Figure 1.** The glucosamine and acetyl glucosamine units that comprise chitosan and chitin. When the degree of deacetylation (DD) is greater than 0.6, the polymer is considered chitosan.

While chitosan films have advantages, such as high anti-microbial and low oxygen-permeability properties [3], their high moisture affinity and relatively poorer mechanical properties compared to common plastics, such as polyethylene terephthalate, low density polyethylene, and polypropylene, limit their packaging applications. The water vapor permeability (WVP) of chitosan films is typically a magnitude higher than that of thermoplastics. Their tensile strength (TS) is in the same range as some plastics; however, their elongation capacities and elastic moduli are lower. A comparison of mechanical and barrier properties of chitosan and plastic films is provided in Table 1.

Improving mechanical properties and reducing hydrophilicity of chitosan films has been attempted by several methods, including (i) composite formation with fatty acids [6,7] and other polysaccharides [8], (ii) grafting hydrophobic compounds [9,10] or phenolics [11,12], and (iii) crosslinking the polymer chains [8]. Composites with hydrophobic compounds, such as beeswax [13], or fatty acids such as stearic [7] and palmitic [13] acid, do not always improve WVP, possibly as a consequence of a decrease in film density [6,13]. Composites with starch [14] and cellulose-derivatives [15] or with proteins such as gelatin [14] have ultimately not shown much improvement in the WVP and TS properties of chitosan films. Crosslinking appears to be a promising method for modification for food-related applications.

**Table 1.** Properties of Chitosan Films and Plastics from Literature.

| Property | Chitosan | Common Plastics |
|---|---|---|
| Water Vapor Permeability (WVP) (g/(m·s·Pa)) | $1\text{--}10 \times 10^{-11}$ [2,3,6,7,15] | $0.01\text{--}1 \times 10^{-11}$ [16–18] |
| Oxygen Gas Permeability (OP) (cm$^3$/(m·s·Pa)) | $1 \times 10^{-15}\text{--}1 \times 10^{-13}$ [3,6,7,15] | $1 \times 10^{-13}\text{--}1 \times 10^{-12}$ [16–18] |
| Tensile Strength (TS) (MPa) | 1–100 [2–4,7,15] | 1–100 [16] |
| Elongation Before Break (EBB) (%) | 1–50 [2–4,7], 100 (with plasticizer) [2,7] | 1–500 [16] |
| Young's Modulus ($E$) (GPa) | 0.1–3 [7,15] | 1–10 |

An important criterion for the selection of a crosslinking agent for food applications is that the compound needs to be non-toxic. This eliminates cytotoxic compounds such as glutaraldehyde and epichlorohydrin, which are commonly used to crosslink chitosan [19,20] for industrial membrane applications such as waste water treatment. Similarly, grafting or crosslinking reactions facilitated by 1-ethyl-3-(3-dimethylaminopropyl)carbodiimide with *N*-hydroxysuccinimide [11] are unsuitable since these compounds are not food grade.

Tannic acid has been studied as a crosslinking agent for chitosan films [21], and has shown improvement in WVP and TS. However, tannic acid is a complex, large polyalcohol compared to typical crosslinkers, making it more difficult to understand its effect on film properties. Genipin, an extract from gardenia and jagua fruit, is another option [22,23], and although reported to have herbal benefits and is used for drug delivery [22], it has not yet been approved by the USA Food and Drug Administration. Citric acid is non-toxic and has been used as a crosslinker for chitosan and other polysaccharides for edible films [8,24–26], textiles [27,28], and hydrogels for drug delivery [29–31]. The reactions between chitosan and citric acid take place in the films at temperatures between 110 to 190 °C [8,32], ranging from a few minutes to several hours [8,32,33], with citric acid concentrations of 5 to 30% (of dry polymer weight) [34], and either with [8,35] or without [26,29,32] a catalyst. Recent work on plasticized chitosan films crosslinked with citric acid (1:1 *w/w* chitosan) reports lower water absorption and WVP, but at the expense of lower TS and mechanical modulus [26]. Composite hemicellulose-chitosan foams show improvement in TS [33] after crosslinking with citric acid.

The research on chitosan edible films crosslinked by citric acid has thus far performed the crosslinking in a homogeneous manner [8,26,32]. That is, with a filmogenic solution containing citric acid in addition to acetic acid. However, this method might not be the most appropriate. A study on the effect of heating chitosan films has shown amidization reactions occur with the acids used in the preparation [36]. However, at 60 °C, citric acid does not appear to react with the amine of chitosan [36] to the extent that acetic acid and propionic acid do. Therefore, in order to avoid competing reactions between acetic and citric acid, acetic acid should be removed prior to the incorporation of the crosslinker into the polymer matrix. Such a multi-step process is referred to as heterogeneous crosslinking, and has been performed with epichlorohydrin [19], glutaraldehyde [19,37,38] and genipin [23] for chitosan.

In this work, the potential crosslinking of chitosan with citric acid using the heterogeneous procedure was investigated. Two crosslinkers were used: citric acid and a model crosslinker, glutaraldehyde (GTA). Additionally, homogeneous and heterogeneous methods were compared. While Fourier transform infrared spectroscopy is an effective technique for studying changes to chemical structure, such as the detection of new amide [33] or ester [24,35] bonds, the method cannot provide distinction between grafting and crosslinking. In this work, crosslinking was examined by studying the viscoelastic properties (storage modulus, $E'$, loss modulus, $E''$, and $\tan\delta$) using dynamic mechanical analysis (DMA) by considering principles of the rubber elasticity theory. This same approach has been used previously for analyzing the crosslinking characteristics of various polysaccharide and protein materials, such as methyl cellulose crosslinked with GTA [39], whey proteins with formaldehyde [40], starch with trisodium trimetaphosphate [41], and chitosan hydrogels crosslinked with genipin [42]. Figure 2 illustrates the postulated covalent crosslinking by conversion of primary amine to amide upon heat treatment of a chitosan film with citric acid, irrespective of film preparation by either the homogeneous or heterogeneous crosslinking methods. Figure 2 also illustrates the ionic crosslinking in a chitosan—citrate film with no heat treatment.

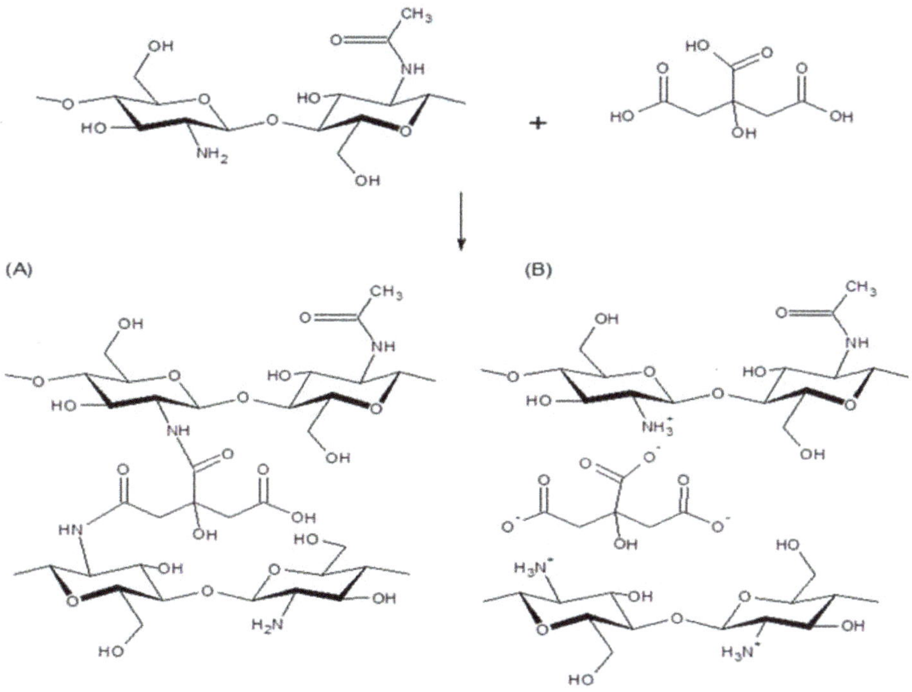

**Figure 2.** (**A**) Postulated covalent crosslinking of chitosan and citric acid with heat treatment and (**B**) electrostatic bonding and ionic crosslinking between the protonated amine and carboxylate ion of a non-heat treated chitosan citrate film.

## 2. Materials and Methods

### 2.1. Materials

Chitosan ($Mv$ = 50–190 kDa, 20–300 cP, 1 wt.% in 1% acetic acid solution at 25 °C, 75 to 85% DD), acetic acid (>99.7%), and glutaraldehyde (Grade II, 25 wt.% aqueous solution) were purchased from Sigma Aldrich (St. Louis, MO, USA). Citric acid (>99.5%) was obtained from Fisher Scientific (Fair Lawn, NJ, USA) and sodium hydroxide (>95%) from EMD Chemicals (Gibbstown, NJ, USA). All reagents were used as received without further modification. Ultra-pure water was used in the film preparation process.

### 2.2. Film Preparation

The following films were prepared: (i) neat films, (ii) neutralized films, (iii) films with different GTA concentration crosslinked homogeneously, (iv) films with different GTA concentration crosslinked heterogeneously, (v) films with citric acid prepared heterogeneously without subsequent heat treatment, and (vi) films with citric acid prepared heterogeneously with subsequent heat treatment. Table 2 lists the film types, their crosslinker content, and their corresponding code names.

Neat chitosan films were made by the solvent casting method with 300 mL of filmogenic solutions of 2% ($w/v$) chitosan in 2% ($v/v$) aqueous acetic acid. After mixing, solutions were filtered through cheesecloth to remove undissolved material and impurities and subsequently degassed using a vacuum aspirator to reduce dissolved gases. The solutions were cast on glass trays (16 × 30 cm) at ambient conditions. Temperature and relative humidity (RH) conditions ranged from 18–23 °C and 20–25%, respectively, and were monitored by a thermo-hygrometer (SMART$^2$, InterTAN Inc., Barrie, ON, Canada). The films required approximately 48 to 60 h to completely dry and form.

Neutralized films were prepared by submerging dried neat films in solutions of 0.2 M NaOH for 30 min, and were then thoroughly rinsed with ultra-pure water until the pH of the diluent reached that of water. Excess water was wiped off the surface, and the wet neutralized films were firmly clamped between a frame and glass plate to maintain shape and dimensions and avoid shrinkage [43], and were then dried in an environmental chamber at 23 °C and 50% RH for 24 h.

Films crosslinked with GTA were prepared via the homogeneous and heterogeneous methods. For homogeneously crosslinked films, denoted as GTA-HOM from now on, the filmogenic solutions were prepared as the neat films; however, prior to casting, a predetermined amount of GTA (3, 6, 12 wt.% of chitosan) was slowly added to the solution. The crosslinking reactions proceeded instantaneously and continued during the drying phase, which took approximately 24 to 36 h. Heterogeneously crosslinked films, referred to as GTA-HET from now on, were prepared by immersing dried neutralized films in 200 mL GTA aqueous solutions for 24 h at ambient conditions, where the GTA (6, 12 wt.% of dried neutralized film) absorbed into the film and reacted with the chitosan. The longer duration for crosslinking in the heterogeneous method is due to slower reaction kinetics controlled by the lower rate of diffusion of GTA into the already formed film [38]. The wet GTA-HET films were dried for 24 h in the environmental chamber (23 °C and 50% RH). The GTA-containing films changed color to an orange-hue, as expected and reported previously [44].

Films containing citric acid (denoted as CA films) were prepared via the heterogeneous method. Dried neutralized films were immersed 200 mL of citric acid aqueous solutions for 5 h at ambient conditions. The concentration of citric acid was 15% ($w/w$ dried neutralized film). The wet CA film was clamped and dried for 24 h in the environmental chamber. The films were partitioned and one piece was heat treated, (denoted as CA-HT films), at 150 °C for 0.5 h in an attempt to induce covalent crosslinking. These heat treatment conditions were chosen based on the information reported in the literature [8,32] and by taking into consideration that citric acid degrades after melting above 160 °C [29]. Based on the DD of the chitosan provided by the supplier, the approximate ratio of [NH$_2$] to [COOH] could vary from 0.88 to 0.99, near stoichiometric ratio.

**Table 2.** Chitosan Film Types.

| Film | Code Name |
| --- | --- |
| Neat | Neat |
| Neutralized | Neutralized |
| Homogeneously crosslinked with 3% (*w/w*) glutaraldehyde (GTA) | GTA-HOM-3 |
| Homogeneously crosslinked with 6% (*w/w*) glutaraldehyde | GTA-HOM-6 |
| Homogeneously crosslinked with 12% (*w/w*) glutaraldehyde | GTA-HOM-12 |
| Heterogeneously crosslinked with 6% (*w/w*) glutaraldehyde | GTA-HET-6 |
| Heterogeneously crosslinked with 12% (*w/w*) glutaraldehyde | GTA-HET-12 |
| Heterogeneously prepared with 15% (*w/w*) citric acid (CA) | CA |
| Heterogeneously prepared with 15% (*w/w*) citric acid, heat treated | CA-HT |

## 2.3. Dynamic Mechanical Analysis

The viscoelastic properties of the films were measured by temperature-ramp experiments in tensile mode using a TA DMA Q800 (TA Instruments, New Castle, DE, USA) following the guidance of ASTM D5206 (Standard Test Method for Plastics: Dynamic Mechanical Properties: In Tension). The specimen dimensions were $5.5 \pm 0.4$ mm (width) $\times$ $10.0 \pm 0.1$ mm (gauge length). A preload force of 1 N was applied during gauge length measurements. The thickness of the films ranged between 80 and 120 μm, and were measured using a digital micrometer (0.002 mm accuracy, Marathon Watch Company, Richmond Hill, ON, Canada).

Films were heated at a constant rate of 3 °C/min, at a fixed frequency of 1 Hz and constant strain of 0.15%. Preliminary strain-ramp tests showed that neat, neutralized, and crosslinked films displayed a linear viscoelastic response above a strain of 0.12% at 30 °C. The films were tested in triplicate, with the exception of GTA films which were tested with single or duplicate measurements. The viscoelastic properties measured by the DMA and used for the analysis included storage modulus, $E'$, loss modulus, $E''$, and $\tan\delta = E''/E'$.

The influence of absorbed water in the films on the viscoelastic properties was also evaluated by preheating the specimens prior to DMA. The specimens were heated at 140 °C for 10 min without any strain, followed by cooling to room temperature, and then conducting the tests as per the above conditions. DMA tests are referred to film specimen with or without preheating.

Peak fitting was performed on the $\tan\delta$ curves. The $\tan\delta$ peaks were fitted with a constant baseline, whose value was chosen as the minimum $\tan\delta$ value for each specific plot. Fitting was performed with OriginPro 8's Peak Analyzer feature, where the final fit was based on the minimization of chi-square and visual inspection of the fitted and experimental curves. The peak center and full width at half maximum (FWHM) values were statistically evaluated using the Least Significant Difference (LSD) test with a significance level of $\alpha = 0.1$.

## 3. Results and Discussion

### 3.1. Description of Viscoelastic Behavior of Neat, Neutralized and CA Films

The plots of non-preheated neat, neutralized, CA and CA-HT films are shown in Figure 3. Their temperature-dependent viscoelastic behaviors are now briefly described. (Note: The ranges of temperatures discussed are based on trends collectively observed from replicate runs.) The $E'$ of neat films decreased relatively linearly from 30 °C to approximately 100 °C, and then either plateaued or reached a local minimum, increased to approximately 120–140 °C, decreased to a local minimum near 160–180 °C and increased upon further heating. The corresponding $E''$ increased from 30 °C and passed through a shoulder peak until it reached a local maximum near 100 °C, decreased to a local minimum near 130–140 °C, and then increased to a maximum at 150–160 °C. The corresponding $\tan\delta$ increased from 30 °C to a small, broad peak near 100 °C along with a smaller secondary peak, reached a minimum between 120–130 °C and increased to a maximum near 160–170 °C.

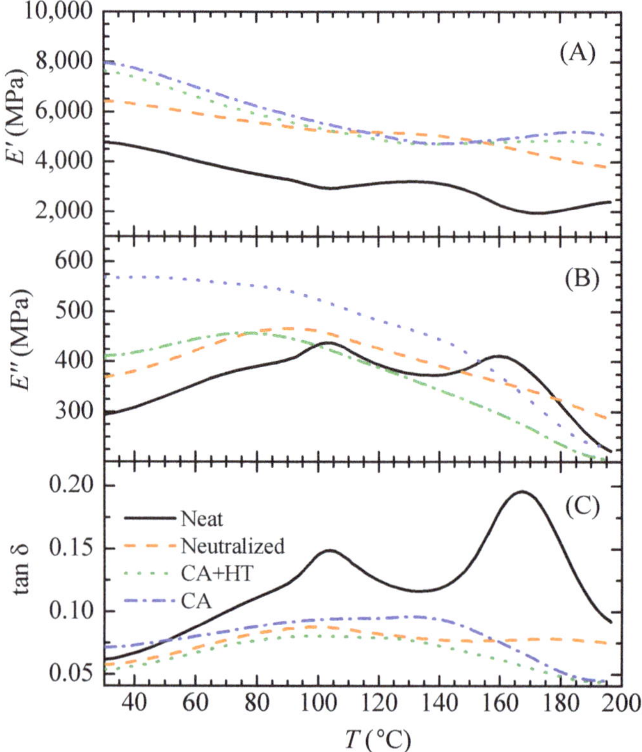

**Figure 3.** The plots of (**A**) storage modulus, (**B**) loss modulus, and (**C**) tanδ against temperature of neat, neutralized, CA, and CA-HT films with no preheating.

The tanδ peak near 100 °C is the water induced relaxation peak arising from desorption and subsequent evaporation of water molecules [45]. The additional secondary-peak, centered around 110 °C, was likely induced by the presence of acetic acid, as it was absent from the scans of neutralized, GTA (Section 3.3) and CA films. It may be related to the $\sigma$-type conduction of acetate ions and protons observed in isochronal dielectric measurements [46] in the temperature range of $-10$ to 150 °C of neutralized and non-neutralized chitosan films.

The $E'$ for neutralized films decreased linearly from 30 °C to approximately 100 °C, then plateaued until approximately 160 °C, and thereafter decreased. The corresponding tanδ increased from 30 °C to a broad maximum near 100 °C, decreased to a local minimum near 140–160 °C before it increased to a weak maximum near 180 °C. For CA and CA-HT films, a sufficient amount of similarity in their DMA scans warrant a joint description and discussion. The $E'$ of the CA and CA-HT films decreased from 30 °C to a minimum near 140 °C, and then increased to a local maximum between 170–190 °C. For $E''$, a local maximum appeared near 80 °C, passed through a shoulder peak near 140 °C and declined thereafter. The corresponding tanδ plots showed an increase up to 100–120 °C at which point the curve either plateaued or displayed a second local maximum between 120–140 °C.

The $E'$ values at 35 °C and 195 °C extracted from the plots of non-preheated neat, neutralized, CA, and CA-HT films are listed in Table 3. The two temperature levels are near ambient conditions, and onset of degradation, respectively. The tanδ peak values and corresponding temperatures are also listed in Table 3 (last four columns). The low temperature peak near 100 °C is designated as 'peak 1' and the high temperature peak near 160 °C is designated as 'peak 2'.

Comparison of the $E'$ values provides an assessment of material stiffness, which is related to crosslinking as per the rubber elasticity theory. Increasing crosslink density, or decreasing the average molecular weight between crosslinks, increases $E'$ in the rubbery-region of a material [47,48]. The tan$\delta$ glass-transition relaxation peak broadens as the distribution of relaxation times increases and, it shifts to higher temperatures as crosslinking increases the glass transition temperature, $T_g$. The drop in $E'$ with temperature across the glass-rubber transition for polysaccharides does not occur over several magnitudes as it does with thermoplastics or rubbers. The applicability of rubber elasticity theory on polysaccharide gels has been questioned as high stiffness of a material creates an internal energy dependence to the overall change in free energy when it is mechanically deformed, in violation of the entropic assumption of the theory [49]. Therefore, a quantitative estimate of crosslink density is avoided here, and only evaluation of $E'$ and tan$\delta$ is used.

From Table 3 it can be seen that: (i) neat films had a lower $E'$ overall, (ii) $E'$ and tan$\delta$ of CA and CA-HT films were not statistically different from one another, and (iii) the $E'$ (195 °C) values of CA and CA-HT films were not statistically different compared to neutralized films. Regarding (i), the results were as expected: neutralized films are stiffer than neat films because of a lower absorbed water content of approximately 3% [50], which results in an increase in stronger inter- and intramolecular hydrogen bonds. Crosslinking by GTA restricts chain mobility, hence the higher $E'$ of GTA-HOM films compared to neat films. For (ii), these results demonstrate that heat treatment followed by 72 h of conditioning did not influence the viscoelastic behavior of films with citric acid. Regarding (iii), this observation shows that citric acid did not influence the properties relative to neutralized films during the approach towards the glass transition as anticipated. An example of such an observation was made with a DMA frequency sweep of a heterogeneously crosslinked chitosan-genipin film studied under physiological conditions that exhibited a storage modulus value more than double that of its neutralized counterpart [42]. Any differences between neutralized and CA films could be structure related, and not necessarily related to crosslinking. For instance, X-ray diffraction (XRD) scans of chitosan films cast with citric acid displayed a more amorphous structure than chitosan acetate films [36]. Since polymers with a higher crystallinity typically display greater storage modulus [47], it might be possible that an increase in amorphous structure caused by citric acid had a counter balancing effect on the increase in rigidity from crosslinking, and thus the difference in $E'$ (195 °C) of neutralized and CA films was not statistically significant. Neutralization does decrease the degree of crystallinity [50], by about 6% [51] compared to neat films.

Comparing the crosslinked and non-crosslinked forms of a material does have some challenges. If the degree of physical entanglements is high it could mask the detection of new covalent crosslinks in the rubbery-plateau region if the degree of covalent junctures is low. Changes to the storage modulus in this region from crosslinking become more observable when tested using lower frequencies [52], and 1 Hz is sufficiently low. While crosslinking is expected to increase $E'$ in the rubbery region, it does not always increase it in the glassy region, as demonstrated by photo-crosslinked polyacrylate membranes [48]. For this reason, $E'$ (35 °C) is considered less of an indicator for crosslinking here, especially as $E'$ (35 °C) of CA and CA-HT are not significantly different.

The crosslinking procedure can influence the degree of crystallinity and the type of bond between the crosslinker and chitosan [19,37]. For example, heterogeneous crosslinking with chitosan-glutaraldehyde does not reduce crystallinity to the extent that the homogeneous procedure does [19], and it is hypothesized that heterogeneous crosslinking mostly occurs at the surface [19] and in the amorphous regions of the polymer matrix [37]. It is similarly expected that citric acid will absorb and react mostly at the surface and amorphous domains of the films.

Preload forces may have some impact on measured viscoelastic properties, especially on softer, biological materials [53]. Strain hardening from preloading can be avoided with strain-rate measurements which negate the use of a trigger force before the measurements. Bartolini et al. [54] studied the viscoelastic response of nano-indented poly(dimethylsiloxane) strips and found that the apparent elastic moduli of previously strained samples were lower than non pre-strained samples.

Here, the 1 N force applied to the specimens during gauge length measurements is unlikely to make a great impact, as it is less than the net force applied to the specimens for 0.15% strain. It is possible that the static force applied during dynamic moduli measurements could affect the different film types to different degrees, thereby creating a bias in the data.

Chitosan films are non-isotropic with a significant degree of variation of density, crystallinity, and chemical structure throughout. This will impact the apparent mechanical moduli values which are determined as a bulk quantity, averaged over the specimen volume. Inhomogeneities within the crosslinked films can be probed using microscale techniques such as nano-indentation, where the indenting load is applied cyclically, either at a constant or increasing load rate [55,56]. This could be considered for assessing differences in heterogeneity between neat, neutralized and CA films.

**Table 3.** Characteristics of the dynamic mechanical analysis (DMA) curves for film specimens with no-preheating (see Table 2 for film code names).

| Film | Parameter | $E'$ (35 °C) | $E'$ (195 °C) | tanδ Peak 1 | | tanδ Peak 2 | |
|---|---|---|---|---|---|---|---|
| | | | | $T$ | tanδ | $T$ | tanδ |
| | | (MPa) | (MPa) | (°C) | | (°C) | |
| Neat | mean | 5071 [a] | 2382 [a] | 102.0 [a] | 0.129 [a] | 164.8 [a] | 0.216 [a] |
| | ± | 432 | 127 | 2.8 | 0.015 | 2.8 | 0.045 |
| | COV * (%) | 8.5 | 5.3 | 2.7 | 11.7 | 1.7 | 21.1 |
| | $n$ | 6 | 6 | 6 | 6 | 6 | 6 |
| Neutralized | mean | 6665 [b] | 4143 [b] | 106.7 [a] | 0.089 [b] | 187.6 [b] | 0.078 [b] |
| | ± | 728 | 334 | 12.9 | 0.003 | 9.8 | 0.001 |
| | COV * (%) | 11 | 8 | 12.1 | 3.4 | 5.2 | 1.7 |
| | $n$ | 5 | 5 | 5 | 5 | 4 | 5 |
| CA | mean | 7588 [c] | 4727 [c] | 106.9 [a] | 0.099 [b] | 130.2 [c] | 0.097 [b,c] |
| | ± | 489 | 349 | 8.9 | 0.016 | 8.4 | 0.026 |
| | COV * (%) | 6.4 | 7.4 | 8.3 | 16.4 | 6.5 | 26.3 |
| | $n$ | 3 | 3 | 3 | 3 | 2 | 2 |
| CA+HT | mean | 7605 [c] | 4479 [b,c] | 104.2 [a] | 0.088 [b] | 131.1 [c] | 0.096 [c] |
| | ± | 871 | 635 | 4.6 | 0.005 | - | - |
| | COV * (%) | 11.5 | 14.2 | 4.4 | 5.5 | - | - |
| | $n$ | 4 | 4 | 4 | 4 | 1 | 1 |
| GTA-HOM-3 | $n = 1$ | 5123 | 2475 | 106.5 | 0.156 | 165.7 | 0.201 |
| GTA-HOM-6 | $n = 1$ | 6222 | 3367 | 96.1 | 0.110 | 168.9 | 0.110 |
| GTA-HOM-12 | $n = 1$ | 5882 | 3160 | 97.3 | 0.090 | 167.0 | 0.091 |

* COV—Coefficient of Variation. Significant statistical difference between means is indicated by different superscripted letters (LSD test, $\alpha = 0.1$).

### 3.2. Effect of Preheating Film Specimens and Thermal Treatment on CA Films

The effects of heat treatment on CA films and preheating on films are now discussed. Scans of preheated neat, neutralized, CA, and CA-HT films are shown in Figure 4. At low temperatures, $E'$ increased relative to non-preheated films by approximately 2000 MPa and the slope of $\Delta E'/\Delta T$ was greater for the majority of the scan. Above 170 °C the storage modulus plots of the non-preheated and preheated scans merged and overlapped (or see Park and Ruckenstein [39] for similar observation with methylcellulose). This is due to the difference in water content becoming less with increasing temperature. Additionally, preheating reduced peak 1 from values in the range of 0.89–0.99 down to 0.52–0.59 for neutralized and CA films, respectively. For the full set of tanδ and $E'$ values of preheated neat, neutralized, CA and CA-HT films, see Table 4.

**Table 4.** Characteristics of DMA curves for Preheated Specimens (see Table 2 for film code names).

| Film | Parameter | $E'$ (35 °C) (MPa) | $E'$ (195 °C) (MPa) | tanδ Peak 1 | | tanδ Peak 2 | |
|---|---|---|---|---|---|---|---|
| | | | | $T$ (°C) | tanδ | $T$ (°C) | tanδ |
| Neat | mean | 7826 [a] | 2289 [a] | - | - | 169.3 [a] | 0.213 [a] |
| | s | 427 | 152 | - | - | 4.2 | 0.034 |
| | COV * (%) | 5.5 | 6.6 | - | - | 2.5 | 16.0 |
| | $n$ | 4 | 4 | 0 | 0 | 4 | 4 |
| Neutralized | mean | 9384 [b] | 4162 [b] | 124.1 [a] | 0.059 [a] | 187.2 [b] | 0.078 [b] |
| | s | 859 | 385 | 3.8 | 0.002 | 5.5 | 0.003 |
| | COV (%) | 6.7 | 6.8 | 2.5 | 3.4 | 2.3 | 2.6 |
| | $n$ | 6 | 6 | 6 | 6 | 6 | 6 |
| CA | mean | 9982 [b] | 4348 [b] | 117.8 [b] | 0.059 [a,b] | 166.5 [a] | 0.061 [c] |
| | s | 220 | 93 | 1.0 | 0.005 | 1.1 | 0.005 |
| | COV (%) | 2.2 | 2.1 | 0.8 | 8.3 | 0.7 | 9.0 |
| | $n$ | 2 | 2 | 2 | 2 | 2 | 2 |
| CA+HT | mean | 9351 [b] | 4612 [b] | 125.8 [a,b] | 0.050 [b] | 165.3 [a] | 0.051 [c] |
| | s | 609 | 244 | 10 | 0.004 | 3.2 | 0.004 |
| | COV (%) | 6.5 | 5.2 | 7.9 | 8.0 | 1.9 | 8.8 |
| | $n$ | 3 | 3 | 3 | 3 | 3 | 3 |
| GTA-HOM-3 | mean | 9119 | 2749 | 97.3 | 0.060 | 175.3 | 0.167 |
| | s | 282.1 | 159.3 | 5.8 | 0.000 | 0.5 | 0.008 |
| | $n$ | 2 | 2 | 2 | 2 | 2 | 2 |
| GTA-HOM-6 | $n = 1$ | 9740 | 3162 | 92.4 | 0.068 | 181.6 | 0.101 |
| GTA-HOM-12 | $n = 1$ | 9413 | 3584 | 77.9 | 0.067 | 191.8 | 0.080 |

* COV—Coefficient of Variation. Significant statistical difference between means is indicated by different superscripted letters (LSD test, $\alpha = 0.1$).

Since the film specimens were heated by ramping in the DMA, and CA-HT films were thermally treated prior to DMA testing, it is important to consider changes to chemistry and chain structure following heat treatment and their subsequent effects on mechanical properties. Infrared spectroscopy studies reveal an increase in the intensity of bands corresponding to secondary amine (–NH–) (amide II) [21] and amide-carbonyl (–N–C=O) (amide I) [21,36] at approximately 1560 and 1650 cm$^{-1}$, respectively, as a consequence of heating. They also show a simultaneous decrease in protonated amine and carboxylate ion peaks at approximately 1515/1615 and 1555 cm$^{-1}$, respectively. As stated earlier, the potential reactions between chitosan amine and carboxyl groups from residual acetic acid is the main motive for neutralizing the films prior to the absorption of citric acid into the matrix: to avoid competition of amidization between the two acids. Further evidence of reaction resulting from conditioning at high temperatures (>100 °C) is reduced film solubility in water [21] and acid aqueous solutions [21,57], which agrees with the effect of amidization as protonation of the amine is necessary for chitosan dissolution.

Some authors report an overall decrease in crystallinity [36] with heating. In this situation, the higher ratio of amorphous to crystalline regions would likely strengthen the appearance of $T_g$ and cause it to shift to higher temperatures and decrease $E'$. If there is such a reduction in crystallinity in preheated or heat treated films here, it is negligible with respect to the error associated with experimentation as the $E'$ (195 °C) of preheated and non-preheated scans are not significantly different. Other studies report a transformation from the 'tendon' (hydrated) to the 'annealed' (anhydrous) crystal structure [57] after heating. Despite all potential changes to physico-chemical and structural properties from preheating or heat treatment that may affect mechanical properties, they are less likely to be detectable in the glassy state by the more pronounced effect of increased rigidity when absorbed water is evaporated out. In short, no effects from changes to crystal structure or crystallinity from heating were observed here.

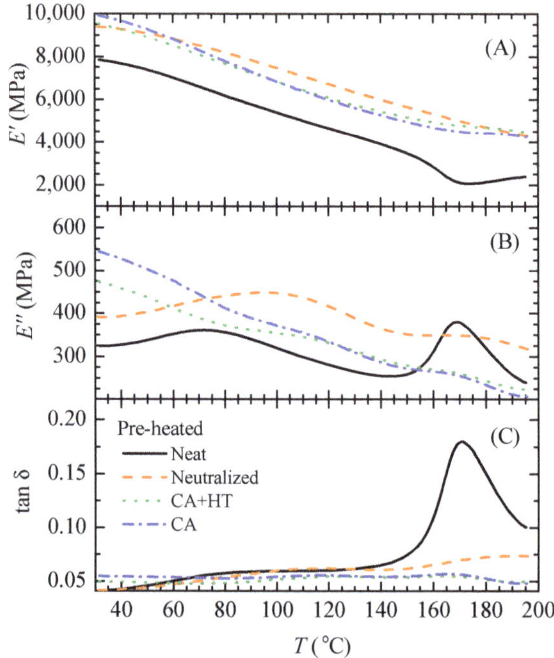

**Figure 4.** The plots of (**A**) storage modulus, (**B**) loss modulus, and (**C**) tanδ against temperature of preheated neat, neutralized, CA, and CA-HT films.

### 3.3. GTA-Crosslinked Films

The tanδ and $E'$ values from the non-preheated and preheated scans of GTA-HOM films are listed in Tables 3 and 4, respectively; see Table 2 for the film names. The plots for preheated scans for GTA-HOM films are shown in Figure 5, and plots of non-preheated scans are shown in Figure A1 in the Appendix A. Comparisons with neat films, and the effect of GTA concentration are now discussed. The film with the lowest concentration, GTA-HOM-3, displayed similar quantitative and qualitative viscoelastic behavior to the neat films. For both non-preheated and preheated conditions, the $E'$ (195 °C) values of GTA-HOM films exceed those of neat films, as expected with crosslinked films. Increasing the concentration from 3 to 12% caused the following changes: (i) a decrease in magnitude of peak 1 (non-preheated scans) as amide/imide formation would reduce H-bonding capacity, (ii) a decrease in tanδ peak 2 magnitude and a shift to higher temperatures (Table 4), and (iii) an overall increase in $E'$. The $E'$, $E''$ and tanδ curves for preheated GTA-HOM-6 and -12 films overlapped reasonably well, indicating marginal differences in the degree of crosslinking above 6% GTA. A systematic increase in $E'$ (195 °C) with an increase in GTA concentration was observed for preheated GTA-HOM films. However, this was not seen in the non-preheated GTA-HOM specimens, which may be related to the stiffness of the polymer chain. Park et al. [39] did not observe significant changes to the glass transition peak in their DMA tanδ scans of methylcellulose-GTA crosslinked hydrogels, and they claimed that the high rigidity of the polysaccharide backbone prevents any indication of changes induced by covalent crosslinking.

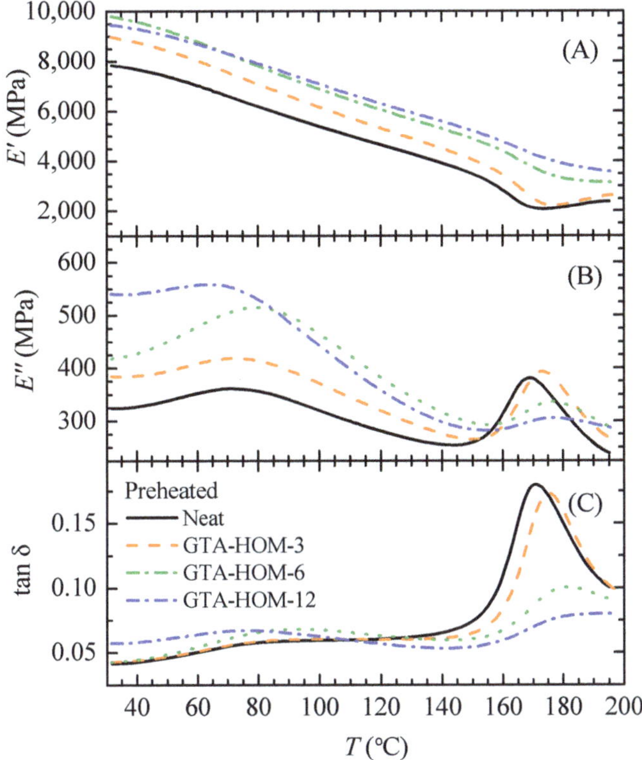

**Figure 5.** The plots of (**A**) storage modulus, (**B**) loss modulus, and (**C**) tanδ against temperature of preheated 3, 6, and 12% GTA-HOM films and a neat chitosan film.

In addition to GTA-HOM films, heterogeneously prepared GTA films were also investigated. (DMA studies of heterogeneously crosslinked chitosan-glutaraldehyde films are seldom reported but are still available in the literature [58].) Figure 6 shows the DMA scans of GTA-HET-6 and GTA-HET-12 films. For these films, the maximum temperature of the scan was increased from 200 to 220 °C to test the temperature-DMA limits of the chitosan films. Despite being more brittle, the $E'$ values of the GTA-HET crosslinked films were less than both the homogenously prepared GTA and neutralized films within the majority of the temperature range recorded, as shown in Figure 6. This would suggest that producing films heterogeneously might have cleaved the polymer chains while simultaneously crosslinking them, mostly at the exterior [19]. New bonds formed with glutaraldehyde may either be imines, or a combination of imine and Michael-type adducts, for heterogeneous versus homogeneous crosslinking, respectively [19]. The difference in bond formation and majority of crosslinking at the exterior of the film might account for the extra brittleness exhibited by GTA-HET films.

The tanδ peak 2 was not visible for the GTA-HET films, but rather a new, broad peak with an onset near 170 °C began to emerge, whose center was out of the measured temperature range. This peak could be more representative of a glass-rubber transition which is speculated to exist within the film degradation range. The degradation of neutralized chitosan films typically begins near 200 °C and reaches a maximum degradation rate near 275 °C [43]. For our neutralized films, only a 1% mass loss was found between 200 and 250 °C and the differential thermogravimetric analysis peak was at 290 °C. It cannot be ruled out that the GTA-HET peak onset at 170 °C could be due to an earlier onset of degradation, as crosslinking with GTA [58] has been found to do. However, in the case of heterogeneously crosslinked chitosan membranes formed by electrospinning, thermogravimetric analysis scans by Correia et al. [59] did not demonstrate differences between neutralized chitosan films and ethanol neutralized-heterogeneously crosslinked GTA membranes, showing that heterogeneous crosslinking with GTA does not necessarily lower the decomposition onset temperature.

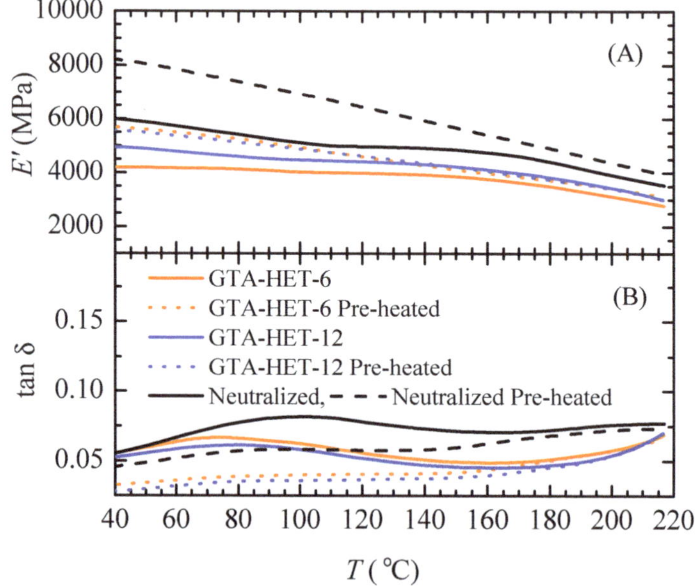


**Figure 6.** The plots of (**A**) storage modulus and (**B**) tanδ against temperature of non-preheated and preheated scans of 6 and 12% GTA-HET films.

### 3.4. Analysis of tanδ and Relation to Crosslinking

To evaluate the relaxation properties of the films further, peak deconvolution was performed on the tanδ curves to obtain a more accurate peak center and to estimate the FWHM (units of temperature). If tanδ peak 2 is related to the glass transition, one would expect peak widening as a consequence of crosslinking and a shift to higher temperatures.

The tanδ peak 2 deconvolution analysis is presented in Table 5. The peak 2 FWHM values of the neutralized films were higher (statistically significant) than those of the CA films for both preheated and non-preheated specimens, contrary to expectations of peak broadening with the addition of citric acid and formation of covalent crosslinks. Furthermore, the FWHM values of preheated CA and CA-HT films were not significantly different, which suggests that thermal treatment likely did not change the bonding type from ionic to covalent, just as the lack of difference in $E'$ values previously indicated, as discussed in Section 3.1. By contrast, the peak width of GTA-HOM films increased from 35 to 59 °C by increasing GTA concentration from 3 to 12%, and peak 2 moved to higher temperatures, according to expectations of an increase in crosslink density.

The height of the tanδ peak 2 diminished from 0.213 ± 0.045 for neat to 0.078 ± 0.01 for neutralized films. The reason for this drop may be due to changes to the chemical nature of the films. Gartner et al. [50] speculated that the origins of this relaxation peak are from electrostatic, ionic interactions between the conjugate base of the solvating acid and the protonated amine. They compared the tanδ properties of neat and neutralized films made from acetic and hydrochloric acid using DMA, and similar tanδ temperature-dependence of their chitosan acetate film was found with this study. Their [15]N nuclear magnetic resonance (NMR) scans [50] provide supporting evidence for the electrostatic interactions by showing shifts in the peak correlated with the amine group of the HCl-prepared neutralized film to a position approximately that observed for the unprotonated chitosan powder, thus suggesting the conversion from $-NH_3^+$ back to $-NH_2$ [50] following treatment with NaOH. Thus, the lower height of tanδ peak 2 for neutralized chitosan films observed in this study makes the hypothesis of ionic effects responsible for the relaxation peak as more plausible (and less likely to be the $T_g$).

Preheating film specimens prior to DMA scans did not significantly change the tanδ peak 2 height or position for neat and neutralized films, as shown in Table 5. By contrast, Sakurai et al. [60] found that preheating their chitosan films at 180 °C caused the tanδ peak at 150 °C to subside compared to non-preheated film, and instead a new peak emerged at 205 °C, which they speculated was closer to the true $T_g$, and thus argued the peak at 150 °C was from a pseudo-stable state of the polymer chains. The removal of plasticization effects from water would increase $T_g$; however, as this was not observed here for $T$ (peak 2) of neat and neutralized films, this casts further doubt on the plausibility that the origin of peak 2 being the glass-rubber transition. Moreover, with preheating of neutralized specimens, peak 2 became more observable, indicating a partial masking effect caused by absorbed water and the water-induced relaxation peak.

To further elucidate the nature of the tanδ peak 2, films containing 15% citric acid were prepared homogeneously, with a 2% acetic acid solution. This film was not subsequently heat treated so the crosslinking was ionic. The similarity of the tanδ curves of the homogeneously and heterogeneously prepared CA films sufficiently demonstrated that the shift of the peak 2 center from 165 to 130 °C with the incorporation of citric acid into the films was independent of the film preparation method. See Figure A2 in the Appendix A. This further supports the notion that the high temperature peak is from ionic effects, as citric acid will be in its conjugate form in a film crosslinked homogeneously. The difference between the viscoelastic properties of a chitosan acetate film and a chitosan citrate film is that preheating has a more significant effect on $T$ (peak 2) for the latter than the former.

**Table 5.** Peak Deconvolution, Fitting Data for tanδ Peak 2 (see also Tables 2–4).

| Film Type | Preheat | Parameter | Baseline | Center (°C) | FWHM | Height |
|---|---|---|---|---|---|---|
| Neat | N | mean | 0.068 [a] | 166.0 [a] | 36.5 [a] | 0.130 [a] |
|  |  | ± | 0.005 | 2.9 | 2.6 | 0.034 |
| Neat | Y | mean | 0.052 [b] | 177.2 [b] | 29.9 [a] | 0.116 [a] |
|  |  | ± | 0.008 | 6.5 | 4.4 | 0.047 |
| Neutralized | N | mean | 0.057 [b] | 188.8 [c] | 105.5 [b] | 0.021 [b,c] |
|  |  | ± | 0.002 | 12.2 | 29.1 | 0.002 |
| Neutralized | Y | mean | 0.029 [c] | 188.5 [c] | 67.9 [c] | 0.048 [c] |
|  |  | ± | 0.006 | 3.4 | 4.8 | 0.008 |
| CA+HT | N | mean | 0.046 [b] | 137.6 [d] | 56.8 [c] | 0.013 [b] |
|  |  | ± | 0.002 | 5.7 | 3.4 | 0.002 |
| CA+HT | Y | mean | 0.045 [b] | 168.3 [a] | 32.2 [a] | 0.006 [b] |
|  |  | ± | 0.006 | 4.1 | 7.6 | 0.004 |
| CA | N | mean | 0.046 [b] | 146.2 [d] | 44.7 [a] | 0.024 [b] |
|  |  | ± | 0.004 | 2.0 | 5.5 | 0.004 |
| CA | Y | mean | 0.045 [b] | 173.3 [a,b] | 44.1 [a] | 0.012 [b] |
|  |  | ± | 0.005 | 13.3 | 27.8 | 0.008 |

Significant statistical difference between means is indicated by different superscripted letters (LSD test, $\alpha = 0.1$). FWHM: full width at half maximum.

## 4. Conclusions

The viscoelastic properties, namely the $E'$, $E''$, and tanδ of chitosan films were investigated to gain insights on the presence of potential covalent crosslinking between citric acid and chitosan. According to the rubbery elasticity theory for crosslinked polymers, the storage modulus is expected to be higher as the polymer approaches and enters into the rubbery region, due to a lower molecular weight between crosslinks. While the effect of crosslinking for the model GTA films was demonstrated with respect to $E'$, no statistical difference was observed for $E'$ between CA, CA-HT, and neutralized chitosan films near 200 °C suggesting that the post thermal treatment of CA films did not induce covalent crosslinking and the interaction between the chitosan amine and CA remained ionic. Furthermore, the difference in $E'$ at temperatures above 170 °C for the preheated and non-preheated film specimens was negligible indicating that moisture did not affect the structure of the chitosan films entering the rubbery region. The tanδ peak at high temperature (peak 2) was also used as an indicator for crosslinking and the glass transition. The tanδ peak 2 of GTA-HOM films shifted to higher temperatures which seemed supportive of an increase in glass transition temperature. However, other changes to tanδ from neutralization and the addition of citric acid indicate that the tanδ peak 2 is more likely related to ionic bonding. Peak fitting analysis of tanδ peak 2 showed that neither the presence of citric acid nor thermal treatment of CA films resulted in the broadening of the peak as would have been expected for an increase in distribution of relaxation times with crosslinking. This was also confirmed by a quantitative and qualitative similarity of the tanδ plots of homogeneously and heterogeneously prepared chitosan-citrate and CA films. The high temperature tanδ relaxation peak shifted from 170 °C for the neat film down to 130 °C for the CA films, irrespective of the CA film preparation method (homogeneous or heterogeneous). Thus, the DMA measurements seemed to confirm that the high temperature relaxation peak near 170 °C is phenomenologically connected with the ionic state of the polymer, and not an indication of the glass transition. However GTA-HET films did show the emergence of a weakly defined peak above 200 °C, which is more likely to be correlated with the glass transition than any of the other films. Most notable is that these DMA tests confirmed that the heterogeneous method of producing chitosan films with citric acid has potential, just as it is with other crosslinkers such as GTA, epichlorohydrin, and genipin. This allows for future considerations on how to properly induce covalent crosslinking with citric acid using the heterogeneous procedure. This could be achieved by optimizing heat treatment conditions or utilizing a phosphate-based catalyst.

**Author Contributions:** J.K. conceived, designed, and performed the experiments with input from C.M. and A.P., who supervised the research. The paper was written by J.K. with contributions from C.M. and A.P.

**Funding:** Financial support from the Natural Sciences and Engineering Research Council (NSERC) of Canada is gratefully acknowledged.

**Conflicts of Interest:** The authors declare no conflicts of interest.

## Appendix A

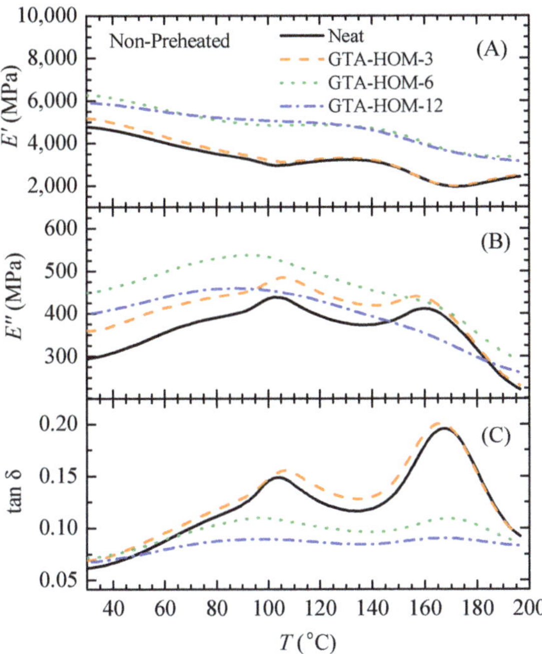

**Figure A1.** The plots of (**A**) storage modulus, (**B**) loss modulus, and (**C**) tanδ against temperature of non-preheated 3, 6, and 12% GTA-HOM films and a neat chitosan film.

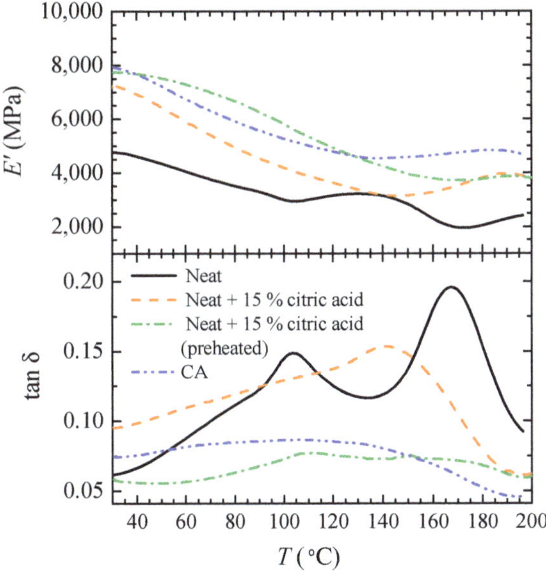

**Figure A2.** The plots of (**A**) storage modulus and (**B**) tanδ against temperature of non-preheated specimens of neat film, a neat film with 15% citric acid, a CA film, and a preheated neat film with 15% citric acid.

## References


1. Wang, H.; Qian, J.; Ding, F. Emerging Chitosan-Based Films for Food Packaging Applications. *J. Agric. Food Chem.* **2018**, *66*, 395–413. [CrossRef] [PubMed]
2. Rhim, J.; Weller, C.; Ham, K. Characteristics of Chitosan Films as Affected by the Type of Solvent Acid. *Food Sci. Food Biotechnol.* **1998**, *7*, 263–268.
3. Park, S.; Marsh, K.; Rhim, J. Characteristics of Different Molecular Weight Chitosan Films Affected by the Type of Organic Solvents. *J. Food Sci.* **2002**, *67*, 194–197. [CrossRef]
4. Blair, H.; Guthrie, J.; Law, T.; Turkington, P. Chitosan and Modified Chitosan Membranes: I. Preparation and Characterisation. *J. Appl. Polym. Sci.* **1987**, *33*, 641–656. [CrossRef]
5. Chen, R.; Lin, J.; Yang, M. Relationships Between the Chain Flexibilities of Chitosan Molecules and the Physical Properties of Their Casted Films. *Carbohydr. Polym.* **1994**, *24*, 41–46. [CrossRef]
6. Wong, D.; Gastineau, F.; Gregorski, K.; Tillin, S.; Pavlath, A. Chitosan-Lipid Films: Microstructure and Surface Energy. *J. Agric. Food Chem.* **1992**, *40*, 540–544. [CrossRef]
7. Srinivasa, P.; Ramesh, M.; Tharanathan, R. Effect of Plasticizers and Fatty Acids on Mechanical and Permeability Characteristics of Chitosan Films. *Food Hydrocoll.* **2007**, *21*, 1113–1122. [CrossRef]
8. Möller, H.; Grelier, S.; Pardon, P.; Coma, V. Antimicrobial and Physicochemical Properties of Chitosan-HPMC-Based Films. *J. Agric. Food Chem.* **2004**, *52*, 6585–6591. [CrossRef] [PubMed]
9. Bordenave, N.; Grelier, S.; Coma, V. Hydrophobization and Antimicrobial Activity of Chitosan and Paper-Based Packaging Material. *Biomacromolecules* **2010**, *11*, 88–96. [CrossRef]
10. Höhne, S.; Frenzel, R.; Heppe, A.; Simon, F. Hydrophobic Chitosan Microparticles: Heterogeneous Phase Reaction with Hydrophobic Carbonyl Reagents. *Biomacromolecules* **2007**, *8*, 2051–2058. [CrossRef]
11. Schreiber, S.; Bozell, J.; Hayes, D.; Zivanovic, S. Introduction of Primary Antioxidant Activity to Chitosan for Application as a Multifunctional Food Packaging Material. *Food Hydrocoll.* **2013**, *33*, 207–214. [CrossRef]
12. Wu, C.; Tian, J.; Li, S.; Wu, T.; Hu, Y.; Chen, S.; Sugawara, T.; Ye, X. Structural Properties of Films and Rheology of Film-Forming Solutions of Chitosan Gallate for Food Packaging. *Carbohydr. Polym.* **2016**, *146*, 10–19. [CrossRef]
13. Yoshida, C.; Oliveira, E.; Franco, T. Chitosan Tailor-Made Films: The Effects of Additives on Barrier and Mechanical Properties. *Packag. Technol. Sci.* **2009**, *22*, 161–170. [CrossRef]
14. Bof, M.; Bordagaray, V.; Locaso, D.; García, M. Chitosan Molecular Weight Effect on Starch-Composite Film Properties. *Food Hydrocoll.* **2015**, *51*, 281–294. [CrossRef]
15. Rubentheren, V.; Ward, T.; Chee, C.; Nair, P.; Salami, E.; Fearday, C. Effects of Heat Treatment on Chitosan Nanocomposite Film Reinforced with Nanocrystalline Cellulose and Tannic Acid. *Carbohydr. Polym.* **2016**, *140*, 202–208. [CrossRef]
16. Jimenez, A.; Fabra, M.J.; Talens, P.; Chiralt, A. Edible and Biodegradable Starch Films: A Review. *Food Bioprocess Technol.* **2012**, *5*, 2058–2076. [CrossRef]
17. Guilbert, S.; Gontard, N.; Gorris, L.G.M. Prolongation of the Shelf-Life of Perishable Food Products using Biodegradable Films and Coatings. *LWT Food Sci. Technol.* **1996**, *29*, 10–17. [CrossRef]
18. Miller, K.S.; Krochta, J.M. Oxygen and Aroma Barrier Properties of Edible Films: A Review. *Trends Food Sci. Technol.* **1997**, *8*, 228–237. [CrossRef]
19. Wan, Y.; Creber, K.; Peppley, B.; Bui, V. Ionic Conductivity and Related Properties of Crosslinked Chitosan Membranes. *J. Appl. Polym. Sci.* **2003**, *89*, 306–317. [CrossRef]
20. Pratt, D.; Wilson, L.; Kozinski, J. Preparation and Sorption Studies of Glutaraldehyde Cross-Linked Chitosan Copolymers. *J. Colloid Interface Sci.* **2013**, *395*, 205–211. [CrossRef]
21. Rivero, S.; García, M.; Pinotti, A. Heat Treatment to Modify the Structural and Physical Properties of Chitosan-Based Films. *J. Agric. Food Chem.* **2012**, *60*, 492–499. [CrossRef] [PubMed]
22. Muzzarelli, R. Genipin-Crosslinked Chitosan Hydrogels as Biomedical and Pharmaceutical Aids. *Carbohydr. Polym.* **2009**, *77*, 1–9. [CrossRef]
23. Pandis, C.; Madeira, S.; Matos, J.; Kyritsis, A.; Mano, J.; Ribelles, J. Chitosan-Silica Hybrid Porous Membranes. *Mater. Sci. Eng. C* **2014**, *42*, 553–561. [CrossRef] [PubMed]
24. Coma, V.; Sebti, I.; Pardon, F.; Pichavant, F.; Deschamps, A. Film Properties from Crosslinking of Cellulosic Derivatives with a Polyfunctional Carboxylic Acid. *Carbohydr. Polym.* **2003**, *51*, 265–271. [CrossRef]

25. Ghanbarzadeh, B.; Almasi, H.; Entezami, A. Improving the Barrier and Mechanical Properties of Corn Starch-Based Edible Films: Effect of Citric Acid and Carboxymethyl Cellulose. *Ind. Crops Prod.* **2011**, *33*, 229–235. [CrossRef]

26. Priyadarshi, R.; Sauraj; Kumar, B.; Negi, Y. Chitosan Film Incorporated with Citric Acid and Glycerol as an Active Packaging Material for Extension of Green Chilli Shelf Life. *Carbohydr. Polym.* **2018**, *195*, 329–338. [CrossRef]

27. Welch, C.; Andrews, B. Ester Crosslinks: A Route to High Performance Nonformaldehyde Finishing of Cotton. *Text. Chem. Color.* **1989**, *21*, 13–17.

28. Gawish, S.; Abo El-Ola, S.; Ramadan, A.; Abou El-Kheir, A. Citric Acid Used as a Crosslinking Agent for the Grafting of Chitosan onto Woolen Fabric. *J. Appl. Polym. Sci.* **2012**, *123*, 3345–3353. [CrossRef]

29. Demitri, C.; Del Sole, R.; Scalera, F.; Sannino, A.; Vasapollo, G.; Maffezzoli, A.; Ambrosio, L.; Nicolais, L. Novel Superabsorbent Cellulose-Based Hydrogels Crosslinked with Citric Acid. *J. Appl. Polym. Sci.* **2008**, *110*, 2453–2460. [CrossRef]

30. Fahmy, H.; Fouda, M. Crosslinking of Alginic Acid/Chitosan Matrices Using Polycarboxylic Acids and Their Utilization for Sodium Diclofenac Release. *Carbohydr. Polym.* **2008**, *73*, 606–611. [CrossRef]

31. Varshosaz, J.; Alinagari, R. Effect of Citric Acid as Cross-Linking Agent on Insulin Loaded Chitosan Microspheres. *Iran. Polym. J.* **2005**, *7*, 647–656.

32. Cui, Z.; Beach, E.; Anastas, P. Modification of Chitosan Films with Environmentally Benign Reagents for Increased Water Resistance. *Green Chem. Lett. Rev.* **2011**, *4*, 35–40. [CrossRef]

33. Salam, A.; Venditti, R.; Pawlak, J.; El-Tahlawy, K. Crosslinked Hemicellulose Citrate-Chitosan Aerogel Foams. *Carbohydr. Polym.* **2011**, *84*, 1221–1229. [CrossRef]

34. Olsson, E.; Hedenqvist, M.; Johansson, C.; Järnström, L. Influence of Citric Acid and Curing on Moisture Sorption, Diffusion and Permeability of Starch Films. *Carbohydr. Polym.* **2013**, *94*, 765–772. [CrossRef]

35. Reddy, N.; Yang, Y. Citric Acid Cross-Linking of Starch Films. *Food Chem.* **2010**, *118*, 702–711. [CrossRef]

36. Ritthidej, G.; Phaechamud, T.; Koizumi, T. Moist Heat Treatment on Physicochemical Change of Chitosan Salt Films. *Int. J. Pharm.* **2002**, *232*, 11–22. [CrossRef]

37. Tual, C.; Espuche, E.; Escoubes, M.; Domard, A. Transport Properties of Chitosan Membranes: Influence of Crosslinking. *J. Polym. Sci. Part B Polym. Phys.* **2000**, *38*, 1521–1529. [CrossRef]

38. Hsien, T.; Rorrer, G. Heterogeneous Cross-Linking of Chitosan Gel Beads: Kinetics, Modeling, and Influence on Cadmium Ion Adsorption Capacity. *Ind. Eng. Chem. Res.* **1997**, *36*, 3631–3638. [CrossRef]

39. Park, J.; Park, J.; Ruckenstein, E. Thermal and Dynamic Mechanical Analysis of PVA/MC Blend Hydrogels. *Polymer* **2001**, *42*, 4271–4280. [CrossRef]

40. Galietta, G.; di Gioia, L.; Guilbert, S.; Cuq, B. Mechanical and Thermomechanical Properties of Films Based on Whey Proteins as Affected by Plasticizer and Crosslinking Agents. *J. Dairy Sci.* **1998**, *81*, 3123–3130. [CrossRef]

41. Demirgöz, D.; Elvira, C.; Mano, J.; Cunha, A.; Piskin, E.; Reis, R. Chemical Modification of Starch Based Biodegradable Polymeric Blends: Effects on Water Uptake, Degradation Behaviour and Mechanical Properties. *Polym. Degrad. Stab.* **2000**, *70*, 161–170. [CrossRef]

42. Silva, S.; Caridade, S.; Mano, J.; Reis, R. Effect of Crosslinking in Chitosan/Aloe Vera-Based Membranes for Biomedical Applications. *Carbohydr. Polym.* **2013**, *98*, 581–588. [CrossRef]

43. Casey, L.; Wilson, L. Investigation of Chitosan-PVA Composite Films and Their Adsorption Properties. *J. Geosci. Environ. Prot.* **2015**, *3*, 78–84. [CrossRef]

44. Roberts, G.; Taylor, K. The Formation of Gels by Reaction of Chitosan with Glutaraldehyde. *Macromol. Chem. Phys.* **1989**, *190*, 951–960. [CrossRef]

45. Pizzoli, M.; Ceccorulli, G.; Scandola, M. Molecular Motions of Chitosan in the Solid State. *Carbohydr. Polym.* **1991**, *222*, 205–213. [CrossRef]

46. Viciosa, M.; Dionisio, M.; Silva, R.; Reis, R.; Mano, J. Molecular Motions in Chitosan Studied by Dielectric Relaxation Spectroscopy. *Biomacromolecules* **2004**, *5*, 2073–2078. [CrossRef]

47. Nielsen, L.; Landel, R. *Mechanical Properties of Polymers and Composites*, 2nd ed.; Marcel Dekker, Inc.: New York, NY, USA, 1994.

48. Krongauz, V. Diffusion in Polymers Dependence on Crosslink Density. *J. Therm. Anal. Calorim.* **2010**, *102*, 435–445. [CrossRef]

49. Mitchell, J. The Rheology of Gels. *J. Texture Stud.* **1980**, *11*, 315–337. [CrossRef]

50. Gartner, C.; López, B.; Sierra, L.; Graf, R.; Spiess, H.; Gaborieau, M. Interplay between Structure and Dynamics in Chitosan Films Investigated with Solid-State NMR, Dynamic Mechanical Analysis, and X-ray Diffraction. *Biomacromolecules* **2011**, *12*, 1380–1386. [CrossRef]
51. Sakurai, K.; Takugi, M.; Takahashi, T. Crystal Structure of Chitosan. I. Unit Cell Parameters. *Sen'i Gakkaishi* **1984**, *40*, 246–253. [CrossRef]
52. Ferry, J. *Viscoelastic Properties of Polymers*, 3rd ed.; John Wiley & Sons Inc.: Hoboken, NJ, USA, 1980.
53. Mattei, G.; Ahluwalia, A. Sample, testing and analysis variables affecting liver mechanical properties: A review. *Acta Biomater.* **2016**, *45*, 60–71. [CrossRef] [PubMed]
54. Bartolini, L.; Iannuzzi, D.; Mattei, G. Comparison of frequency and strain-rate domain mechanical characterization. *Sci. Rep.* **2018**, *8*, 13697. [CrossRef] [PubMed]
55. Oyen, M.L. Nanoindentation of biological and biomimetic materials. *Exp. Technol.* **2013**, *37*, 73–87. [CrossRef]
56. van Hoorn, H.; Kurniawan, N.A.; Koenderink, G.H.; Iannuzzi, D. Local dynamic mechanical analysis for heterogeneous soft matter using ferrule-top indentation. *Soft Matter.* **2016**, *12*, 3066–3073. [CrossRef] [PubMed]
57. Lim, L.; Khor, E.; Ling, C. Effects of Dry Heat and Saturated Steam on the Physical Properties of Chitosan. *J. Biomed. Mater. Res.* **1999**, *48*, 111–116. [CrossRef]
58. Neto, C.; Giacometti, J.; Job, A.; Ferreira, F.; Fonseca, J. Thermal Analysis of Chitosan Based Networks. *Carbohydr. Polym.* **2005**, *62*, 97–103. [CrossRef]
59. Correia, D.; Gámiz-González, M.; Botelho, G.; Vidaurre, A.; Gomez Ribelles, J.; Lanceros-Mendez, A.; Sencadas, V. Effect of Neutralization and Cross-Linking on the Thermal Degradation of Chitosan Electrospun Membranes. *J. Therm. Anal. Calorim.* **2014**, *117*, 123–130. [CrossRef]
60. Sakurai, K.; Maegawa, T.; Takahashi, T. Glass Transition Temperature of Chitosan and Miscibility of Chitosan/Poly(N-Vinyl Pyrrolidone) Blends. *Polymer* **2000**, *41*, 7051–7056. [CrossRef]

 *processes*

*Review*

# On the Use of Starch in Emulsion Polymerizations

Shidan Cummings [1], Yujie Zhang [1], Niels Smeets [2], Michael Cunningham [3] and Marc A. Dubé [1,*]

[1] Department of Chemical and Biological Engineering, Centre for Catalysis Research and Innovation, University of Ottawa, 161 Louis Pasteur Private, Ottawa, ON K1N 6N5, Canada; shidan.cummings19@gmail.com (S.C.); zhangyujie0603@gmail.com (Y.Z.)
[2] Department of Chemical Engineering, McMaster University, 1280 Main Street West, Hamilton, ON L8S 4L7, Canada; nmbsmeets@gmail.com
[3] Department of Chemical Engineering, Queens University, 19 Division Street, Kingston, ON K7L 3N6, Canada; michael.cunningham@queensu.ca
* Correspondence: marc.dube@uottawa.ca

Received: 17 January 2019; Accepted: 28 February 2019; Published: 6 March 2019

**Abstract:** The substitution of petroleum-based synthetic polymers in latex formulations with sustainable and/or bio-based sources has increasingly been a focus of both academic and industrial research. Emulsion polymerization already provides a more sustainable way to produce polymers for coatings and adhesives, because it is a water-based process. It can be made even more attractive as a green alternative with the addition of starch, a renewable material that has proven to be extremely useful as a filler, stabilizer, property modifier and macromer. This work provides a critical review of attempts to modify and incorporate various types of starch in emulsion polymerizations. This review focusses on the method of initiation, grafting mechanisms, starch feeding strategies and the characterization methods. It provides a needed guide for those looking to modify starch in an emulsion polymerization to achieve a target grafting performance or to incorporate starch in latex formulations for the replacement of synthetic polymers.

**Keywords:** Starch; graft; polymerization; emulsion; polysaccharide

---

## 1. Introduction

To achieve a sustainable society, we must intensify responsible agriculture and industry and reduce our greenhouse gas (predominantly $CO_2$) emissions and use of non-biodegradable and single-use materials. The polymer industry has a major role to play in this shift in practice [1]. A bio-based economy focusses on extracting useful organic materials from renewable sources such as starch and other saccharides. It is expected that the proportion of plastic products that are bio-based in the market will increase each year and that up to 20% of all chemicals will be bio-based by 2020 [2]. The bioplastics market had an annual growth rate of 40% worldwide and 50% in Europe from 2003 to 2007 and is currently growing at 20–30% per year [3]. Starch-based polymers accounted for 70% of the bio-based market in 2003 (equivalent to 25,000 tons), increasing to 114,000 tons in 2007 and is projected to increase to 810,000 tons in 2020 [4]. In 2009, 75% of the starch-based plastics market was located in Europe [3]. There is a wide range of technologies used for the production of starch-based products, including polymerization reactions, blending and thermoforming and that range from those at the research and development stage to fully commercialized processes. Growth has been slower than that of biofuels, however, due to a lack of government incentives [2].

Emulsion polymerization, as opposed to bulk and solution polymerization, is a method that provides enhanced heat transfer by using water as a dispersion medium, allows for high molecular weight products at simultaneously high reaction rates, enhanced control over polymer morphology and structure, as well as resulting in a final latex with a viscosity independent of molecular weight [5,6]. Paints and coatings are the largest uses of latexes, with 60% being for decorative uses and 10% for industrial applications. Other films are applied for laminates, as construction adhesives or pressure sensitive adhesives. Perhaps most famously, the first large volume commercial product from an emulsion polymerization was styrene butadiene latex rubber as a replacement for natural rubber during World War II [7]. Latexes have also been used to improve the strength and durability of asphalt and concrete, as coatings for textiles, films for packaging, pellets or gels for moulding, as additives for printing inks, as templates for nanostructured optoelectronic materials and as the key ingredient in toner, waxes, polymer alloys, caulks, sealants, engine fluids, pigments and carpet backing [8–11]. Niche applications include drug delivery, whereby the active molecule is embedded into or immobilized onto the latex polymer particle, calibration material for analytical equipment, as coatings for magnetic recording media and for purification of biological formulations [7,12]

From a green chemistry perspective, it would be ideal to have the entire polymer formulation sustainably sourced and produced. One strategy is to ensure that monomers (feedstock for polymerizations) are synthesized from non-petroleum-based feedstock. Fillers and property modifiers should also be sourced from natural and sustainable feedstock where possible. Biopolymers such as polysaccharides offer naturally occurring and sustainably grown materials to act as fillers or macromers as a substitute for synthetic, petroleum-based monomers in polymer formulations. Polysaccharides have been extensively studied and are among the most abundant, versatile, inexpensive and commercially utilized biopolymers [5,13–32].

Starch is a polysaccharide material consisting of amylose and amylopectin arranged in a granular structure with both amorphous and crystalline domains. Starches have found widespread application in many industries for use as food additives, pigment binders in paper production and rheology modifiers for pharmaceutical and textile industries [17,33,34]. Starch can be suspended or, after its granular structure is removed, dissolved in water and its properties are tuneable by physical or chemical modification. When combined with other materials, modified starches impart a wide range of properties to a polymer composite depending on the starch type, its pre-treatment and functionalization [15,28,34–43]. As a filler or reactive co-polymer, starch can be used to displace reasonable amounts of synthetic material, reducing the products' dependence on petroleum based feedstock, making the overall process more sustainable. Overcoming certain undesirable properties of starch (i.e., tensile strength, hardness, hygroscopicity), as well as compatibility issues between raw or modified starch and synthetic polymers, is essential for the effective use of starch in industrial latex formulations.

While starch is used extensively as a substrate for graft polymerizations, its use in polymer latexes is an emerging field of interest [44,45]. Because of the many challenges inherent in using starch in water-based formulations in combination with hydrophobic materials, innovative strategies are required. This work focuses on the use of starch as a filler, colloidal stabilizer and reactive polymer in emulsion polymerizations, as well as the feeding strategies that have been employed to incorporate various modified starches into the latex, the effect of reaction variables on grafting performance and final latex properties and finally, on the analysis and characterization methods used to determine grafting performance and other properties. In this review we highlight instances where elevated rates of incorporation of starch and especially of encapsulation, have been reported.

## 2. Polymer Latex

### 2.1. Emulsion Polymerization

Emulsion polymerization is a form of heterogeneous free radical chain polymerization, where hydrophobic polymer particles are formed in an aqueous dispersion medium [46]. It satisfies, to a high degree, the 12 principles of green chemistry: as in solution polymerization, the use of small amounts of free radical initiator overcomes the need for excess stoichiometric reagents (principle 9); greater control over molecular weight and final polymer properties at high conversion, relative to solution polymerization, reduces the amount of off-spec and waste materials (principle 1); the use of water as the reaction medium removes the need for solvent use and subsequent evaporation, thus enhancing process safety and reducing energy costs (principles 3, 4 and 5); and lower viscosity profiles lead to enhanced heat transfer and therefore, lower energy requirements (principle 11) [1,12]. The addition of sustainably produced and sourced monomers and fillers (principles 4, 7 and 10) further enhances the "greenness" of the process by displacing typically petroleum-based monomers, giving the overall formulation a smaller environmental footprint.

The mechanism of emulsion polymerization typically begins with the formation of micelles (aggregates of surfactant with their hydrophobic tails oriented inwards) in the aqueous reaction media, which occurs when the surfactant molecules reach their critical micelle concentration (CMC). A significant amount of surfactant remains dissolved in the water phase. When a hydrophobic monomer is added, the monomer partitions into three loci: (1) large monomer droplets that are stabilized by the excess surfactant, (2) micelles and (3) dissolved in the aqueous phase (as a small percentage). Water-soluble initiator is added to the water phase where it slowly dissociates into free radicals during the polymerization. The free radicals react with the monomer dissolved in the aqueous phase to form water-soluble oligomeric radicals. When the oligomeric radicals reach a critical chain length to render them sufficiently hydrophobic, they enter the monomer-swollen micelles, thus nucleating a polymer particle. Due to the large surface area occupied by the micelles compared to the monomer droplets, the nucleation of micelles is the dominant particle nucleation mechanism. The entry of oligomeric radicals into the micelles is called heterogeneous or micellar nucleation and is dominant when surfactant is present above its CMC. Homogeneous particle nucleation can also occur (especially when the surfactant concentration is below its CMC) where the low molecular weight oligomers reach their insolubility limit and precipitate. They are then stabilized by surfactant and become polymer particles. Other ingredients such as buffer, chain transfer agent and crosslinker can be used to modify the reaction kinetics and latex properties. Final latexes can be used as produced or with the addition of post-polymerization additives.

The polymerization can be further described by three reaction intervals (I, II, III) (Figure 1). Interval I comprises nucleation of polymer particles and sees an increase particle number (and consequently the polymerization rate) over time. By the end of this interval (5–15 wt.% monomer conversion), monomer-swollen micelles are no longer present, having been nucleated or the surfactant distributed and absorbed onto the growing polymer particles; the number of particles (and consequently polymerization rate) becomes constant. During interval II, a thermodynamic equilibrium is maintained that keeps the monomer concentration in the polymer particles constant by the continuous transport of monomer from the monomer droplets to the particles. Thus, the polymerization rate remains constant (and no new particles are nucleated), while the polymer particles increase in size and the monomer droplets are depleted. The end of interval II is signalled by the disappearance of the monomer droplets. Interval III begins anywhere between 40–80 wt.% monomer conversion. With the particle number remaining constant and a lack of new monomer added to the system, the polymer concentration in each particle increases and the polymerization rate decreases until full conversion is reached. Final latex polymer particles are typically 50–300 nm in diameter, with a latex viscosity ranging from 30–1000 cp and solids content from 30–70 wt.%. Several particle morphologies are possible such as core-shell, occluded and moon-like. The morphology is dependent on the hydrophobicity and reactivity of the monomers, the glass transition temperature ($T_g$) of the resultant polymers, the presence of fillers and macromers, the distribution of free radicals within the composite polymer particle and the degree of polymerization between components. Polymerization conditions such as formulation, semi-batch feed strategy, initiation method and temperature are also factors that affect morphology [7,47–50].

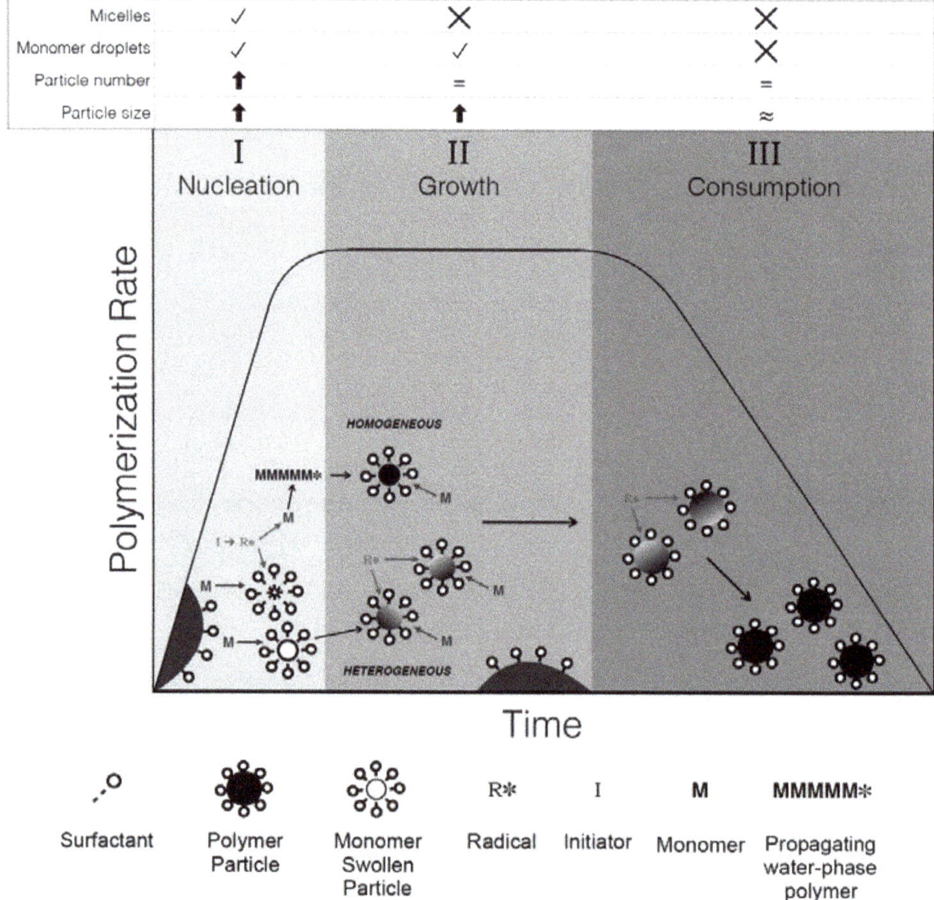



**Figure 1.** Mechanisms of the three intervals of an emulsion polymerization.

## 2.2. Film Formation

The mechanism of film formation is generally understood to consist of three stages: drying, deformation and coalescence (Figure 2) [47,51–53]. Drying begins once the latex is evenly spread upon a substrate, after which water evaporates at a constant rate until a polymer volume fraction of 0.6–0.75 is reached. During this stage, the particles form a coordinated array, the specific orientation of which depends on ionic strength, viscosity and other factors. Deformation occurs as the evaporation rate slows and particles become closely packed; this occurs above the minimum film formation temperature (MFFT). A thin film may first form at the surface, causing further evaporation to occur by diffusion of water through this polymer "skin." Particles then contact each other and start to deform to polyhedral structures through air-water, water-polymer and polymer-air interfacial tensions, osmotic forces and surface adhesive forces. Finally, polymer chains diffuse across particle boundaries to reduce surface energy and the spheres coalesce into a continuous film. The type of surfactant used to stabilize the particles can affect their mobility and the subsequent film's uniformity. The use of additives that modify the density and surface properties of the particles, as well as fillers that alter the zeta-potential with hairy layers or introduce layers of polymer with varying glass transition temperatures, will most likely have an effect not only on the deformation and coalescence of the particles but also on the homogeneity of the final film, as well as its sensitivity to environmental factors.

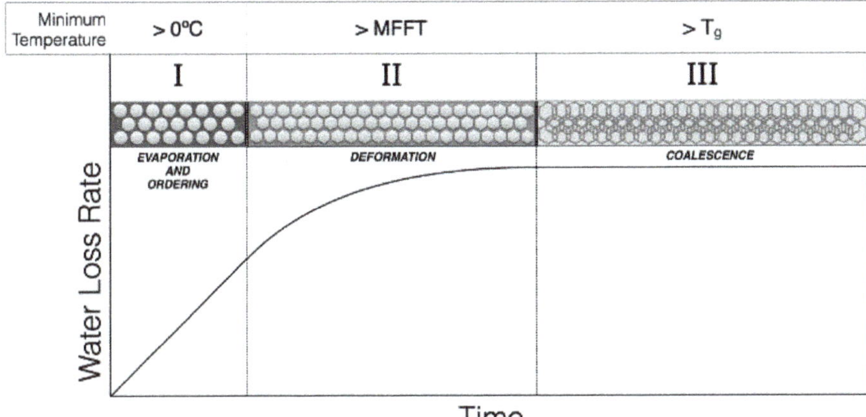

**Figure 2.** Film formation mechanism of polymer latex.

## 3. Starch Properties and Modification

### *3.1. Structure and Properties*

Starch is considered a low cost feedstock that offers an economically viable alternative to petroleum sourced materials (e.g., synthetic monomers) [4]. The two largest producers of maize starch worldwide are the United States of America and China. The use of starch as a source of non-food products such as ethanol has increased prices and raised concerns over whether or not to divert food resources to fuels and materials [54].

Two types of polysaccharides make up starch, amylose and amylopectin. Amylose is linear with anhydroglucose repeat units connected by α-1,4-D-glucoside bonds, while amylopectin is branched and contains α-1,4-D-glucoside bonded glucose repeat units with short branches of α-1,4-D-glucopyranose bonded polysaccharide chains at α-1,6 branch points (every ~22 repeat units) (Figure 3) [55].

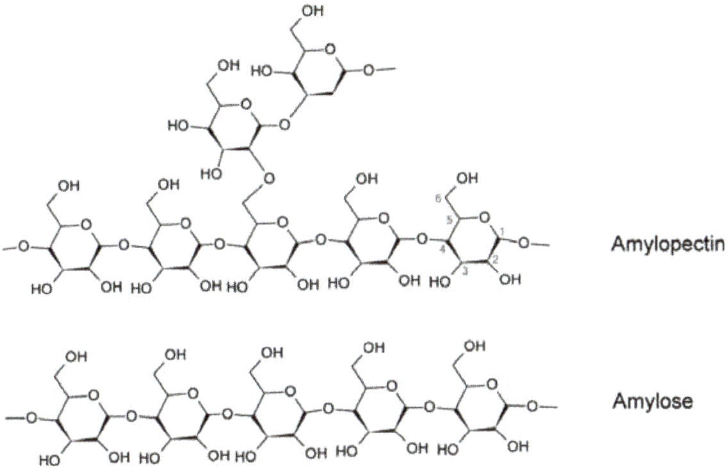

**Figure 3.** Structure of starch with carbon atoms labelled.

The molecular weights of amylose and amylopectin are dependent on the botanical source but are generally in the range of ~$10^5$–$10^6$ g/mol and ~$10^7$–$10^8$ g/mol, respectively [56]. These polymers are naturally occurring and are principally extracted from corn (maize), potato, wheat and rice feedstock. Starch is nontoxic and biocompatible, making it safe for human contact and food applications, while its biodegradable properties make it appealing for sustainable materials [18,57]. The granule and particle size of starch depends on its source (e.g., maize, wheat, potato), number of contaminants (e.g., protein) and morphology (e.g., crystallinity). Starch is first extracted from its source in the form of microscopic granules (2–100 μm) with rings (120–500 nm) of "blocklets" (20–50 nm), each containing alternating layered lamella (9 nm) of crystalline and amorphous regions [5,58,59]. The granules are often colloidally stable in solution, as are other physically modified forms of starch, so long as this stability is not affected by complexing of associable amylose and amylopectin [60]. The crystallinity of extracted starch ranges from 19% in high amylose starch and 39% in waxy maize (low amylose), to 51% in rice starch [59]. Native starch granules have low cold water solubility, are sensitive/responsive to changes in pH, moisture, temperature and mechanical stress [61,62]. Typically, at temperatures above 60 °C, native starch undergoes irreversible gelatinization which begins with swelling of the granules (30 and 100 times by volume for maize and potato, respectively) followed by dissolution into the continuous aqueous phase, forming a viscous solution [17,60]. Once these solutions cool, the linear amylose chains orient themselves into double-helices, while amylopectin crystallizes by attraction of the short branch chains, all of which forms a network that separates from the continuous phase in a process called retrogradation [60,63]. Amylose and amylopectin carry no charge and contain an abundance of hydroxyl functional groups, making them a prime candidate for the addition of customized moieties or for grafting reactions [59].

Starches (particularly native and cooked) suffer from limited solubility, high hydrophilicity, high viscosity, moisture sensitivity and brittleness and thus require some form of physical and chemical modification to make them applicable in polymer composite formulations. At the same time, starch can be viewed as a useful additive to synthetic polymer formulations as a property modifier [26] (e.g., increase oxygen and toluene and decrease water barrier resistance [64]) or to increase the biodegradability [29] and/or sustainability of the final product.

*3.2. Physical and Chemical Treatment*

Native starch granules (2–100 μm) can be used in polymerizations but are typically first swollen and gelatinized through cooking, which removes the granular structure [5,65,66]. "Native" starch is defined herein as starch that is in the granular form and has not undergone a change in molecular weight, crystallinity, composition or functionality (modification of moieties). Distinction will be made between granules, which are not cooked and gelatinized starch, which has been swollen and become soluble in water. Several mechanical and chemical treatment methods exist that modify the properties of starch but do not introduce functionality to the starch structure. These methods, summarized in Table 1, are often used to extract a specific type or molecular weight of amylose or amylopectin from the granules, to affect granule particle size, degree of crystallinity or overall morphology. Many of these methods are combined for multiple effects. Often treatment is followed by or coupled with, chemical functionalization that introduces reactive or functional moieties, which compatibilizes the starch for a particular polymer formulation or application. A popular and effective strategy for achieving high loadings of starch into polymer latexes at acceptable viscosities has been to severely reduce the amylose and amylopectin molecular weights.

The use of crystalline and amorphous nanoparticles has recently garnered significant attention [5,24,26]. SNCs (Starch nano-crystals) are produced by the removal (usually through hydrolysis) of the amorphous regions of native starch, thus exposing the crystalline platelets (>40% crystallinity). These nano-crystals are typically <100 nm in size, degrade at 251 °C and have a broad distribution of viscosity in solution. SNPs (Starch nano-particles), on the other hand, are more amorphous particles where the crystalline regions have been destroyed (<40% crystallinity), typically by mechanical treatment. Regenerated starch nanoparticles (RSNPs) are near completely amorphous (96–98%) due to treatment via extrusion [5]. These amorphous nano-particles range between 100–500 nm, degrade at 285–294 °C and produce a lower viscosity in solution than native starch.

**Table 1.** Summary of some starch physical and chemical modification pathways.

| Method | Pathways | Primary Modification[a] | Product |
|---|---|---|---|
| **Physical** | | | |
| Mechanical extrusion [67] | Heat treatment and mechanical work | Molecular weight | Amorphous pre-gelatinized starch |
| Microfluidization [68] | / | / | SNP[b] |
| Electric field [69] | / | Crystallinity | Amorphous pre-gelatinized starch |
| High pressure [70,71] | Hydrostatic processing Supercritical medium Homogenization Microfluidization | Crystallinity | Amorphous gelatinized starch |
| Irradiation [14,69] | UV λ γ Microwave | Molecular weight | Lower molecular weight starch |
| Ball milling [72] | / | Crystallinity Particle size | Amorphous partially gelatinized starch granules |
| Ultrasonication [73] | 24kHz treatment of starch aqueous solution | Crystallinity | Partially amorphous starch granules |
| Microwave [74] | Treatment of starch paste | Crystallinity | Partially amorphous starch granules |
| Thermal degradation [75] | Treatment of dried starch | Molecular weight | Lower molecular weight starch |
| Re-crystallization [76] | Hydrothermal Heat-moisture treatment | Crystallinity | Crystalline "resistant" starch SNC[c] |
| **Chemical** | | | |
| Acid Hydrolysis [77] | Acid | Crystallinity Branching | SNC Debranched starch |
| Emulsion precipitation | Butanol complexing co-crystallization [78] | Crystallinity | SNC |
| | Non-solvent precipitation crosslinking [79] | Degree of crosslinking | SNP |
| Reactive extrusion [80] | Water-phase crosslinking | Crystallinity Molecular weight Crosslink density | SNP |
| Alkaline Treatment [72] | NaOH | Molecular weight | Low amylose content, lower molecular weight starch |
| **Enzymatic** | | | |
| Hydrolysis [81] | Enzymatic | Crystallinity Branching | SNC Debranched starch |
| Re-crystallization [76] | Enzymatic/melt/crystallization | Crystallinity | SNC |

[a] Particle size is reduced in all methods. [b] Starch nano-particle (SNP): mostly amorphous, in some cases completely so. [c] Starch nano-crystal (SNC): mostly crystalline, with the amorphous regions degraded and removed to differing degrees.

## 3.3. Functionalization

Modification of starch has been performed through functionalization by small molecule chemistry (addition of moieties) or by "grafting from" or "grafting to" approaches. The abundance of hydroxyl groups, the reactivity of which is C-6 > C-2 > C-3 [82], provide sites for addition (the likelihood of each hydroxyl to be functionalized also depends on the specific reaction mechanism). However, certain mechanisms allow for the formation of a bond directly to one of the carbons on the starch backbone. Because of its hydrophilicity, starch typically requires hydrophobization or functionalization to become compatible or reactive with hydrophobic synthetic polymers [59]. The literature provides examples of starch modified by small molecule chemistry with a wide variety of moieties (Figures 4–6), including anhydrides [13,16,61,83–85], esters [16,61,82,83], ethers [16,61,82] and other cationic and anionic functional groups [59,61,82,83,86,87].

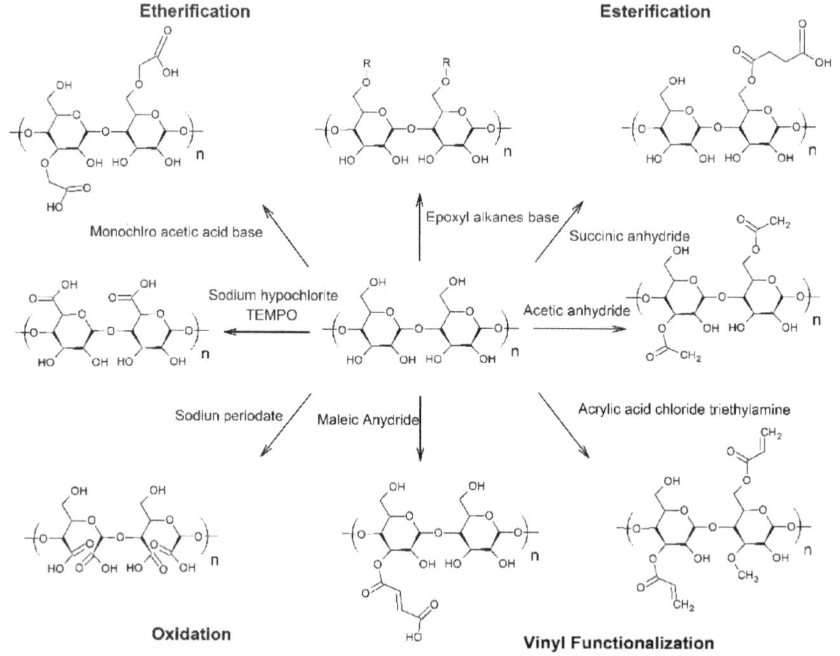

Figure 4. Some examples of starch modification by small molecule chemistry.

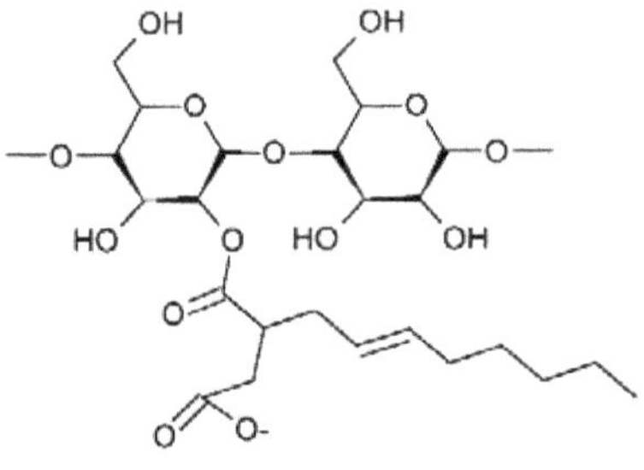

Figure 5. Example of anionic functionalized starch.

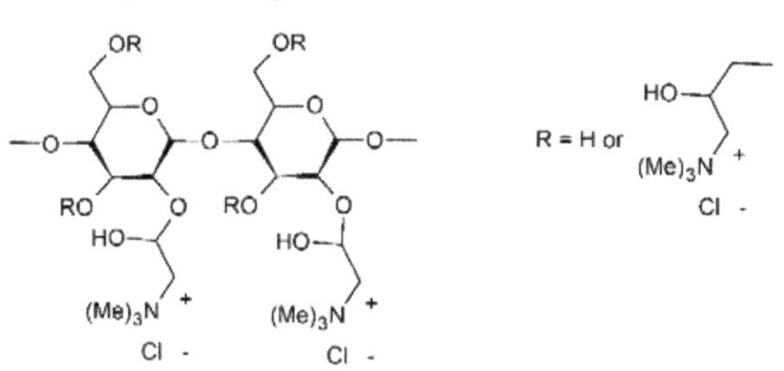

Figure 6. Example of cationic functionalized starch.

## 4. Measuring Grafting Performance

To determine the extent of grafting and the effect of varying formulations and procedures in achieving the desired particle morphology and composition, there are five types of results that are typically sought (Table 2). After graft polymerization has finished, most analytical methods use solvent extraction to separate homopolymer and/or unreacted starch from the graft polymer prior to analysis. Grafting performance can be expressed by grafting efficiency (Equation (1)), percent grafting through add-on (Equation (2)) or ratio (Equation (3)) or the molecular weight of the grafts (Equation (4)). In cases where the incorporation (embedding) of starch in the synthetic polymer matrix is desired, performance may be reported as the wt.% of starch that cannot be removed from the graft polymer by washing with polar solvent. Incorporation wt.% does not imply grafting but rather the success of the polymerization in removing starch from the continuous water phase and into the synthetic polymer solids. Encapsulation describes a near completely covered starch particle, where a core-shell morphology exists with the starch in the core.

**Table 2.** Summary of measurements for grafting performance.

| Variable/Characteristic | Results | Extraction Methods | Analysis Methods |
|---|---|---|---|
| Grafting efficiency | wt.% of grafted polymer in total formulation | Soxhlet extraction | FTIR $^{13}$C-NMR $^{1}$H-NMR Gravimetry |
| Percent Grafting (Ratio) | wt.% ratio of synthetic grafts to the starch to which they are bonded | Centrifugal solvent extraction | |
| Percent Grafting (Add-on) | wt.% of synthetic polymer in total grafted polymer | Hydrolysis | |
| Homopolymerization ratio | wt.% of synthetic monomer that resulted only in homopolymer | | |
| Molecular weight of grafts | Weight or number average g/mol of the grafted synthetic polymer chains | Enzyme or acid hydrolysis of starch within extracted graft polymer will release the grafted synthetic chains | GPC |

$$\text{Grafting ratio (GR)} = (\text{weight of synthetic polymer} \, / \, \text{weight of starch}) \times 100 \tag{1}$$

$$\text{Add-on (AO)} = (\text{weight of synthetic polymer grafted} \, / \, \text{total weight of grafted polymer}) \times 100 \tag{2}$$

$$\text{Grafting efficiency (GE)} = \text{weight of grafted polymer} \, / \, (\text{weight of grafted polymer} + \text{weight of homopolymer}) \times 100 \tag{3}$$

$$\text{AGU/graft} = M_n \text{ of grafted polymer} \times (\text{wt.\% of starch in graft copolymer} \, / \, \text{wt.\% of grafted polymer}) \times (\text{MW of repeat unit})^{-1} \tag{4}$$

Depending on the desired measure of performance for including starch in an emulsion polymerization, either the unreacted starch or ungrafted synthetic polymer will need to be separated for analysis of the remaining graft polymer. In some instances, both will need to be removed. Either can be achieved by solvent extraction, assuming that an appropriate solvent or non-solvent can be identified. Centrifugation is often used to improve separation and is combined with washing of the dried centrifugation precipitate with the same solvent or with a final separation step with a starch non-solvent such as methanol. A Soxhlet extractor is commonly used to separate a mixture of hydrophobically opposite components. In all procedures used to isolate the grafted polymers, there will always be a portion of the grafts that are removed in the separation stage. Some grafts contain mostly starch and are not sufficiently modified (hydrophobic) to separate out with the bulk of the synthetic polymer. Additionally, the presence of hydrophilic polymer (dependent on the formulation) in the decanted solvent can affect gravimetric calculations. In all cases however, the percentage of starch in the final extracted graft polymer compared to the solvent will be underestimated, meaning final results can be considered as a minimum level of performance.

NMR can be used to determine the presence of synthetic grafts or functionality on starch by identifying either the decrease in starch hydroxyl or backbone hydrogen atoms or the presence of synthetic hydrogen or carbon atoms. Solid state $^{13}$C-NMR and $^{1}$H-NMR do not typically reveal high resolution spectra and it is recommended using cross polarized magic angle spinning (CP-MAS). The use of IR to determine grafting in an unpurified latex has limitations as homopolymer and unreacted starch will appear on the resulting spectra [88]. In-line IR analysis to track the change in starch hydroxyl or backbone hydrogens in an emulsion polymerization can yield extremely useful information relating to the rate of grafting and reaction kinetics. Gravimetric methods such as centrifugation, precipitation and solvent extraction can be used to separate the starch graft polymer from either the unreacted starch or the synthetic homopolymer, providing quantitative results on grafting performance.

### 4.1. Characterization of Polymer Products

Once the synthetic homopolymer or unreacted starch is separated from the graft polymer, any of the three samples can be characterized (Table 3).

**Table 3.** Common characterization methods for each component of a starch graft polymer.

| Sample Type | Characteristic | Method | Typical Goal of Study |
|---|---|---|---|
| Graft polymer | Composition | Gravimetry IR/NMR | Achieve target grafting performance or create a new hybrid starch-synthetic particle. |
| | Morphology Crystallinity | (cryo)TEM/SEM XRD | |
| | | Titration | |
| | Degree of substitution (max of 3.0) | Elemental analysis $^1$H-NMR | |
| | | TGA-IR | |
| | MW of grafted chains | GPC | |
| Homopolymer | Molecular weight Degree of branching | GPC HPLC | Effect of reaction variables, formulation or procedure on kinetics of homopolymerization in the presence of starch. |
| Unreacted starch | Morphology Crystallinity Particle size/density Molecular weight | TEM/SEM XRD Flow fractionation DLS GPC | Effect of reaction variables, formulation or procedure on structure of starch in the presence of initiator and vinyl monomer. |

XPS and elemental analysis are alternative methods to NMR and IR to identify and quantify the concentration of chemical bonds and specific functional groups, respectively. Although strongly affected by the properties of the reaction medium (e.g., turbidity), Raman spectroscopy can be used in-line to monitor the signal of a particular bond, similar to in-line FTIR. The location of particles within an emulsion can be observed in-line by confocal laser scanning microscopy (CLSM). Care must be taken in the preparation of starch based samples for TEM/SEM as the water soluble starches can settle on the latex particles during evaporation and create misleading artefacts [89]. The changes in crystallinity of starch towards that of the synthetic polymer or vice versa, provides strong evidence of grafting [43,90–93]. The molecular weight of the synthetic polymer grafts can inform us about the effect of reaction conditions and monomer composition on grafting kinetics and can be determined by GPC. The branching of grafted polymer chains decreases the mobility of the grafted polymer, imparts inhibition of the movement of small molecules towards the backbone of the starch and affects the organization of the polymer molecules (e.g., retrogradation). Various methods including $^1$H-NMR and $^{13}$C-NMR [94–96], HPLC [97] and GPC [98] techniques have been used to quantify the degree or extent of branching of polymers. Asymmetric flow field-flow fractionation, coupled with a DLS device, can quantifiably detect both synthetic latex particles and more swollen and transparent starch particles by first separating them by size prior to sending them to a detector. Other methods used to determine the particle size of grafted polymers include hydrodynamic chromatography (HDC) using a PSD analyser [99], dynamic image analysis (particles in the micron range) and disc centrifugation (>75 nm). The use of DLS exclusively is not recommended due to the high degree of swelling of some starch particles, giving them a refractive index similar to water. Differential centrifugal sedimentation can be applied to measure the density of nanoparticles when the particle size is known from one of these methods.

### 4.2. Implicit Proof of Grafting

The characterization methods described previously allow for explicit analysis of grafting performance. To explain the effect of a grafting performance on a particular final property, additional characterization is often undertaken on the latexes or dried films/resin (Tables 4 and 5). Unless otherwise stated, information in this section is taken from the thesis by Cummings [89].

**Table 4.** Summary of characterization methods for determination of latex properties.

| Characteristic | Analysis Method |
| --- | --- |
| Viscosity | Viscometry |
| Particle size | Asymmetric flow field-flow fractionation coupled with DLS |
| Birefringence | Angle dependent optical reflectometry |
| pH | Electrical potential |
| Conductivity | |
| Zeta Potential | Electrophoretic light-scattering |
| Film formation | Minimum film formation temperature (MFFT) |
| | Visual |
| Shelf life | Visual |
| Grit | Gravimetry |

**Table 5.** Summary of characterization methods for determination of film properties.

| Characteristic | Analysis Method |
| --- | --- |
| **Hydrophobicity** | Contact angle |
| Adhesive Properties | Tack, peel and shear |
| Opacity | Spectrophotometry |
| Water whitening | Spectrophotometry |
| Gel content | Solvent swelling, extraction, gravimetry |
| Solubility | Solvent swelling, extraction, gravimetry |
| Tensile strength | Instron tensile strength |
| Elasticity | |
| Water/oxygen barrier | Water and oxygen permeation |
| Glass transition temperature | DSC |
| Melting temperature | |
| Thermal stability | TGA |
| Biodegradability | Enzyme, culture, composting |

#### 4.2.1. Latex Properties

Viscosity is typically increased by the presence of starch in the aqueous phase of polymer latex and in some cases can cause coagulation. Even in cases where the starch is not completely encapsulated by the synthetic polymer, latex viscosity (100–1000 cp) can be similar to standard formulations. Molecular weight reduction methods have served to decrease the viscosity of starch dispersions [100–102]. Although high viscosity can persist after functionalization and grafting, it is possible to reduce the viscosity by reducing the likelihood of retrogradation, hydrogen bonding and persistent water soluble starches. Crystalline regions of starch contribute to its birefringence properties, which could be weakened or eliminated by deformation and modification [103–105]. Starch loaded latex can still exhibit birefringence in the form of blue or green hues upon cooling to room temperature. Due to the abundance of hydroxyl groups on starch molecules, the pH of the water phase can cause or prevent deprotonation, as well as alter the speed at which granules gelatinize or particles disperse. Although modified starches can exhibit strong zeta potential, a much smaller value is seen with starch-synthetic hybrid latexes. Although this should cause instability, the superior steric stabilization properties of starch prevent this [99,106,107]. The homogeneity of films formed from polymer latex containing starch is a key indicator for the presence of water phase compounds and the compatibility of the latex with the substrate. The increased viscosity and hydrogen bonding potential of starch loaded latexes can improve film formation (depending on the substrate), while any exposed starch once the film is dried can alter the composites sensitivity to humidity and temperature. The long term stability of latexes, also known as the shelf life, is affected both by the electrostatic or steric stability of the individual latex particles, as well as the presence of destabilizing or phase separation inducing molecules in the water phase. Starch loaded latex carry the risk of destabilization and phase separation in the cases of retrogradation or persistent non-grafted (hydrophilic) water soluble starch. The presence of grit (large impurities in the final latex) can indicate poor dispersion of the added starch particles/molecules, the burning of the starch due to accumulation on reactor walls or the general coagulation of ungrafted and unreacted starch material.

### 4.2.2. Film Properties

Even at high levels of incorporation, the presence of small amounts of non-grafted starch on the surface of a polymer film will make it more hydrophilic and increase the contact angle with water droplets. Generally, starch will reduce the adhesive properties of polymer films, especially in dry environments, however shear strength typically increases due to the rigidity, hydrogen bonding and potential crosslinking effect of starch particles. Since the degree of water whitening is dependent on the polarity of the polymer and the diffusivity of water through the matrix, this characteristic can be affected by the presence of starch [108]. The hydrophilicity of starch and the hydrophobicity of the synthetic polymer, combined with crosslinking that may occur during polymerization and intermolecular interactions (hydrogen bonding), all affect the gel and soluble content of the final graft polymer [109]. Starch content increases the polymer film's sensitivity to humidity and polar plasticizers, all of which reduce the tensile strength. In dry environments, the loading of starch in synthetic polymer formulations tends to increase the tensile strength [41,76,110,111]. Starch is hygroscopic and therefore SNPs do not naturally have good moisture barrier properties. SNCs however, due to their crystalline structure, impart much better moisture and gas barrier properties to polymer composites, as long as deformation of gelatinization of the crystalline regions does not occur. In general, starch has extremely good oxygen barrier properties under low moisture conditions [5,64,76]. The precise functionalization, synthetic graft polymer formulation and final composite particle morphology all have a large effect on final film barrier properties. Starch grafted polymers exhibit a change in $T_g$ towards that of native starch, except in cases where the synthetic polymer can plasticize and disrupt the structural hydrogen bonding of the starch chains [15,88]. In some cases, the $T_g$ remains unaffected by synthetic grafts and in others (where grafting is low) two transitions are observed (synthetic polymer and starch). Typically, starch grafted polymers exhibit thermal stability up to 200–400 °C, with the properties of both starch and the synthetic polymer influencing the decomposition temperature [5,15,75,112,113]. Degradation of a polymer can be photoinduced, thermal, chemical or biological, while usually beginning with a reduction in molecular weight followed by digestion by microorganisms. The presence of biodegradable starch in polymer formulations warrants testing of the final product's biodegradation potential. Testing typically begins with enzyme and culture experiments, followed by composting conditions in a reactor [29,31]. Starch is readily biodegradable and typically serves to increase the biodegradation of synthetic formulations, so long as there is a mechanism for radiation, water, enzymes and microorganisms to make contact with the starch chains [114–116].

## 5. Grafting Mechanisms

Most polymerizations involving starch have been initiated through radical chain mechanisms. Both native and functionalized starch have been "grafted from" or "grafted to" using polymerization techniques (Figures 7 and 8) [15]. A "grafting from" approach involves generating a radical on the starch backbone or hydroxyl groups through a covalently bound initiating group, which subsequently reacts with a monomer to initiate the growth of a polymer chain. Alternatively, the chain-end of a pre-formed polymeric/oligomeric radical can react with the starch backbone, hydroxyl groups or other added functionality, "grafting to" the starch molecule. Various initiators, monomers and semi-batch feed strategies have been used to include starch into synthetic polymer latexes [117,118].

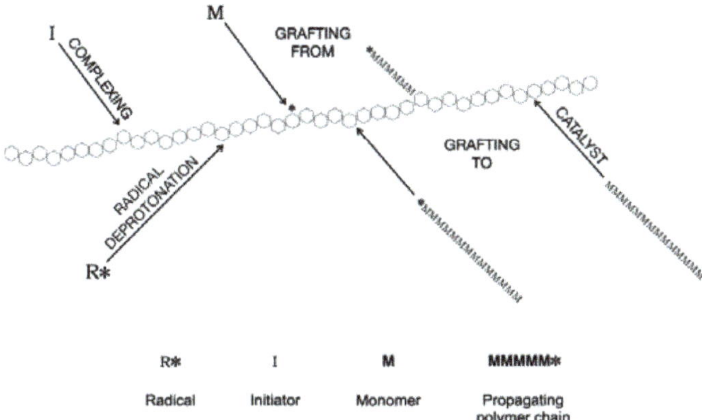


**Figure 7.** Visualization of the "grafting to" and "grafting from" mechanisms. "Grafting to" occurs when propagating or fully reacted polymer chains bond to the starch chains. "Grafting from" describes initiator complexing or radical deprotonation of the anhydroglucose repeat unit, which creates a radical from which monomer can polymerize.

$$St(H) + R^{\cdot} \rightarrow St^{\cdot} + RH \qquad\qquad \text{Initiation}$$

$$St^{\cdot} + M \rightarrow St - M^{\cdot} \qquad\qquad \text{Propagation}$$

$$St - M^{\cdot} + nM \rightarrow St - M_{(n+1)}$$

$$St - M_n^{\cdot} + M(H) \rightarrow St - M_n H + M^{\cdot} \qquad \text{Chain transfer to monomer}$$

$$St - M_n^{\cdot} + St - M_m^{\cdot} \rightarrow St - M_{(n+m)} - St \qquad \text{Termination}$$

$$St - M_n^{\cdot} + M_m^{\cdot} \rightarrow St - M_{(n+m)}$$

where St = starch, R$^{\cdot}$ = radical, M = monomer, M$_n$ and M$_m$ = polymer

**Figure 8.** Initiation, propagation, chain transfer and termination mechanisms from the grafting of vinyl monomer to the starch backbone. Adapted from [119].

Attempts to graft or otherwise incorporate starch into synthetic polymers can be categorized based on their mechanism of grafting, which is governed by the initiator system employed. Free radical, controlled/living radical, ring opening and condensation polymerizations have been used to graft synthetic monomer from a starch [45,66,117]. Various "grafting to" formulations are prepared via coupling agents, radical coupling, activated esters or acids, acyl halides and reversible addition-fragmentation chain transfer (RAFT) click chemistry [66].

There are several examples where grafting of starch has been conducted by atom-transfer radical polymerization (ATRP) and RAFT [117]. ATRP has been used to graft starch to MMA, AA, 2-hydroxyethyl acrylate (HEA) and other monomers mostly in heterogeneous polymerizations in organic solvents. In some cases, extremely high graft percentages were reported (up to 1800%, 18x more synthetic mass than starch mass in the graft polymer). RAFT has been used to graft VAc to acid-functionalized, esterified and other starches. Both ATRP and RAFT have the advantages of producing very low amounts of homopolymer while offering control of the molecular weights of the synthetic graft chains [117].

Starch hydroxyl groups can initiate ROP of cyclic esters such as lactones and lactides in the presence of a catalyst (metal, anionic, basic organocatalysis, ionic liquid) [45,66]. Polycondensation polymerizations are stepwise reactions between bifunctional or polyfunctional monomers with liberation of small molecules such as water and have been used to graft starch [66,120,121].

Grafting via free radical polymerization occurs by hydrogen abstraction from starch to an initiator-derived radical, propagation of the starch macroinitiator with vinyl monomers and termination of the graft chains by either an initiator-derived radical, starch coupling or disproportionation. Hydrogen abstraction can either occur with the C-H on the starch backbone or from the hydroxyl groups. Unless initiation is taking place through complexation of starch with the initiating species, it is possible that both types of hydrogen abstraction could occur. Initiator-derived radicals can also directly initiate polymerization, resulting in the production of homopolymer in addition to graft polymer. In the case of persulfates, the grafting efficiency (GE) with starch is often low [66]. Examples of exceptions to this low efficiency are styrene and acrylic acid grafting, where the hydrophilicity of acrylic acid gives higher compatibility with starch and the slower polymerization reaction rate of styrene favours grafting. Ferric initiators gives slightly better grafting efficiencies but is generally lower then cerium ion initiators [66]. Ceric and persulfate initiators are the most widely used for starch grafting and while persulfates typically result in large proportions of homopolymer, ceric initiation produces the best GE, capable of reaching 85–100% with acrylonitrile, acrylamide, methyl methacrylate and styrene [32,45,66]. The molecular weight of graft chains from all methods are found to be in the range of $1–13.5 \times 10^5$ g/mol [66]. A common trend with grafting performance (GE and GP) has been a parabolic dependence on polymerization parameters leading to a maximum or optimum condition. An observed decrease in GE beyond a maximum temperature is most likely caused by chain transfer and increased termination of graft chains and with reaction time because of increased local viscosity and polymer concentration around the starch molecules.

The most common types of free radical initiators used to graft synthetic material from starch in emulsion polymerizations can be categorized into how decomposition to produce radicals is induced; thermal decomposition (persulfate, azo), redox (ferrous/peroxide, thiosulfate/persulfate, ceric ion/acid, manganese/acid) or irradiation. The species resulting from the decomposition of persulfate, ferrous and photo initiators typically produce radicals by abstraction of the C2 or C3 hydrogen or a hydroxyl hydrogen, while those deriving from manganese and cerium initiators produce a radical typically by forming a complex with the anhydroglucose unit (Figure 9). A summary of grafting results from starch with each initiator is presented below (Table 6). Preference is given to emulsion polymerization mechanisms, unless studies are limited, in which case applicable cases from solution polymerizations are presented. For emulsion polymerizations, a variety of synthetic surfactants have been employed, including those that are non-ionic, cationic and anionic. Unless otherwise reported, GE is similar between studies when comparing other grafting performance variables.

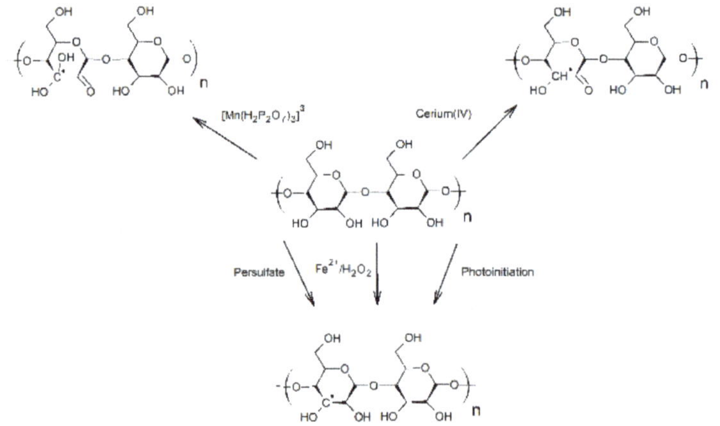

**Figure 9.** Mechanisms for initiation of starch graft polymerization.

**Table 6.** Summary of grafting results for different initiation methods in starch graft polymerization.

| Initiator | Maximum Grafting Efficiency (%) | Maximum Percentage Grafting[a] (%) | Monomers |
|---|---|---|---|
| | | | *Solution* |
| | 64 | 30 | Acrylic acid (AA) [122] |
| | 9 | GR = 17 | Vinyl acetate (VAc) [123] |
| | 84 | AO = 60 GR = 150 | Acrylamide (AM)/itaconic acid (IA) [124] |
| | 69 | - | Methyl methacrylate (MMA) [125] |
| | 65 | - | Ethyl methacrylate (EMA) [125] |
| | 60 | - | Butyl methacrylate (BMA) [125] |
| | 79 | 84 | Butyl acrylate (BA)/VAc [126] |
| | | | *Emulsion* |
| | 37–74 | 104–301 | BA/St [88,100,127] |
| Persulfate | - | - | 2-ethylhexyl acrylate (EHA) [101,128] |
| | - | - | Methacrylic acid (MAA) [101] |
| | 68 | 49 | BA [102,128] |
| | - | - | AA [101,128] |
| | - | - | BMA [129–131] |
| | 59–60 | PG = 60, GR = 57 | Styrene (St) [43,132,133] |
| | - | - | MMA [134,135] |
| | - | 350 | BA/MMA/diacetone acrylamide (DAAM) [136] |
| | - | - | BA/MMA/AA/ethyl acrylate (EA) [90] |
| | 52 | - | VAc [137] |
| | - | - | EA [135] |
| | | | *Solution* |
| | - | - | Acrylonitrile (AN) [138] |
| AIBN | 5 | 24 | VAc [139] |
| (Azobisisobutyronitrile) | 8 | 73 | MMA [140] |
| | 57 | 27 | AA [122] |
| | | | *Emulsion* |
| | - | - | St [141] |
| | | | *Solution* |
| | 99 | - | N-Methylol acrylamide (NMA) [142] |
| | - | - | BA [143] |
| | 86 | - | Methyl acrylate (MA)/NMA [142] |
| Thiosulfate + persulfate | 68 | - | MA [142] |
| | - | GR = 30 | MAA [144,145] |
| | 41 | - | VAc [137] |
| | | | *Emulsion* |
| | - | - | - |
| | | | *Solution* |
| | - | GR = 42 | AN [146–148] |
| | 99 | 50 | Glycidyl methacrylate (GMA) [149] |
| | 69 | 10 | BA [150] |
| | 65–95 | 75–110 | MMA [151–153] |
| | - | - | MA [154] |
| | - | - | EA [154] |
| Ceric | - | - | MAA [154] |
| | 33 | - | N-tert-butyl acrylamide (BAM) [155] |
| | 81 | GR = 162 | AM [156,157] |
| | 37 | 20 | VAc [123] |
| | | | *Emulsion* |
| | - | - | AN/isoprene [158] |
| | - | AO = 45 | MA [159] |
| | 65 | - | MMA [160] |
| | | | *Solution* |
| | 96 | AO = 45–49 | AN [21,22,161,162] |
| | 8.5–100 | GR = 120 AO = 36–253 | MMA [23,93,99,163–165] |
| | 59 | AO = 19 | AM [165] |
| | <9 | AO = 38 | VAc [166] |
| Manganese | - | - | AA [111] |
| | | | *Emulsion* |
| | 30–50 | - | St/AN [167] |
| | 30–50 | - | St/EA [167] |
| | 77 | 52 | St/MMA [168] |
| | 90 | 200 | St [167,169] |

**Table 6.** *Cont.*

| Initiator | Maximum Grafting Efficiency (%) | Maximum Percentage Grafting[a] (%) | Monomers |
|---|---|---|---|
| | | | *Solution* |
| | 96 | - | AN [138] |
| | 98 | AO = 54 | MMA [138] |
| | 85 | - | MA [170] |
| | 85 | AO = 20 | MAA [171] |
| Ferrous | 53 | AO = 18 | DMAEMA/AM [172] |
| | 44 | 17 | AA [173] |
| | 49 | 35 | St/BA [34] |
| | 44 | 20 | St/MMA [34] |
| | | | *Emulsion* |
| | 36 | 59 | St/BA [91] |
| | | | *Solution* |
| | - | 260 | MMA [174] |
| | - | 190 | AN [174] |
| | - | 30 | St [174] |
| | 72 | 91−907 | AM [174−176] |
| Photoinitiator | - | ~350 | AA [174] |
| | - | 420 | MAA [174] |
| | | | *Emulsion* |
| | 13 | - | Styrene [177] |
| | 81 | - | BA [177] |
| | 30 | - | MMA [178] |
| | | | *Solution* |
| | - | 135 | MA [179] |
| Irradiation | >99 | 158−243 | AA [180,181] |
| | ~85 | 200 | AM [182] |
| | | | *Emulsion* |
| | - | - | - |

[a] If not specified as add-on (AO) or grafting ratio (GR), values were reported in the article as "percent grafting (PG)," which usually represents the grafting ratio.

## 5.1. Persulfates

Sulphate radicals generated from ammonium persulfate (APS), potassium persulfate (KPS) or a thiosulfate/persulfate redox pair, have been used for the grafting of starch with vinyl monomers according to a "grafting from" mechanism through abstraction of the CH or OH hydrogen of the anhydroglucose unit (Figure 10). The hydrogen is abstracted either by a sulphate or hydroxyl radical and the formed starch radical provides a site for vinyl monomer addition and propagation. Grafting efficiency is limited with persulfate initiation relative to other more selective methods due to the favourability of homopolymerization of the synthetic monomer over radical formation on the starch anhydroglucose repeat unit.

$$S_2O_8^{-2} \rightarrow 2SO^{\cdot 4-}$$

$$2SO^{\cdot 4-} + H_2O \rightarrow HSO^{4-} + HO^{\cdot}$$

$$2OH^{\cdot} \rightarrow HOOH$$

$$HO^{\cdot} + HOOH \rightarrow H_2O + HO_2^{\cdot}$$

$$S_2O_8 + HO_2 \rightarrow HSO^{4-} + SO^{4-} + O_2$$

$$St(OH) + R^{\cdot} \rightarrow St(O^{\cdot}) + RH$$

$$St(O^{\cdot}) + CH_2 = CH-R \rightarrow St(O-CH_2-HC^{\cdot}-R)$$

or

$$St(H) + R^{\cdot} \rightarrow St^{\cdot} + RH$$

$$St^{\cdot} + CH_2 = CH-R \rightarrow St-CH_2-HC^{\cdot}-R$$

where St = starch, R$^{\cdot}$ = radical species such as HO$^{\cdot}$ and HO$^{\cdot}_2$

**Figure 10.** Mechanism of vinyl monomer "grafting from" starch by persulfate initiation. Adapted from [15].

Graft polymerizations of acrylic acid onto hydrolysed starch in solution produced higher PG and GE than identical reactions with AIBN and benzoyl peroxide, presumably due to lower overall reaction rate and thus less homopolymerization [122]. An induction time for grafting was generally observed in grafting polymerizations with KPS, while overall conversion induction time increased with higher starch content. Grafting of VAc onto granular and gelatinized starch exhibited a larger induction time than ceric ammonium nitrate (CAN) due to lack of selectivity of KPS towards hydrogen abstraction from starch. KPS served to add stability when using granular starch due to the addition of charged sulphate groups on the starch surface, since non-gelatinized native starch does not exhibit colloidal stability. A mechanism was proposed for KPS initiation where homopolymerization dominated the beginning of the reaction, at which point chain transfer from polymer radicals initiated graft polymerization on the starch molecules. The VAc polymer would eventually embed itself into the core of the graft polymer while the starch provided stabilization. Increase in starch:monomer ratio served to decrease all grafting performance as well as overall conversion [123,124,136].

Increased homopolymerization, termination of grafted polymeric chains or chain transfer to monomer above 60 °C decreased GE significantly in emulsion polymerizations. An increase in large moieties that cause steric hindrance also decreased GE [125]. Generally GE followed a parabolic trend with time caused by generation and then depletion of starch radicals [126]. A parabolic trend was also seen between GE and initiator concentration [126].

Several methods can be employed to enhance the grafting performance of emulsion polymerization with KPS. Pre-treatment (mixing of starch and KPS in solution prior to polymerization) increased GP and GE due to the increased formation of radicals on the starch molecules as well as a decrease in molecular weight [100,127]. The induction time often seen with persulfate initiators can be reduced with a pre-treatment [102]. A reduction in viscosity was also observed when starch molecular weight was reduced, either via pre-polymerization or in situ. This at times meant the difference between stability and coagulation [101,106]. Alternatively, hydrophobization of starch at any molecular weight increased GE [137]. Gelatinized starch, whether functionalized or not, exhibited stabilization properties and was able to be grafted with several monomers in the absence of surfactant [127,128]. Native and hydrophobized SNPs and SNCs also exhibited excellent stabilization properties in batch emulsion polymerizations [43,129–132]. In the presence or absence of synthetic surfactant, starch overwhelmingly resided on the surface of the final latex particles, thermodynamically favouring the polar water molecules. It is believed that starch provided stabilization as well as radical grafting sites [135]. Styrene/BA was grafted to gelatinized oxidized cassava starch where it was found that the impact of variables on GE in ascending order was initiator < monomer < temperature < starch, with a maximum efficiency of 42% and on PG was starch < initiator < temperature < monomer, with a maximum value of 301% [88]. More recently, a seeded semi-batch emulsion polymerization was conducted to prepare pressure sensitive adhesives using BA, MMA and AA. SNPs were first crosslinked with sodium trimetaphosphate (STMP), then functionalized with a sugar-based monomer (vinyl group) and finally reacted with butyl vinyl ether (BVE) to add hydrophobization. After in situ modification, the SNP hybrid particles were grafted with the primary monomer feed and KPS initiator, resulting in a core-shell latex particle with the starch completely encapsulated [183].

Enhancement by ultrasound irradiation served to increase the diffusion and collisions between starch, initiator and monomer during polymerization, increasing grafting [102]. Persulfates can also be combined with thiosulfate in a redox pair initiating system to produce generally high GE (>40%) [137,142–145]. Alternatively, a potassium persulfate/amine redox system or a dual initiator system such as with benzoyl peroxide or acetone sodium bisulphate can be conducted at comparable grafting efficiencies to KPS alone [184–186]. Persulfates are commonly used to graft synthetic monomers to starches in emulsion polymerization. While selectivity towards starch continues to limit GE and reaction rate, the above mentioned strategies have worked around this issue to produce useful latexes with starch content.

## 5.2. Ceric ions

Ceric ions (Ce (IV)) produced from ceric ammonium nitrate and ceric ammonium sulphate can be used to produce starch grafted latexes with high GE relative to other free radical methods [15]. The mechanism of ceric ion initiation is such that the ions form a complex with either the C6 carbon, C6 oxygen, C2-C3 glycol or the C1 hemiacetal at the reducing end of the starch chains (Figure 11) [15]. The most probable grafting site is the C2-C3 glycol [187,188]. The complex then decomposes by abstracting a hydrogen or through the ring opening of the anhydroglucose unit, resulting in a Ce (III) ion and a radical on the starch backbone. Less likely is the chain cleaving mechanism, where the Ce (IV) ion forms a radical on the C1 carbon or the adjacent oxygen from the ether group on a non-reducing end anhydroglucose unit (Figure 12) [32]. Alternatively, a reducing agent other than starch can be used to initiate vinyl monomer polymerization, such as a thiol, glycol, aldehyde or alcohol. Latex formed by the use of heavy metal initiators such as cerium may have a coloured hue.

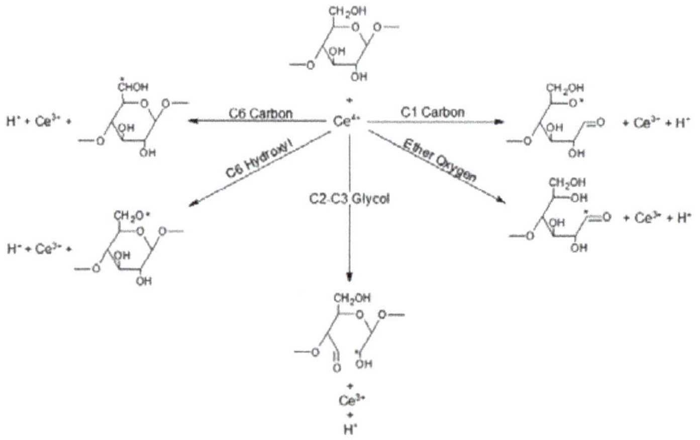

**Figure 11.** Possible mechanisms of Ceric ion initiation of starch. Adapted from [188].

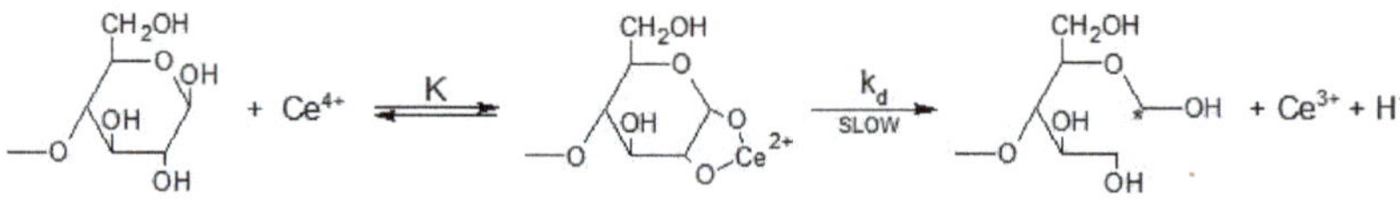

**Figure 12.** Ceric ion initiation of C1 hemiacetal at reducing end of starch chain. Adapted from [187].

The advantage of using Ce (IV) ions is that they do not react directly with vinyl groups on monomers, avoiding or greatly limiting the amount of homopolymer produced and, consequently, achieving high GE. Homopolymer can still be formed, however, for example through chain transfer to monomer or polymer. A disadvantage of this type of initiator in aqueous media is that an acid is often required to avoid hydrolysis of cerium ammonium nitrate to $[Ce\text{-}O\text{-}Ce]^{6+}$, which has little to no reactivity [15].

When using ceric initiator to graft AN to granular wheat starch, gelatinization had no effect on grafting performance. This is contrary to the effect of gelatinization in KPS systems, most likely due to the greater need for diffusion of monomer and initiator into the starch matrix for an initiator with low selectivity [147,148]. A "depleted initiator" method was employed with an MMA system where $Ce^{4+}$ that was not adsorbed to starch was removed. This led to higher grafting efficiencies than the usual continuous method [153]. When AM and N,N'-methylene-bisacrylamide (MBA) were grafted to gelatinized starches of varying amylose content, GE, AO and PG were highest with 50% amylose content and lowest with waxy starch (very low amylose %). It was thought that this occurred due to the more complete gelatinization of starches with higher amylose content compared to those with less, which require a higher cooking temperature. Additionally, the branched nature of the higher molecular weight amylopectin reduced the diffusivity and mobility of the polymer chains, resulting in higher viscosity and resistant chain growth [156].

In emulsion polymerizations, if a monomer is desired that does not react with starch (e.g., isoprene) it can be combined with a monomer that does readily graft in order to incorporate the unreactive monomer into the graft polymer. Gugliemelli et al. used this approach to prepare poly (isoprene-co-acrylonitrile) starch grafts for vulcanized rubber, while studying the effect of monomer ratio and other reaction variables on grafting performance. A disadvantage of this approach is that the unreactive monomer may retard the grafting rate and subsequently the overall polymerization rate (depending on the initiator system employed) [158].

A seeded emulsion polymerization strategy was used with a PMMA latex seed that was subsequently swelled with MMA and semi-batch fed CAN with DTAB surfactant in the presence of oligosaccharides (low molecular weight polysaccharides from enzymatically degraded starch). BA and styrene were sufficiently hydrophobic to make the oligosaccharides amphiphilic with only a few monomer units. Due to this hydrophobicity, there was not a high enough concentration of BA or styrene in the aqueous phase to give a high rate of propagation with the hydrophilic initiator. Coagulum was far lower when MMA was used (1.5%) compared to BA and styrene (20%). The reason given was that the oligosaccharides were grafted with MMA until they became sufficiently surface active to adsorb onto the monomer swollen latex particles, after which the PMMA portion of the graft polymer continued to polymerize inside the particle [99]. Debranched cationic starch was grafted to: (1) pre-synthesized and monomer swollen poly (MMA) particles (GE = 8.5%); (2) MMA monomer in an emulsion polymerization (GE = 24%), and; (3) MMA monomer in the presence of poly (N-isopropylacrylamide). The second feed strategy led to slow particle nucleation and hence a broad particle distribution and low GE [160].

Ozonolyzed amylopectin was grafted with cerium and persulfate/glucose for enhanced activity. Seeded polymer particles were stabilized by the amylopectin and the polymerization continued, where growing latex particles formed within the amylopectin scaffold, which underwent controlled coalescence. The adsorbed amylopectin remained partly exposed, leaving stabilizing chains on the particle surface. A unique use of energy-dispersive spectroscopy (EDS) showed the presence of starch throughout the latex particle, essentially confirming near complete encapsulation [189,190]. Another effective dual initiation system with ceric ions and KPS was used to graft semi-batch fed AM to a DADMAC and gelatinized starch batch charge. GE was 89% while PG was 173%. Higher monomer concentration increased PG to a limit, beyond which radical transfer from starch macroradicals to monomer may have increased, causing a decreased grafting rate. The coupled initiator system was more effective than either initiator alone due to the KPS regenerating depleted $Ce^{3+}$ ions back into $Ce^{4+}$ ions that could further initiate radicals on starch [65].

Due to its preference for starch radical formation over monomer initiation, high GE can be achieved with ceric initiators through relatively straightforward strategies [32,149,151–153]. Since homopolymerization is not a great concern, focus can be given to optimizing the interaction of starch, monomer and surfactant. Ceric initiators are optimal for attempting to encapsulate starch within latex particles due to the isolation of polymerization to the starch backbone rather than the grafting reaction competing with water phase or micellar polymerization. Starch can be grafted in the aqueous phase until hydrophobic enough to adsorb onto nucleating polymer particles, increasing the chances of incorporation [99].

### 5.3. Manganic ions

Manganese initiation using manganic ($Mn^{4+}$) pyrophosphate or potassium permanganate ($Mn^{3+}$) most likely proceeds either by reaction with an acid (e.g., oxalic, sulphuric) to form a radical species, which abstracts a hydrogen from the starch hydroxyl groups (Figure 13) or by the formation of a complex with the starch similar to that of ceric initiators, followed by hydrogen abstraction and radical formation on the starch backbone (Figure 14) [15,32,187].

$$Mn^{4+} + H_2SO_4 \rightarrow SO^{\cdot\,4-} + Mn^{3+} + 2H^+$$

$$Mn^{4+} + H_2O \rightarrow Mn^{3+} + H^+ + HO^{\cdot}$$

**Figure 13.** Radical generation in a potassium permanganate and sulphuric acid system. Adapted from [15].

$$St - OH + Mn^{4+} \rightarrow St - O^{\cdot} + Mn^{3+} + H^+$$

$$St - OH + Mn^{3+} \rightarrow St - O^{\cdot} + Mn^{2+} + H^+$$

where St = starch

**Figure 14.** Direct radical formation of manganese initiators and starch. Adapted from [15].

The rate determining step of manganic grafting of monomer to starch was found to be the cleaving of the glycol group of the anhydroglucose repeat unit, which was dependent on the pyrophosphate and $Mn^{3+}$ concentration, acidity and glycol (crosslinker) concentration [21]. An increase in either acid concentration or $Mn^{3+}$ concentration caused a decrease in molecular weight of the grafts, while AO was parabolically dependent on $Mn^{3+}$ concentration. The increased number of grafts caused an increase in monomer conversion [21]. When AN was grafted from gelatinized potato starch, a higher agitation rate was necessary compared to granular starch due to the higher viscosity of the gel phase and thus lower GE from diffusion limitations of the monomer and initiator. Monomer conversion (77–84%), AO (48–49%) and the molecular weight of grafts (33,000–430,000) all increased as the starch pre-treatment temperature was increased from 30–85 °C [22]. Grafting efficiencies were generally higher with AN than MMA due to increased rates of termination with MMA [23]. When a methanol:water mixture was used as the solvent, the GE linearly increased to 100% as methanol:water ratio increased to 50:1 (v:v). This behaviour was attributed to the increased solubility of MMA in methanol [163].

Manganese initiators have been used with styrene binary systems including AN [167], EA [167] and MMA [168], where the highest performance was observed for pure styrene (GE = 90, PG = 200) [167,169].

### 5.4. Iron ions

With a ferrous salt and hydrogen peroxide redox system, a hydroxyl radical is first produced by the decomposition of hydrogen peroxide by $Fe^{2+}$ and then a radical is subsequently formed on the starch backbone by hydrogen abstraction (Figure 15). This initiation system, unlike that of ceric and manganese initiators, is able to polymerize vinyl monomers using hydroxyl free radicals [15,34,91,123,137,191].

$$H_2O_2 + Fe^{2+} \rightarrow HO^{\cdot} + OH^{\cdot} + Fe^{3+}$$

$$HO^{\cdot} + CH_2 = CH - R_1 \rightarrow CH_2 - HC^{\cdot} - R_1$$

$$HO^{\cdot} + St(H) \rightarrow St^{\cdot} + H_2O$$

$$CH_2 - HC^{\cdot} - R_2 + St(H) \rightarrow St^{\cdot} + CH_2 - CH_2 - R_2$$

$$St^{\cdot} + CH_2 = CH - R_1 \rightarrow CH_2 - HC^{\cdot} - St$$

$$CH_2 - HC^{\cdot} - St + CH_2 = CH - R_2 \rightarrow Starch\ Graft\ Copolymer$$

where St = starch, $R_1$ = any monomer group, $R_2$ = any polymer group

**Figure 15.** $Fe^{2+}/H_2O_2$ initiation of vinyl monomer or starch. Adapted from [15].

AN was grafted more readily to native starch granules than MMA due to its increased hydrophilicity [138]. Grafting efficiency increased as peroxide concentration increased, however degradation of starch began to occur. Oxidized starch was generally more susceptible to grafting than native starch [138]. MAA was grafted to granular starch and it was found that at the termination of the polymerization, the rate of grafting (>85% conversion) was much faster than that of homopolymerization (~10% conversion). Pre-gelatinized starch achieved only half the conversion of granular starch. Explanations for this were given as increased intrinsic reactivity of the starch macroradical, the stability of the propagating free radical species and high local monomer concentration within the starch granule due to association of MAA with the starch hydroxyls [171]. There was a negligible effect of increasing $Fe^{2+}$ concentration above a minimum, with a brownish hue detected at excess amounts. Longer reaction times are more effective due to added time to homogenize the polymerization as the grafting reaction happens quickly with ferrous initiator pairs and sometimes generates regions of heterogeneity [173].

Semi-batch (monomer and hydrogen peroxide feed) surfactant free emulsion polymerization was used to graft styrene and BA to gelatinized cationic acetylated starch. It was found that conversion increased with $H_2O_2$ concentration as well as increased to a maximum with monomer concentration. Grafting efficiency and PG had parabolic dependency on $H_2O_2$ concentration and while both variables increased with monomer concentration, GE decreased after a maximum. Grafting efficiency and PG were 35% and 59% respectively [91].

*5.5. AIBN*

Azobisisobutyronitrile (AIBN) is a hydrophobic initiator that dissociates into two 2-cyanoprop-2-yl radicals capable of initiating polymerization of vinyl monomers (Figure 16). These radicals are capable of abstracting hydrogen from a hydroxyl group on the anhydroglucose units, thus creating a radical on the starch backbone [11,15,122,141,192].

$$R_2N - N = N - NR_2 \rightarrow 2R_2N^{\cdot} + N_2$$
$$2R_2N^{\cdot} + CH_2 = CH\text{-}R_1 \rightarrow CH_2 - HC^{\cdot}\text{-}R_1$$
$$2R_2N^{\cdot} + St(H) \rightarrow St^{\cdot} + H_2O$$
$$CH_2 - HC^{\cdot}\text{-}R_2 + St(H) \rightarrow St^{\cdot} + CH_2 - CH_2\text{-}R_2$$
$$St^{\cdot} + CH_2 = CH\text{-}R_1 \rightarrow CH_2 - HC^{\cdot}\text{-}St$$

where St = starch, $R_1$ = any monomer group,

$R_2$ = any group of an azo initiator

**Figure 16.** Asobisisobutryonitrile (AIBN) initiation of vinyl monomer or starch. Adapted from [11].

The grafting of AA onto hydrolysed potato starch was conducted using AIBN as initiator in aqueous solution. Monomer conversion approached 100% as expected but GE and PG were limited to 57% and 27% respectively, due to the high amount of homopolymer produced and lack of grafting sites produced on the starch molecules [122]. MMA was grafted to gelatinized starch to achieve PG = 73% and GE = 8% and it was concluded that AIBN radicals were not reactive enough to abstract a hydrogen from starch. Any grafting that occurred came from "grafting to" of polymeric radicals, which were available dependent on the concentration of relatively hydrophobic monomer present, which the radicals were reactive enough to polymerize [140]. Similar observations were made when grafting VAc at PG = 24% and GE = 5% [139]. A starch-based bromide chain transfer agent was grafted with AIBN to vinyl acetate via RAFT polymerization for use in biomedical and other applications. The structure, composition and molecular weight of the grafted polymer was successfully controlled and well defined [193].

Due to the hydrophobicity of AIBN, it is more suited to mini emulsion or suspension polymerization than emulsion polymerization [192,194]. A notable exception was the use of OSA-acetic anhydride (AC) modified SNPs in a seeded emulsion polymerization with styrene. The SNPs were found to first act as Pickering stabilizers of the styrene but then to transition under higher temperature to a core-shell seed with styrene in the core of the starch particle. The polymerization proceeded at the interphase of the swollen SNP seed, with the AIBN hydrophobic initiator initiating creating oligomer radicals that would propagate inward to the core of the particles. The results were an apparent core-shell morphology with most of the starch in the core of the particle [141].

*5.6. Other Initiators*

Other initiators suitable for grafting from starch include anionic polymerization by alkoxides, metal complexes such as cobalt or chromium (similar to manganese mechanism), copper acetylacetonate-trichloroacetic acid, potassium peroxyvanadate, hydrogen peroxide/iron sulphate/thiourea dioxide, potassium bromate/thiourea dioxide, alkyl hydroperoxide/amino group, vanadium mercaptosuccinic acid, horseradish peroxidase/hydrogen peroxide/2,4 pentane dione, ozone and enzymes [15,32,45,66].

The enzyme horseradish peroxidase (HRP), an oxidoreductase catalyst extracted from the roots of horseradish, has been used on several occasions to perform grafting reactions with starch. BA (GE = 19%) [195], MA (GE = 45%) [196], AM (GE = 66%) [197] and DMDAAC (GE = 99%) [198], have all been grafted from starch in solution polymerization in the presence of hydrogen peroxide as the oxidant and a reductant (e.g., 2,4-pentanedione). Although HRP has been used in emulsion polymerizations, there is no reported work grafting starch using such an approach [199]. This presents a worthwhile avenue to be pursued due to the moderate reaction temperatures and high grafting efficiency associated with HRP, in addition to the fact that it is a bio-catalyst that is renewable.

Benzoyl peroxide was used as the sole initiator for the emulsion polymerization of styrene/maleic anhydride with crosslinked SNPs and hydroxyethyl cellulose as stabilizers. An inverse emulsion was first prepared where SNPs were mixed with water in an organic phase consisting of maleic anhydride, styrene and benzoyl peroxide pre-polymerized to 25% conversion. The SNPs were shown to act as Pickering stabilizers of the water droplets. Subsequently, the inverse emulsion was transferred to an aqueous dispersion of hydroxyethyl cellulose, where the cellulose stabilized a region of the organic media around several smaller SNP stabilized water droplets and the polymerization was continued to completion. Although grafting performance was not reported, FTIR of the purified starch grafted polymer was shown as evidence of esterification between hydroxyl groups and maleic anhydride [107].

*5.7. Irradiation*

Microwave, UV and high energy (gamma and electron) irradiation have become increasingly popular for grafting synthetic monomer to polysaccharides without the need for additional reagents. High energy ionizing initiation can occur through: (1) a direct method where monomer and polysaccharide are simultaneously initiated; (2) a peroxidation method where hydroperoxide or peroxide functional groups are formed by reaction of the polysaccharide and air, which later form free radical groups under heat; or (3) a pre-irradiation method where the radiation forms free radicals on the polysaccharide in an inert atmosphere, prior to being mixed with monomer [200]. It has been found that high irradiation doses or reaction times can lead to starch degradation [174]. Exposing a photoinitiator (e.g., dimethoxy-2-phenylacetophenone (DMPA), benzoin methyl ether (BME), benzophenone (BP)) to a certain wavelength of radiation causes generation of a radical that can attack vinyl monomer or starch [201]. Irradiation methods can also be used to enhance other initiation methods [202,203]. The precise grafting performance and kinetics of each monomer/initiator pair has to be evaluated separately as their behaviour cannot necessarily be assumed to apply universally [174].

Emulsion polymerization was used to graft styrene, BA and MMA to amylopectin with sodium dodecylbenzene sulfonate as surfactant [177,178]. Using the direct irradiation method, surfactant, water, monomer, starch and initiator were mixed and lightly stirred, purged with $N_2$ and subsequently illuminated. A pre-illumination method was applied where monomer, surfactant and water were added to an addition funnel and purged to form a pre-emulsion, while the starch dispersion was illuminated. Subsequently the pre-emulsion was then added to the reaction vessel with the pre-illuminated starch dispersion. For the direct method, styrene and BA resulted in 50% and 87.8% conversion at ~0% and 11.5% GE respectively. Using pre-illumination, styrene and BA yielded 37.5% and 89.1% conversion and 12.5% and 81.4% GE respectively, a significant improvement over simultaneous illumination. For BA, which has a faster polymerization rate, pre-illumination has a greater effect on GE due to the higher chance of grafting over homopolymerization when the radicals are formed only on the starch backbone.

Irradiative initiation methods present clear benefits over other initiators due to good control over polymerization rate and the potential for radical formation on the starch backbone without the need for addition of potentially harmful chemicals such as initiator, heavy metal salts and acids. So long as degradation is avoided, irradiation can be combined with other initiation methods to enhance GE.

## 6. Strategies for Starch Incorporation in Emulsion Polymerization

From the numerous studies described above, certain conclusions can be drawn regarding latex formulations and strategies as well as the dependence of grafting performance variables on various reaction parameters. These strategies are summarized below as a guide for using starches in emulsion polymerization.

### 6.1. Initiator

The appropriate choice of initiator depends on the constraints of a particular system with regard to industrial scalability and final product specifications (mostly GE and PG). Persulfates are the most widely used initiators in industry and there is abundant experience in handling and using these initiators in emulsion polymerizations. If homopolymer formation is not of concern, this is the clear direction to pursue [45]. Ceric initiators can be active at room temperature, do not polymerize vinyl monomers directly and are nearly completely selective towards starch initiation [66]. The need for acid in the polymerization and the presence of heavy metals and colour in the final product may be deterrents for its use [15]. Synergistic effects of multiple initiator systems can be attractive and some creativity in feed strategy can limit the need for a high concentration of heavy metal initiators while achieving improved grafting performance over persulfates alone [65,202,204,205]. The combination of irradiation with persulfate or ceric initiators can more than double the GE [203]. In the case where an initiator can initiate homopolymerization, a pre-treatment step where initiator and starch mix is able to generate starch macroradicals prior to the addition of monomer, as well as reduce the starch molecular weight, which can assist in increasing GE [100,102]. An alternative strategy is to modify the starch through small molecule chemistry to add covalently bonded vinyl groups that can participate in initiation [183,206,207]. These vinyl groups must be incapable of reacting with each other (e.g., maleic anhydrides) or else microgel networks of starch particles will form, while they must readily react with the monomer formulation chosen [89,189].

Although degradation can be caused by prolonged reaction times or irradiation strength, radiation initiation provides a simple method for graft polymerization, provided one has access to the appropriate equipment at scale [174,179]. The degradation of starch by any initiating method can result in lower molecular weight starch radicals, which may improve grafting and lower viscosity. This degradation could also increase the amount of soluble starch that migrates to the water phase, which may derail attempts at achieving a particular final latex particle morphology or stability (such as the degradation of initially well-defined SNPs or SNCs).

The type of initiator used can be expected to have an effect on the properties of the final latex (e.g., viscosity) due to differences in initiation and subsequently, on nucleation mechanisms, amount of homopolymer and water phase polymer, degradation of starch and grafting performance. Generally, the effect of temperature on GE is universally parabolic with the efficiency increasing with temperature due to increased radicals being created on the starch molecules, with subsequent decreases due to higher rates of termination and chain transfer reactions. A similar trend can be expected with reaction time due to consumption of monomer and initiator as well as diffusion limits from longer grafted chains, preventing reaction components from contacting the starch. The temperature of a reaction should be high enough to initiate radical formation but not so high as to produce excess homopolymer formation [125]. Several kinetic models have been prepared for specific systems [32]. One should recall that if the initiator, monomer or starch type change, model parameters will need to be recalculated.

*6.2. Monomer*

Choice of monomer is made by matching the resulting polymer to the properties of the desired product. The glass transition temperature, mechanical and surface properties and functionality of moieties of the final product are among determining factors when selecting the appropriate monomer (s). In cases of starch grafted polymers, additional attention must be paid to the interaction of the monomer (and resulting polymer) and starch molecules, perhaps the most important compatibility issue when targeting a particular particle morphology or level of grafting performance. Monomer hydrophilicity and reactivity are the primary factors defining this interaction. If a monomer is too hydrophilic it may cause excess water phase homopolymer formation, while extreme hydrophobicity may reduce contact between the monomer and hydrophilic starch. A way around this issue is to functionalize the starch with hydrophobic groups prior to polymerization, although this could add undesired steps to an industrial process and reduce the overall mass of bio-sourced material in the starch pre-cursor material [129,130].

If monomers are required that do not readily graft with starch, they can be coupled with initiating monomers, although this tends to increase the induction time common with certain initiators (such as persulfates), as well as decrease the overall reaction rate [158]. Monomers with sterically hindered vinyl groups can affect reactivity with starch and subsequently the GE [125]. The GE of the same monomers used with persulfates, cerium and manganese initiators vary greatly, suggesting further that such observations cannot be applied universally.

The application of mixed monomer/starch batch seed generation, as well as the use of high shear mixing narrowed the particle size distribution (PSD) and improved the GE [137,160,189,190,208]. Semi-batch feeding of monomer or a pre-emulsion to achieve monomer starved conditions could be favourable for starch incorporation into latex particles, especially in the case where homopolymerization is a concern.

*6.3. Surfactant/Stabilizer*

Emulsion polymerizations usually use a stabilizer or surfactant to produce well-defined polymer latex particles. These surfactants are low molecular weight synthetic molecules with a hydrophobic tail and charged hydrophilic head. Alternatively, non-charged steric stabilizers can be used, as well as solid spherical particles (Pickering emulsions) or a combination of all three types.

The vast majority of starch types (except uncooked granules) exhibit stabilization properties in emulsion polymerizations. Carbohydrate based surfactants are obviously tailored to be effective stabilizers as direct replacements for typical surfactants such as SDS [209]. SNPs and SNCs are particularly effective at stabilizing latex particles without the need for synthetic surfactants [107,129–131,192,210]. Although starch particle sizes are often >1 μm, exceptions exist where they are more appropriately sized at <500 nm [106,131,207,211]. Starch particle stabilizers should be smaller than the latex particles they are stabilizing, which typically range from 50–300 nm. The composition of the continuous phase as well as the hydrophobicity of the monomer and polymer affect the particle size and distribution due to the hydrophilicity and strong hydrogen bonding potential of the starch.

Some strategies for improving stabilization could include crosslinking the starch once it has stabilized the particles, functionalizing the starch with cationic/anionic charged groups and including synthetic surfactants to be cooperating stabilizers or to induce a lower starch CMC [207,212,213]. The electric charge of the latex particles, surfactant, monomer and starch should be considered. Surface tension measurements can shed light on the behaviour of surfactants and stabilizers and their interaction with other molecules. Different forms of starch have been adsorbed onto polymer particles once sufficiently grafted with hydrophobic monomer, after which the particles grow and incorporate the starch into its structure, leaving strands of starch chains on the particle surface for stabilization [99].

The use of synthetic surfactant in starch graft polymerizations can improve GE, reduce viscosity, increase storage stability by reducing retrogradation and prevent amylose aggregation in the water phase [214,215]. Adverse effects of excess surface active molecules can arise from their migration to the surface of the subsequently formed polymer films, thus affecting adhesive and barrier properties [47,215]. Reports of successful grafting in surfactant free systems are regularly reported, while cases of no grafting in such systems also exist [128].

Octenyl succinic anhydride (OSA) functionalized starch has been proven to be an extremely effective stabilizer (low CMC and good particle size control) at particle diameters from 280 nm to 100 μm and concentrations of 1–45 wt.% [199]. Non-functionalized waxy SNPs have stabilized BMA latex particles as small as 135 nm at 10 wt.% concentration, while SNCs have stabilized particles as small as 121 nm at 12 wt.% concentration [129,131]. SNPs have been shown to preferentially stabilize smaller polymer latex particles rather than monomer droplets [129]. The higher molecular weight SNPs are superior at stabilizing oily particles compared to other starches and may have better long term stability that synthetic surfactants [216,217]. In contrast, free amylose derives its stabilization properties from the formation of 3D networks between particles, contributing steric stabilization to resist coagulation [217,218]. The hydroxyl groups of starch have a pKa > 12, indicating a reasonably strong resistance to alkalinity, however one can expect more deprotonation as the pH increases to this point. The presence of salts and buffer can also influence the stability of latex, with or without starch, although latex with low concentrations of water soluble starch are highly salt resistant [38,217,219,220]. Although zeta potential provides an established method to test the stability of latex particles, most starch stabilized latexes exhibit weak zeta potentials (0 to −5 mV) while maintaining good long term stability due to steric hindrance.

*6.4. Starch Type*

As previously explained, there are several types of starch that can be used to either modify polymer latex properties or for the reduction of synthetic polymer mass. These include granular and gelatinized non-modified starch, as well as non-functionalized and functionalized gelatinized starch, amylose-rich and amylopectin-rich starch, ultra-low molecular weight (ULMW) polysaccharides, SNPs and SNCs. Gelatinization is typically achieved for any starch type by pre-treating in water at 70–95 °C. The temperature of gelatinization influences grafting performance [22,161,162,172].

It is common to degrade starch either by peroxides, acid, persulfates or by enzymatic treatment in order to increase the likelihood of high encapsulation within the latex particles. In fact, apart from vinyl and hydrophobic functionalization, this strategy is the only one to result in significant encapsulation. This may be due to decreased negative effects on viscosity and reaction kinetics, reduced termination of growing graft chains and the speed at which the molecule can become amphiphilic through grafting of the hydrophobic monomer. As with any polar polymer, a decrease in its molecular weight will increase its solubility in non-polar solvents. SNPs with a DS of 0.07 of OSA and 0.158 of AC are sufficient for a very hydrophobic monomer like styrene to swell the core of the SNPs and polymerize, forming varying final morphologies (as observed by SEM). So long as the monomer feed ratio was kept low, the synthetic polymer remained bound to the SNP at full conversion (the synthetic monomer/polymer migrated during the polymerization in all cases) [208].

Contrary to hydrophobically modified SNPs, grafting onto SNCs was restricted to the surface of the particles only, due to the high degree of crystallinity preventing diffusion of the hydrophobic monomer into the core of the SNC [221–223].

*6.5. Encapsulation*

As mentioned earlier, it could be desirable to encapsulate all or most of the starch material into the latex particles, to maintain the properties of the synthetic standard. Apart from degradation or hydrophobization, vinyl functionalization can transform the starch into a reactive polymer (or macromer) capable of participating in free radical polymerization. This strategy has been pursued with bifunctional groups as well as with glucose surfmer (reactive surfactant) functionalized SNPs [183,206,207,224].

Enzymatically degraded starch was reported to be encapsulated within pMMA latex particles either by early adsorption of the amphiphilic graft polymers onto the nucleated latex particles and subsequent particle growth or by the crosslinking of grafted starches and subsequent monomer swelling and polymerization [160]. The authors demonstrated this by analysing the starch and graft polymer by newly developed and apparent error-free solution $^1$H-NMR and solid $^{13}$C CP-MAS quantification methods, as well as TEM.

OSA-AC modified SNPs were found to first form Pickering emulsions with styrene but then transitioned under higher temperature to a core-shell emulsion with styrene in the core of the starch particle. Upon addition of hydrophobic initiator an apparent core-shell latex particle morphology was formed containing the majority of the starch in the core of the particle, with small amounts of starch chains extending into the shell [141].

Another approach reported by Cummings et al., consisted of RSNPs functionalized with vinyl functionalized glucose-based surfmer. When used in an emulsion polymerization of BA/MMA, these functionalized RSNPs produced 10–40 wt.% incorporation of starch in the polymer latex [206,207]. Subsequently, this functionalization was combined with an additional internal crosslinking of the RSNP with sodium meta triphosphate (SMTP) to reduce the amount of water soluble starch and increase the crosslink density of the RSNPs prior to polymerization. This modification improved incorporation greatly, where the starch was reported to be fully encapsulated in the latex particle [183]. Furthermore, the use of a tie-layer monomer (e.g., butyl vinyl ether) with low reactivity to preferentially graft to the starch vinyl groups, served to hydrophobize the RSNPs prior to the primary monomer feed. The result was successful encapsulation and core-shell morphology of the SNPs within the final latex particles. Although the maximum starch content reported to date was 17 wt.%, this work is significant in that it follows a relatively simple emulsion polymerization procedure with a very common initiator (KPS) to produce completely encapsulated particles [183].

Latexes with low encapsulation can still appear white at low conversion, while those with high encapsulation can still undergo significant increases in viscosity during the polymerization (even when only small amounts of water phase starch is present).

*6.6. Starch Loading and Solids Content*

Challenges persist in regard to the preparation of latexes with high starch loading and solids content and manageable viscosity and grafting (ideally encapsulation). Achievement of such an industrially relevant formulation that could impart some beneficial property modification while displacing a large amount of synthetic (petroleum sourced) material represents a large step forward in applying the 12 principles of green chemistry to the production of synthetic-starch graft polymers.

Solids content limitations typically arise from an increase in viscosity caused by persistent water phase starches. These starches additionally interfere with the diffusion of initiator and monomer to the polymer particles (causing a decrease in reaction rate, potential monomer droplet initiation and subsequent coagulation) and reduce the stability of the latex particles at high or complete conversion [183,206,207].

Starch loading is limited by similar effects but mostly by the decrease in polymerization rate of initiator-starch initiation compared to that of the initiator-monomer, the persistence of water phase starches affecting viscosity and poor GE leading to a product with limited usefulness [183,206,207]. The slower initiator-starch initiation is caused by the higher dissociation energy of the starch hydroxyl hydrogen compared to the π-bond of the monomer's vinyl group, resulting in the faster homopolymerization reaction being preferred when both monomer and starch are present. Achieving a particular loading is usually not of concern in cases where starch is being used as a surfactant/stabilizer, in contrast to when it is used as a filler or synthetic polymer replacement.

Surveying the achieved solids contents and starch loadings in the literature, the highest solids content achieved was ~60 wt.% when using VAc to graft onto waxy maize dextrin (ULMW starch) at a starch loading of 25 wt.% [137]. Also using dextrin, the highest loading of 93 wt.% was achieved at 5 wt.% solids [225]. A degraded amylose was loaded at 70 wt.% with VAc at 45 wt.% solids [214]. Beyond low molecular weight starches, 50 wt.% loading and solids was achieved with native starch being used as a filler and stabilizer with styrene and BA with KPS [226]. For modified starches, high solids and loadings are rarer, with cationic acetylated starch being loaded at 20 wt.% with styrene and BA at 52 wt.% solids with ferrous initiator [91] and SNCs being loaded up to 50 wt.% with styrene at low 6 wt.% solids with KPS [92]. OSA-AC modified SNPs that achieved excellent encapsulation were performed at only 9 wt.% loading at 6 wt.% solids with AIBN [141]. Further, other OSA modified SNPs were used at 2 wt% loading and 53 wt.% solids with KPS [130]. It is increasingly difficult to apply efficient encapsulation strategies at high loadings and solids contents. Brune and Eben-Worlée achieved 30–50 wt.% loading at 30–70 wt.% solids and although copolymerization was assumed, no grafting performance was reported [224]. Our group has achieved partial incorporation of waxy (20 wt.%) and dent (40 wt.%) RSNPs modified with a vinyl functional glucose surfmer. Waxy RSNPs were included at 40 wt.% (20 wt% incorporation) and 50 wt.% (10 wt.% incorporation) loadings and 40 wt.% solids, while dent RSNPs had a limit of 25 wt.% loading at 40 wt.% solids (40 wt.% incorporation) [206,207]. A positive achievement of these latexes is their low viscosity (<250 cp), reasonable particle size (<300 nm), good film formation and good long-term stability.

## 7. Starch-Based Latex Applications

### 7.1. Starch

With its abundance, biodegradability and potential for modification and functionalization, starch has been widely used in industry and for academic study. In fact, a third of the biodegradable polymers being commercially produced are starch based [114]. Despite this, starch is readily plasticized by water, has poor barrier resistance to water vapor and is overly rigid, making it difficult for use in many coating and film applications. Researchers have used starch in composites to impart unique properties and have shown it is applicable for the food, cosmetics, pharmaceuticals, paper making and tire industries [76,227].

Two of the most popular commercialized starch products are Mater-Bi$^{TM}$ (carbon black replacement for tires) and Eco-Sphere$^{TM}$ (bio-based latex for coatings and paper binder) [59]. Due to its stabilizing properties, starch can be used in emulsions for cosmetics, pharmaceuticals, paper products, textiles, packaging and industrial latexes for paints and coatings [228,229]. While the most common use of starch products are for packaging applications, it holds promise for drug delivery and other niche applications [3]. In the food industry, starch can replace fats while maintaining similar rheological properties [14,227,230]. SNPs have been used to reinforce packaging for foodstuffs. In the pharmaceutical industry, starch has been extensively investigated as a drug carrier, although further work must be done to determine the dangers of starch based particles in the blood stream [15,227,231]. These particles are a low cost alternative as an adsorbent for wastewater treatment. In comparison to native starch, SNPs have exhibited improved properties as a paper binder [14,227,230]. Metal implants can conceivably be replaced by starch based biodegradable material. SNC have good reinforcing properties for composites and latexes such as natural rubber and various synthetics [26].

Commercialized starch blends include Bioplast®, NovonTM, Biopar®, Baialene®, Solanyl®, Ecobras, Biolice, Cornpole, Pentafood, Pentaform, Plantic and products Trellis Earth among others [110,232]. Due to the requirements for vegetable packaging, this provides a new area to be explored. Current starch based products include thermoformed trays and cups, injection moulded cutlery, plates and bottles, as fibres for agricultural products and clothing, as thickeners and as additives in adhesives [4,233]. Some commercial leaders in the field include Novamont, Rodenburg, Biotec, Biop, Limagrain, Livan, EcoSynthetix, Cereplast and Plantic, with the bulk of the industry existing in Europe [114]. The future of commercialized starch products is sure to include a high growth rate in innovation and commercialization, especially in the field of starch graft polymer composites for a wide array of previously unexplored applications, including for uses as consumer plastic bags, straws, tapes, films, wraps and paints [114].

*7.2. Starch-g-synthetic Latex*

As mentioned previously, although starch is a useful material in and of itself, its performance can be greatly enhanced by customizing its functional moieties and adding grafted polymer chains. This will allow previously unexplored applications to be tested, as well as open the possibility of replacing purely synthetic formulations with increasing amounts of renewable starch material. Starch material modified with synthetic polymers have shown potential to be used in biomedical and pharmaceutical fields, as an adsorbent for wastewater treatment and in polymer composites as a property modifier.

Other applications for starch graft polymer latexes are as flocculants [15,66], resins in ion exchange filtration [15], superabsorbent/hydrogel for aqueous solutions (water treatment) [15,32], sizing agents for textiles [32,66,117] and in various papermaking applications [15,32,45,66]. Biomedical applications include drug delivery, enzyme immobilization, biosensing and catalysis [15,45,66,117,234]. Biodegradable plastics (including for tissue engineering and implants) can contain a significant amount of starch and continue to have desired properties [15,45,234]. Additionally, there are several oilfield applications where starch graft polymers have been used [15].

## 8. Conclusions and Future Perspectives

The grafting of vinyl monomers from granular, native and functionalized amylose, amylopectin, SNCs and SNPs has been extensively explored in the literature. Initiation methods investigated include persulfates, irradiation, azo and ferrous, ceric and manganese redox initiators. Undesired homopolymer formation during the grafting polymerization is often a challenge, as is proper encapsulation of the starch material. Controlled radical polymerization strategies such as ATRP, RAFT and more recently nitroxide-mediated radical polymerization (NMP), have been investigated for polysaccharide modification and grafting and have the advantage of avoiding homopolymer formation. The bio-catalyst HRP has also garnered recent attention for the grafting of starch and shows promise as a "green" pathway for the production of starch grafted polymers. The use of emulsion polymerizations to produce starch incorporated latexes for coatings and other applications of polymer films shows promise for the enhancement of properties, the reduction of synthetic (petroleum-based) monomers and the further improvement of overall process sustainability. Challenges persist due to the difficulty in maintaining latex stability in the presence of high molecular weight starches and the persistence of water phase starches in the final latex. Although some strategies exist in the literature as proofs of concepts, more work must be done to develop procedures for full encapsulation of starch within the latex particles (thermodynamically unfavourable) for a range of monomers. The resulting core-shell particles would be expected to exhibit the properties of the vinyl monomer rather than that of the encapsulated starch. This may also increase the starch loading limit before undesired effects begin to exhibit themselves during the polymerization. Some of these strategies have included hydrophobization or vinyl functionalization of starches (of mainly SNPs and SNCs) to make them more compatible with vinyl monomer and initiator or the use of sequential polymerization stages to first create an amphiphilic grafted starch seed that can then be encapsulated with semi-batch fed monomer or a pre-emulsion. Another recent approach has been to include carbohydrate-based monomer in polymerizations to increase the bio-content of the final product [235]. Starch is generally cheaper than petroleum by-products and has the potential to be sustainably sourced and biodegradable, making it extremely attractive in an industry looking to reduce its environmental footprint. Even so, the commercialization of starch loaded latexes must also accelerate to increase public acceptance and support of starch-based products, government funding of academic research and industrial start-ups and more comprehensive analysis of the effectiveness of these products and the environmental impact of their full life cycle. We have presented in this work a survey of the attempts to include starch in polymerizations, specifically those carried out through the more sustainable emulsion polymerization process, including those based on granular, native, functionalized and nano-sized starch materials. Strategies for future works were presented and a comprehensive analysis of characterization methods and applications of starch grafted latexes and films was provided.

**Author Contributions:** Conceptualization, S.C., M.D. and N.S.; methodology, S.C., M.D. and N.S.; software, S.C.; validation, M.D., N.S., M.C. and Y.Z.; formal analysis, S.C.; investigation, S.C.; resources, M.D. and S.C.; data curation, S.C.; writing—original draft preparation, S.C.; writing—review and editing, S.C., M.D., N.S. and M.C.; visualization, S.C., M.D., M.C., N.S. and Y.Z.; supervision, M.D., M.C. and N.S.; project administration, M.D.; funding acquisition, M.D. and M.C.

**Funding:** This research was funded by the Natural Sciences and Engineering Research Council (NSERC) of Canada, grant number CRDPJ 476580-14.

**Acknowledgments:** EcoSynthetix, Inc. (Burlington, ON) is acknowledged for their financial support of the research.

**Conflicts of Interest:** The authors declare no conflict of interest.

## Abbreviations

| | |
|---|---|
| $^1$H-NMR | proton nuclear magnetic resonance |
| $^{13}$C-NMR | carbon 13 nuclear magnetic resonance |
| AA | acrylic acid |
| AC | acetic anhydride |
| AIBN | azobisisobutyronitrile |
| AGU | anhydroglucose units |
| AM | acrylamide |
| AN | acrylonitrile |
| AO | add-on |
| APS | ammonium persulfate |
| ATRP | atom-transfer radical polymerization |
| BA | butyl acrylate |
| BAM | N-tert-butyl acrylamide |
| BMA | butyl methacrylate |
| BME | benzoin methyl ether |
| BP | benzophenone |
| CAN | ceric ammonium nitrate |
| CLSM | confocal laser scanning microscopy |
| CMC | critical micelle concentration |
| CP-MAS | cross polarized magic angle spinning |
| DADMAC/DMDAAC | diallyl dimethyl ammonium chloride |
| DAAM | diacetone acrylamide |
| DLS | dynamic light scattering |
| DMA | dynamic mechanical analysis |
| DMAEMA | 2-(dimethylamino)ethyl methacrylate |
| DMPA | dimethoxy-2-phenylacetophenone |
| DTAB | dodecyltrimethylammonium bromide |
| EA | ethyl acrylate |
| EMA | ethyl methacrylate |
| FTIR | Fourier-transform infrared |
| GE | grafting efficiency |
| GMA | glycidyl methacrylate |
| GPC | gel permeation chromatography |
| GR | grafting ratio |
| HDC | hydrodynamic chromatography |

| | |
|---|---|
| HPLC | high performance liquid chromatography |
| IA | itaconic acid |
| KPS | potassium persulfate |
| MA | methyl acrylate |
| MAA | methacrylic acid |
| MBA | N,N′-methylene-bisacrylamide |
| MFFT | minimum film formation temperature |
| MMA | methyl methacrylate |
| MW | molecular weight |
| NMA | N-methylol acrylamide |
| OSA | octenyl succinic anhydride |
| PG | percent grafting |
| PSD | particle size distribution |
| RAFT | reversible addition fragmentation chain transfer |
| RSNP | regenerated starch nano-particle |
| SEM | scanning electron microscopy |
| SMTP | sodium meta triphosphate |
| SNC | starch nano-crystal |
| SNP | starch nano-particle |
| St | styrene |
| TEM | transmission electron microscopy |
| TGA | thermogravimetric analysis |
| VAc | vinyl acetate |
| ULMW | ultra-low molecular weight |
| XRD | x-ray diffraction |

## References


1. Zhang, Y.; Dubé, M.A. Green Emulsion Polymerization Technology. *Adv. Polym. Sci.* **2017**, 1–36.
2. Marques, S.; Moreno, A.D.; Ballesteros, M.; Gírio, F. *Biomass and Green Chemistry*, 1st ed.; Vaz, S., Jr., Ed.; Springer: Cham, Switzerland, 2018; pp. 69–94.
3. Laycock, B.G.; Halley, P.J. *Starch Polymers*, 1st ed.; Halley, P.J., Averous, L., Eds.; Elsevier: New York, NY, USA, 2014; pp. 381–419.
4. Tadini, C.C. *Starch-Based Materials in Food Packaging: Processing, Characterization and Applications*, 1st ed.; Barbosa, S.E., Garcia, M.A., Castillo, L., Lopez, O.V., Vilar, M., Eds.; Academic Press: London, UK, 2017; pp. 19–36.
5. Le Corre, D.; Dufresne, A. Preparation of Starch Nanoparticles. *Biopolym. Nanocomposites Process. Prop. Appl.* **2013**, 153–180.
6. Belgacem, M.N.; Gandini, A.; Silvestre, A.; Galbis, J.A.; García-Martín, M.G.; Benvegnu, T.; Plusquellec, D.; Lemiègre, L.; Pizzi, A.; Gellerstedt, G.; et al. *Monomers, Polymers and Composites From Renewable Resources*, 1st ed.; Belgacem, M.N., Gandini, A., Eds.; Elsevier: New York, NY, USA, 2008.
7. Hansen, F.K.; van Herk, A.M.; Gilbert, R.G.; Leiza, J.R.; Meuldijk, J.; Charleux, B.; Monteiro, M.J.; Heuts, H.; Mercado, Y.R.; Akhmastkaya, E.; et al. *Chemistry and Technology of Emulsion*, 2nd ed.; van Herk, A.M., Ed.; John Wiley & Sons: West Sussex, UK, 2013.
8. Anderson, C.D.; Daniels, E.S. *Rapra Rev* and Latex Applications. *Rapra Rev. Rep.* **2003**, *14*, 1–146.
9. Guyot, A.; Chu, F.; Schneider, M.; Graillat, C.; McKenna, T.F. High Solids Content Latexes. *Prog. Polym. Sci.* **2002**, *27*, 1573–1615. [CrossRef]
10. Rudin, A.; Choi, P. *The Elements of Polymer Science & Engineering*, 3rd ed.; Elsevier: New York, NY, USA, 2013; pp. 427–447.
11. Odian, G. *Principles of Polymerization*, 4th ed.; John Wiley & Sons: Hoboken, NJ, USA, 2004; pp. 198–349.
12. Dubé, M.A.; Salehpour, S. Applying the Principles of Green Chemistry to Polymer Production Technology. *Macromol. React. Eng.* **2014**, *8*, 7–28. [CrossRef]
13. Ačkar, Đ.; Babić, J.; Jozinović, A.; Miličević, B.; Jokić, S.; Miličević, R.; Rajič, M.; Šubarić, D. Starch Modification by Organic Acids and Their Derivatives: A Review. *Molecules* **2015**, *20*, 19554–19570. [CrossRef] [PubMed]

14. Kim, H.Y.; Park, S.S.; Lim, S.T. Preparation, Characterization and Utilization of Starch Nanoparticles. *Colloids Surf. B Biointerfaces* **2015**, *126*, 607–620. [CrossRef] [PubMed]

15. Jyothi, A.N.; Carvalho, A.J.F. *Polysaccharide Based Graft Copolymers*, 1st ed.; Kalia, S., Sabaa, M., Eds.; Springer: Berlin/Heidelberg, Germany, 2013; pp. 59–110.

16. Mollaahmad, M.A. Sustainable Fillers for Paper. Masters's Thesis, Luleå University of Technology, Luleå, Sweden, 2008.

17. Le Corre, D. Starch Nanocrystals: Preparation and Application to Bio-Based Flexible Packaging. Ph.D. Thesis, Université De Grenoble, Grenoble, France, 2011.

18. Dudhani, A.R. *Starch-Based Polymeric Materials and Nanocomposites*, 1st ed.; Ahmed, J., Tiwari, B.K., Imam, S.H., Rao, M.A., Eds.; CRC Press: Boca Raton, FL, USA, 2012; pp. 361–372.

19. Angellier, H.; Molina-Boisseau, S.; Dufresne, A. Mechanical Properties of Waxy Maize Starch Nanocrystal Reinforced Natural Rubber. *Macromolecules* **2005**, *38*, 9161–9170. [CrossRef]

20. Dennenberg, R.J.; Abbott, T.P. Rapid Analysis of Starch Graft Copolymers. *J. Polym. Sci. Polym. Lett. Ed.* **1976**, *14*, 693–696. [CrossRef]

21. Mehrotra, R.; Ranby, B. Graft Copolymerization Onto Starch. II. Grafting of Acrylonitrile to Granular Native Potato Starch by Manganic Pyrophosphate Initiation. Effect of Reaction Conditions on Grafting Parameters. *J. Appl. Polym. Sci.* **1977**, *21*, 3407–3415. [CrossRef]

22. Mehrotra, R.; Ranby, B. Graft Copolymerization Onto Starch. III. Grafting of Acrylonitrile to Gelatinized Potato Starch by Manganic Pyrophosphate Initiation. *J. Appl. Polym. Sci.* **1978**, *22*, 2991–3001. [CrossRef]

23. Mehrotra, R.; Ranby, B. Graft Copolymerization Onto Starch. IV. Grafting of Methyl Methacrylate to Granular Native Potato Starch by Manganic Pyrophosphate Initiation. *J. Appl. Polym. Sci.* **1978**, *22*, 3003–3010. [CrossRef]

24. Dufresne, A. *Polysaccharides: Bioactivity and Biotechnology*, 1st ed.; Ramawat, K.G., Merillon, J.-M., Eds.; Springer International Publishing: Basel, Switzerland, 2015; pp. 417–449.

25. Heinze, T.; Koschella, A. Carboxymethyl Ethers of Cellulose and Starch—A Review. *Macromol. Symp.* **2005**, *223*, 13–39. [CrossRef]

26. Le Corre, D.; Angellier-Coussy, H. Preparation and Application of Starch Nanoparticles for Nanocomposites: A Review. *React. Funct. Polym.* **2014**, *85*, 97–120. [CrossRef]

27. Dufresne, A.; Castaño, J. Polysaccharide Nanomaterial Reinforced Starch Nanocomposites: A Review. *Starch/Stärke* **2016**, *68*, 1–19. [CrossRef]

28. Carvalho, O.; Avérous, L.; Tadini, C. Mechanical Properties of Cassava Starch-Based Nano-Biocomposites. In Proceedings of the 11th International Confress of Engineering and Food, Athens, Greece, 22–26 May; Saravacos, G., Ed.; Elsevier Procedia: New York, NY, USA, 2011; pp. 111–112.

29. Leja, K.; Lewandowicz, G. Polymer Biodegradation and Biodegradable Polymers—A Review. *Polish J. Environ. Stud.* **2010**, *19*, 255–266.

30. Mose, B.R.; Maranga, S.M. A Review on Starch Based Nanocomposites for Bioplastic Materials. *J. Mater. Sci. Eng. B* **2011**, *1*, 239–245.

31. Lu, D.R.; Xiao, C.M.; Xu, S.J. Starch-Based Completely Biodegradable Polymer Materials. *Express Polym. Lett.* **2009**, *3*, 366–375. [CrossRef]

32. Athawale, V.D.; Rathi, S.C. Graft Polymerization: Starch as A Model Substrate. *J. Macromol. Sci. Part C Polym. Rev.* **1999**, *39*, 445–480. [CrossRef]

33. Santana, Á.L.; Angela, M.; Meireles, A. New Starches Are the Trend for Industry Applications: A Review. *Food Public Heal.* **2014**, *4*, 229–241. [CrossRef]

34. Meshram, M.W.; Patil, V.V.; Mhaske, S.T.; Thorat, B.N. Graft Copolymers of Starch and Its Application In Textiles. *Carbohydr. Polym.* **2009**, *75*, 71–78. [CrossRef]

35. Li, M.C.; Ge, X.; Cho, U.R. Emulsion Grafting Vinyl Monomers Onto Starch for Reinforcement of Styrene-Butadiene Rubber. *Macromol. Res.* **2013**, *21*, 519–528. [CrossRef]

36. van den Oever, M.; Molenveld, K. Replacing Fossil Based Plastic Performance Products by Bio-Based Plastic Products-Technical Feasibility. *New Biotechnol.* **2017**, *37*, 48–59. [CrossRef] [PubMed]

37. Glavas, L. Starch and Protein Based Wood Adhesives. Masters's Thesis, Kungluga Tekniska Högskolan, Stockholm, Sweden, 2011.

38. Eutamene, M.; Benbakhti, A.; Khodja, M.; Jada, A. Preparation and Aqueous Properties of Starch-Grafted Polyacrylamide Copolymers. *Starch/Staerke* **2009**, *61*, 81–91. [CrossRef]

39. Shin, J.Y.; Jones, N.; Lee, D.I.; Fleming, P.D.; Joyce, M.K.; Dejong, R.; Bloembergen, S. Rheological Properties of Starch Latex Dispersions and Starch Latex-Containing Coating Colors. In Proceedings of the"Growing the Future", Proceedings of the Paper Conference and Trade Show, TAPPI Press, Peachtree Corners, GA, USA, 22–25 April 2012; pp. 1–25.

40. Misman, M.A.; Azura, A.R.; Hamid, Z.A.A. The Physical and Degradation Properties of Starch-Graft-Acrylonitrile/Carboxylated Nitrile Butadiene Rubber Latex Films. *Carbohydr. Polym.* **2015**, *128*, 1–10. [CrossRef] [PubMed]

41. Zhang, J.F.; Sun, X. Mechanical Properties of Poly(Lactic Acid)/Starch Composites Compatibilized by Maleic Anhydride. *Biomacromolecules* **2004**, *5*, 1446–1451. [CrossRef] [PubMed]

42. Xiao, C.M.; Tan, J.; Xue, G.N. Synthesis and Properties of Starch-g-Poly(Maleic Anhydride-Co-Vinyl Acetate). *Express Polym. Lett.* **2010**, *4*, 9–16. [CrossRef]

43. Wang, C.; Pan, Z.; Wu, M.; Zhao, P. Effect of Reaction Conditions on Grafting Ratio and Properties of Starch Nanocrystals-g-Polystyrene. *J. Appl. Polym. Sci.* **2014**, *131*, 1–7. [CrossRef]

44. Noordergraaf, I.-W.; Fourie, T.K.; Raffa, P. Free-Radical Graft Polymerization Onto Starch as a Tool to Tune Properties in Relation to Potential Applications. A Review. *Processes* **2018**, *6*, 31. [CrossRef]

45. Smeets, N.M.B.; Imbrogno, S.; Bloembergen, S. Carbohydrate Functionalized Hybrid Latex Particles. *Carbohydr. Polym.* **2017**, *173*, 233–252. [CrossRef] [PubMed]

46. Asua, J.M. Emulsion Polymerization: From Fundamental Mechanisms to Process Developments. *J. Polym. Sci. Part A Polym. Chem.* **2004**, *42*, 1025–1041. [CrossRef]

47. Steward, P.A.; Hearn, J.; Wilkinson, M.C. Overview of Polymer Latex Film for mation and Properties. *Adv. Colloid Interface Sci.* **2000**, *86*, 195–267. [CrossRef]

48. Okubo, M.; Yamada, A.; Matsumoto, T. Estimation of Morphology of Composite Polymer Emulsion Particles by the Soap Titration Method. *J. Polym. Sci. Polym. Chem. Ed.* **1980**, *18*, 3219–3228. [CrossRef]

49. Okubo, M.; Katsuta, Y.; Matsumoto, T. Studies on Suspension and Emulsion. LI. Peculiar Morphology of Composite Polymer Particles Produced by Seeded Emulsion Polymerization. *J. Polym. Sci. Polym. Lett. Ed.* **1982**, *20*, 45–51. [CrossRef]

50. Sundberg, D.C.; Durant, Y.G. Latex Particle Morphology, Fundamental Aspects: A Review. *Polym. React. Eng.* **2003**, *11*, 379–432. [CrossRef]

51. Singh, K.B. Understanding Film for mation Mechanism In Latex Dispersions. Ph.D. Thesis, Indian Institute of Technology Bombay, Mumbai, India, 2008.

52. Dalnoki-Veress, K.; Dutcher, J.R.; for rest, J.A. Dynamics and Pattern for mation in Thin Polymer Films. *Phys. Can.* **2003**, *59*, 75–84.

53. Keddie, J.L. Film for mation of Latex. *Mater. Sci. Eng.* **1997**, *21*, 101–170. [CrossRef]

54. Ranum, P.; Peña-Rosas, J.P.; Garcia-Casal, M.N. Global Maize Production, Utilization, and Consumption. *Ann. N. Y. Acad. Sci.* **2014**, *1312*, 105–112. [CrossRef] [PubMed]

55. Rudin, A.; Choi, P. *The Elements of Polymer Science & Engineering*, 3rd ed.; Elsevier: New York, NY, USA, 2013; pp. 521–535.

56. Olsson, E. Effects of Citric Acid on Starch-Based Barrier Coatings. Ph.D. Thesis, Karlstad University, Karlstad, Sweden, 2013.

57. Li, M.; Witt, T.; Xie, F.; Warren, F.J.; Halley, P.J.; Gilbert, R.G. Biodegradation of Starch Films: The Roles of Molecular and Crystalline Structure. *Carbohydr. Polym.* **2015**, *122*, 115–122. [CrossRef] [PubMed]

58. Gallant, D.J.; Bouchet, B.; Baldwin, P.M. Microscopy of Starch: Evidence of A New Level of Granule Organization. *Carbohydr. Polym.* **1997**, *32*, 177–191. [CrossRef]

59. Le Corre, D.; Bras, J.; Dufresne, A. Starch Nanoparticles: A Review. *Biomacromolecules* **2010**, *11*, 1139–1153. [CrossRef] [PubMed]

60. Singh, N.; Singh, J.; Kaur, L.; Sodhi, N.S.; Gill, B.S. Morphological, Thermal and Rheological Properties of Starches From Different Botanical Sources. *Food Chem.* **2003**, *81*, 219–231. [CrossRef]

61. Namazi, H.; Fathi, F.; Heydari, A. *The Delivery of Nanoparticles*, 1st ed.; Hashim, A.A., Ed.; Intech: London, UK, 2012; pp. 149–184.

62. Daniel-da-Silva, A.L.; Trindade, T. *Advances in Nanocomposite Technology*, 1st ed.; Hashim, A.A., Ed.; Intech: London, UK, 2011; pp. 275–291.

63. Davidson, R.L. *Handbook of Water-Soluble Gums and Resins*, 1st ed.; Davidson, R.L., Ed.; McGraw-Hill: New York, NY, USA, 1980.

64. Rindlav-Westling, A.; Stading, M.; Hermansson, A.-M.; Gatenholm, P. Structure, Mechanical and Barrier Properties of Amylose and Amylopectin Films. *Carbohydr. Polym.* **1998**, *36*, 217–224. [CrossRef]
65. Lu, S.; Lin, S.; Yao, K. Study on The Synthesis and Application of Starch-Graft-Poly(AM-Co-DADMAC) by Using A Complex Initiation System of CS-KPS. *Starch/Stärke* **2004**, *56*, 138–143. [CrossRef]
66. Meimoun, J.; Wiatz, V.; Saint-Loup, R.; Parcq, J.; Favrelle, A.; Bonnet, F.; Zinck, P. Modification of Starch by Graft Copolymerization. *Starch/Stärke* **2018**, *70*, 1–23. [CrossRef]
67. Liu, W.C.; Halley, P.J.; Gilbert, R.G. Mechanism of Degradation of Starch, a Highly Branched Polymer, during Extrusion. *Macromolecules* **2010**, *43*, 2855–2864. [CrossRef]
68. Liu, D.; Wu, Q.; Chen, H.; Chang, P.R. Transitional Properties of Starch Colloid with Particle Size Reduction from Micro- to Nanometer. *J. Colloid Interface Sci.* **2009**, *339*, 117–124. [CrossRef] [PubMed]
69. Shah, U.; Naqash, F.; Gani, A.; Masoodi, F.A. Art and Science behind Modified Starch Edible Films and Coatings: A Review. *Compr. Rev. Food Sci. Food Saf.* **2016**, *15*, 568–580. [CrossRef]
70. Zhu, F. Composition, Structure, Physicochemical Properties, and Modifications of Cassava Starch. *Carbohydr. Polym.* **2015**, *122*, 456–480. [CrossRef] [PubMed]
71. Shi, A.M.; Li, D.; Wang, L.J.; Li, B.Z.; Adhikari, B. Preparation of Starch-Based Nanoparticles through High-Pressure Homogenization and Miniemulsion Cross-Linking: Influence of Various Process Parameters on Particle Size and Stability. *Carbohydr. Polym.* **2011**, *83*, 1604–1610. [CrossRef]
72. Li, W.; Cao, F.; Fan, J.; Ouyang, S.; Luo, Q.; Zheng, J.; Zhang, G. Physically Modified Common Buckweat Starch and Their Physicochemical and Structural Properties. *Food Hydrocoll.* **2014**, *40*. [CrossRef]
73. Bel Haaj, S.; Thielemans, W.; Magnin, A.; Boufi, S. Starch Nanocrystals and Starch Nanoparticles from Waxy Maize as Nanoreinforcement: A Comparative Study. *Carbohydr. Polym.* **2016**, *143*, 310–317. [CrossRef] [PubMed]
74. Zondag, M.D. Effect of Microwave Heat-Moisture and Annealing Treatments on Buckwheat Starch Characteristics. Masters's Thesis, University of Wisconsin, Madison, WI, USA, 2003.
75. Liu, X.; Wang, Y.; Yu, L.; Tong, Z.; Chen, L.; Liu, H.; Li, X. Thermal Degradation and Stability of Starch under Different Processing Conditions. *Starch/Staerke* **2013**, *65*, 48–60. [CrossRef]
76. Šárka, E.; Dvořáček, V. New Processing and Applications of Waxy Starch (A Review). *J. Food Eng.* **2017**, *206*, 77–87. [CrossRef]
77. Kim, H.Y.; Lee, J.H.; Kim, J.Y.; Lim, W.J.; Lim, S.T. Characterization of Nanoparticles Prepared by Acid Hydrolysis of Various Starches. *Starch/Stärke* **2012**, *64*, 367–373. [CrossRef]
78. Kim, J.Y.; Yoon, J.W.; Lim, S.T. Formation and Isolation of Nanocrystal Complexes between Dextrins and N-Butanol. *Carbohydr. Polym.* **2009**, *78*, 626–632. [CrossRef]
79. Saari, H.; Fuentes, C.; Sjöö, M.; Rayner, M.; Wahlgren, M. Production of Starch Nanoparticles by Dissolution and Non-Solvent Precipitation for Use in Food-Grade Pickering Emulsions. *Carbohydr. Polym.* **2017**, *157*, 558–566. [CrossRef] [PubMed]
80. Song, D.; Thio, Y.S.; Deng, Y. Starch Nanoparticle for mation Via Reactive Extrusion and Related Mechanism Study. *Carbohydr. Polym.* **2011**, *85*, 208–214. [CrossRef]
81. Kim, J.Y.; Park, D.J.; Lim, S.T. Fragmentation of Waxy Rice Starch Granules by Enzymatic Hydrolysis. *Cereal Chem.* **2008**, *85*, 182–187. [CrossRef]
82. Song, D. Starch Crosslinking for Cellulose Fiber Modification and Starch Nanoparticle for Mation. Ph.D. Thesis, Georgia Institute of Technology, Atlanta, GA, USA, 2011.
83. Lin, N.; Huang, J.; Chang, P.R.; Anderson, D.P.; Yu, J. Preparation, Modification, and Application of Starch Nanocrystals In Nanomaterials: A Review. *J. Nanomater.* **2011**, *2011*, 1–13. [CrossRef]
84. Hui, R.; Qi-he, C.; Ming-liang, F.; Qiong, X.; Guo-qing, H. Preparation and Properties of Octenyl Succinic Anhydride Modified Potato Starch. *Food Chem.* **2009**, *114*, 81–86. [CrossRef]
85. Rzayev, Z.M.O. Graft Copolymers of Maleic Anhydride and Its Isostructural Analogues: High Performance Engineering Materials. *Int. Rev. Chem. Eng.* **2011**, *3*, 1689–1699.
86. Xu, Y.; Ding, W.; Liu, J.; Li, Y.; Kennedy, J.F.; Gu, Q.; Shao, S. Preparation and Characterization of Organic-Soluble Acetylated Starch Nanocrystals. *Carbohydr. Polym.* **2010**, *80*, 1078–1084. [CrossRef]
87. Chakraborty, S.; Sahoo, B.; Teraoka, I.; Miller, L.M.; Gross, R.A. Enzyme-Catalyzed Regioselective Modification of Starch Nanoparticles. *Macromolecules* **2005**, *38*, 61–68. [CrossRef]
88. Cheng, S.; Zhao, W.; Wu, Y. Optimization of Synthesis and Characterization of Oxidized Starch-Graft-Poly(Styrene-Butyl Acrylate) Latex for Paper Coating. *Starch/Stärke* **2015**, *67*, 493–501. [CrossRef]

89. Cummings, S. The Incorporation of Vinyl Modified Regenerated Starch Nanoparticles in Emulsion Polymerizations. Ph.D. Thesis, University of Ottawa, Ottawa, ON, Canada, 2017.

90. Li, M.; Mun, Y.; Cho, U.R. Synthesis of Environmental-Friendly Starch-Acrylic Coating Sols by Emulsion Polymerization. *Elastom. Compos.* **2010**, *45*, 272–279.

91. Mou, J.; Li, X.; Wang, H.; Fei, G.; Liu, Q. Preparation, Characterization, and Water Resistance of Cationic Acetylated Starch-g-Poly(Styrene-Butyl Acrylate) Surfactant-Free Emulsion. *Starch/Stärke* **2012**, *64*, 826–834. [CrossRef]

92. Song, S.; Wang, C.; Pan, Z.; Wang, X. Preparation and Characterization of Amphiphilic Starch Nanocrystals. *J. Appl. Polym. Sci.* **2008**, *107*, 418–422. [CrossRef]

93. Gao, J.-P.; Tian, R.-C.; Yu, J.-G.; Duan, M.-L. Graft Copolymers of Methy Methacrylate onto Canna Starch Using Manganic Pyrophosphate as an Initiator. *J. Appl. Polym. Sci.* **1994**, *53*, 1091–1102. [CrossRef]

94. Wyatt, V.T.; Strahan, G.D. Degree of Branching in Hyperbranched Poly(Glycerol-Co-Diacid)s Synthesized in Toluene. *Polymers* **2012**, *4*, 396–407. [CrossRef]

95. Nilsson, G.S.; Bergquist, K.-E.; Nilsson, U.; Gorton, L. Determination of the Degree of Branching in Normal and Amylopectin Tye Potato Starch with 1H-NMR: Improved Resolution and Two-Dimensional Spectroscopy. *Starch/Stärke* **1996**, *10*, 352–357. [CrossRef]

96. Zhu, X.; Chen, L.; Chen, Y.; Yan, D. Using 2D NMR to Determine the Degree of Branching of Complicated Hyperbranched Polymers. *Sci. China Ser. B Chem.* **2008**, *51*, 1057–1065. [CrossRef]

97. Markoski, L.J.; Thompson, J.L.; Moore, J.S. Indirect Method for Determining Degree of Branching in Hyperbranched Polymers. *Macromolecules* **2002**, *35*, 1599–1603. [CrossRef]

98. Liu, J.M.; Chang, C.S.; Tsiang, R.C.C. Method of Determining the Degree of Branching from the Gel Permeation Chromatogram for Star-Shaped SBS Thermoplastic Block Copolymers. *J. Polym. Sci. Part A Polym. Chem.* **1997**, *35*, 3393–3401. [CrossRef]

99. Mange, S.; Dever, C.; De Bruyn, H.; Gaborieau, M.; Castignolles, P.; Gilbert, R.G. Grafting of Oligosaccharides Onto Synthetic Polymer Colloids. *Biomacromolecules* **2007**, *8*, 1816–1823. [CrossRef] [PubMed]

100. Baijun, W.; Hui, X.; Yonghong, Z. Study on Graft Emulsion Copolymerization of Styrene/Butyl Acrylate Onto Starch. *Spec. Petrochem.* **2005**, *2*.

101. Hwang, J.H.; Ryu, H.; Cho, U.R. A Study on Starch-Acrylic Graft Copolymerization by Emulsion Polymerization. *Elastomer* **2008**, *43*, 221–229.

102. Chu, H.J.; Wei, H.L.; Zhu, J. Ultrasound Enhanced Radical Graft Polymerization of Starch and Butyl Acrylate. *Chem. Eng. Process. Process Intensif.* **2015**, *90*, 1–5. [CrossRef]

103. Peng, B.L.; Dhar, N.; Liu, H.L.; Tam, K.C. Chemistry and Applications of Nanocrystalline Cellulose and its Derivatives: A Nanotechnology Perspective. *Can. J. Chem. Eng.* **2011**, *89*, 1191–1206. [CrossRef]

104. Kweon, M.; Slade, L.; Levine, H. Role of Glassy and Crystalline Transitions in the Responses of Corn Starches to Heat and High Pressure Treatments: Prediction of Solute-Induced Barostability from Solute-Induced Thermostability. *Carbohydr. Polym.* **2008**, *72*, 293–299. [CrossRef]

105. Cranston, E.D.; Gray, D.G. Birefringence in Spin-Coated Films Containing Cellulose Nanocrystals. *Colloids Surfaces A Physicochem. Eng. Asp.* **2008**, *325*, 44–51. [CrossRef]

106. Cheng, S.; Zhang, Y.; Wu, Y. Preparation and Characterization of Enzymatically Degraded Starch-g-Poly(Styrene-Co-Butyl Acrylate) Latex for Paper Coating. *Polym. Plast. Technol. Eng.* **2014**, *53*, 1811–1816. [CrossRef]

107. Nikfarjam, N.; Taheri Qazvini, N.; Deng, Y. Cross-Linked Starch Nanoparticles Stabilized Pickering Emulsion Polymerization of Styrene In W/O/W System. *Colloid Polym. Sci.* **2014**, *292*, 599–612. [CrossRef]

108. Jiang, B.; Tsavalas, J.G.; Sundberg, D.C. Water Whitening of Polymer Films: Mechanistic Studies and Comparisons Between Water and Solvent Borne Films. *Prog. Org. Coat.* **2017**, *105*, 56–66. [CrossRef]

109. Rodehed, C.; Rånby, B. Structure and Molecular Properties of Saponified Starch–Graft-polyacrylonitrile. *J. Appl. Polym. Sci.* **1986**, *32*, 3323–3333. [CrossRef]

110. Glenn, G.M.; Orts, W.; Imam, S.; Chiou, B.; Wood, D.F. *Starch Polymers*, 1st ed.; Halley, P., Averous, L., Eds.; Elsevier: New York, NY, USA, 2014; pp. 421–452.

111. Mostafa, K.M. Graft Polymerization of Acrylic Acid Onto Starch Using Potassium Permanganate Acid (Redox System). *J. Appl. Polym. Sci.* **1995**, *56*, 263–269. [CrossRef]

112. Cheng, S.; Xu, J.; Wu, Y. Preparation and Characterization of Oxidized Starch-graft-Poly(styrene-butyl acrylate) Latex Via Emulsion Polymerization. *J. Polym. Eng.* **2014**, *34*, 611–616. [CrossRef]

113. Jiang, S.; Dai, L.; Xiong, L.; Sun, Q. Preparation and Characterization of Octenyl Succinic Anhydride Modified Taro Starch Nanoparticles. *PLoS ONE* **2016**, *11*, e0150043. [CrossRef] [PubMed]

114. Ortega-Toro, R.; Bonilla, J.; Talens, P.; Chiralt, A. *Starch-Based Materials in Food Packaging: Processing, Characterization and Applications*, 1st ed.; Villar, M.A., Barbosa, S.E., Garcia, A.M., Castillo, L.A., López, O.V., Eds.; Academic Press: London, UK, 2017; pp. 257–312.

115. Ulu, A.; Koytepe, S.; Ates, B. Design of Starch Functionalized Biodegradable P(MMA-co-MMA) as Carrier Matrix for l-asparaginase Immobilization. *Carbohydr. Polym.* **2016**, *153*, 559–572. [CrossRef] [PubMed]

116. Ulu, A.; Koytepe, S.; Ates, B. Synthesis and Characterization of Biodegradable pHEMA-starc Composites for Immobilization of l-asparaginase. *Polym. Bull.* **2016**, *73*, 1891. [CrossRef]

117. Garcia-Valdez, O.; Champagne, P.; Cunningham, M.F. Graft Modification of Natural Polysaccharides via Reversible Deactivation Radical Polymerization. *Prog. Polym. Sci.* **2018**, *76*, 151–173. [CrossRef]

118. Glasing, J.; Champagne, P.; Cunningham, M.F. Graft Modification of Chitosan, Cellulose and Alginate Using Reversible Deactivation Radical Polymerization (RDRP). *Curr. Opin. Green Sustain. Chem.* **2016**, *2*, 15–21. [CrossRef]

119. Chen, Q.; Yu, H.; Wang, L.; Ul Abdin, Z.; Chen, Y.; Wang, J.; Zhou, W.; Yang, X.; Khan, R.U.; Zhang, H.; et al. Recent Progress in Chemical Modification of Starch and Its Applications. *RSC Adv.* **2015**, *5*, 67459–67474. [CrossRef]

120. Lu, D.; Duan, P.; Guo, R.; Yang, L.; Zhang, H. Synthesis and Characterization of Starch Graft Biodegradable Polyester and Polyesteramide by Direct Polycondensation. *Polym. Plast. Technol. Eng.* **2013**, *52*, 200–205. [CrossRef]

121. Maiti, S.; Ranjit, S.; Sa, B. Polysaccharide-Based Graft Copolymers in Controlled Drug Delivery. *Int. J.* **2010**, *2*, 1350–1358.

122. Djordjevic, S.; Nikolic, L.; Kovacevic, S.; Miljkovic, M.; Djordjevic, D. Graft Copolymerization of Acrylic Acid Onto Hydrolyzed Potato Starch Using Various Initiators. *Period. Polytech. Chem. Eng.* **2013**, *57*, 55–61. [CrossRef]

123. Lai, S.M.; Don, T.M.; Liu, Y.H.; Chiu, W.Y. Graft Polymerization of Vinyl Acetate Onto Granular Starch: Comparison on The Potassium Persulfate and Ceric Ammonium Nitrate Initiated System. *J. Appl. Polym. Sci.* **2006**, *102*, 3017–3027. [CrossRef]

124. Lanthong, P.; Nuisin, R.; Kiatkamjornwong, S. Graft Copolymerization, Characterization, and Degradation of Cassava Starch-g-Acrylamide/Itaconic Acid Superabsorbents. *Carbohydr. Polym.* **2006**, *66*, 229–245. [CrossRef]

125. Shi, Z.; Reddy, N.; Shen, L.; Hou, X.; Yang, Y. Effects of Monomers and Homopolymer Contents on the Dry and Wet Tensile Properties of Starch Films Grafted with Various Methacrylates. *J. Agric. Food Chem.* **2014**, *62*, 4668–4676. [CrossRef] [PubMed]

126. Wu, Y.B.; Lv, C.F.; Han, M.N. Synthesis and Performance Study of Polybasic Starch Graft Copolymerization Function Materials. *Adv. Mater. Res.* **2009**, *79–82*, 43–46. [CrossRef]

127. Xue, Y.; Cheng, S.; We, Y. Study on Preparation of Graft Copolymer Emulsion of Zymolytic Starch-Styrene/Butyl Acrylate. *Appl. Chem. Ind.* **2012**, *2*.

128. Lee, M.; Ryu, H.; Cho, U.R. A Study on The Synthesis of Starch-Acrylic Polymer by Emulsion Polymerization. *Polymer* **2010**, *34*, 58–62. [CrossRef]

129. Haaj, S.B.; Bou, S. Starch Nanoparticles Produced Via Ultrasonication As A Sustainable Stabilizer In Pickering Emulsion Polymerization. *RSC Adv.* **2014**, *4*, 42638–42646. [CrossRef]

130. Pei, X.; Tan, Y.; Xu, K.; Liu, C.; Lu, C.; Wang, P. Pickering Polymerization of Styrene Stabilized by Starch-Based Nanospheres. *Polym. Chem.* **2016**, *7*, 3325–3333. [CrossRef]

131. Haaj, S.B.; Thielemans, W.; Magnin, A.; Boufi, S. Starch Nanocrystal Stabilized Pickering Emulsion Polymerization for Nanocomposites with Improved Performance. *ACS Appl. Mater. Interfaces* **2014**, *6*, 8263–8273. [CrossRef] [PubMed]

132. Angellier, H.; Choisnard, L.; Molina-Boisseau, S.; Ozil, P.; Dufresne, A. Optimization of the Preparation of Aqueous Suspensions of Waxy Maize Starch Nanocrystals Using a Response Surface Methodology. *Biomacromolecules* **2004**, *5*, 1545–1551. [CrossRef] [PubMed]

133. Namazi, H.; Dadkhah, A.; Mosadegh, M. New Biopolymer Nanocomposite of Starch-Graft Polystyrene/Montmorillonite Clay Prepared Through Emulsion Polymerization Method. *J. Polym. Environ.* **2012**, *20*, 794–800. [CrossRef]

134. Ulu, A.; Koytepe, S.; Ates, B. Synthesis and Characterization of PMMA Composites Activated With Starch for Immobilization of L-Asparaginase. *J. Appl. Polym. Sci.* **2016**, *133*, 1–11. [CrossRef]

135. Xu, J.; Long, L.; Hu, H. Preparation of Starch-Based Styrene Acrylate Emulsion Used As Surface-Treatment Agent for Decorative Base Paper. *J. Polym. Eng.* **2013**, *33*, 323–330. [CrossRef]

136. Wang, R.-M.; Wang, X.-W.; Guo, J.-F.; He, Y.-F.; Jiang, M.-L. Crosslinkable Potato Starch-Based Graft Copolymer Emulsion for Humidity Controlling Coatings. *Mater. Res.* **2013**, *16*, 1246–1253. [CrossRef]

137. Terpstra, R. Potato Starch Stabilized Synthetic Latexes. Ph.D. Thesis, University of Groningen, Groningen, The Netherlands, 2015.

138. Brockway, C.E.; Seaberg, P.A. Grafting of Polyacrylonitrile to Granular Corn Starch. *J. Polym. Sci. Part A-1* **1967**, *5*, 1313–1326. [CrossRef]

139. Misra, B.N.; Dogra, R.; Kaur, I.; Sood, D. Grafting onto Starch. II. Graft Copolymerization of Vinyl Acetate on to Starch by Radical Initiator. *J. Polym. Sci. Polym. Chem. Ed.* **1980**, *18*, 341–344. [CrossRef]

140. Misra, B.N.; Dogra, R. Grafting onto Starch. IV. Graft Copolymerization of Methyl Methacrylate by Use of AIBN as Radical Initiator. *J. Macromol. Sci.* **2006**, *14*, 763–770. [CrossRef]

141. Pei, X.; Zhai, K.; Tan, Y.; Xu, K.; Lu, C.; Wang, P.; Wang, T.; Chen, C.; Tao, Y.; Dai, L.; et al. Synthesis of Monodisperse Starch-Polystyrene Core-Shell Nanoparticles Via Seeded Emulsion Polymerization without Stabilizer. *Polymer* **2016**, *108*, 78–86. [CrossRef]

142. Abo-Shosha, M.H.; Ibrahim, N.A. Synthesis and Characterization of Starch-Glycidyl Methacrylate-Acrylic Acid Cation Exchange Composites. *Starch/Stärke* **1992**, *44*, 466–471. [CrossRef]

143. Wu, G.; Limei, Z. Study on KPS-Na2S2O3 Initiated Graft Copolymerization of Butyl Acrylate Onto Corn Starch. *Petrochemical Technol.* **1993**, *5*.

144. Hebeish, A.; Beliakova, M.K.; Bayazeed, A. Improved Synthesis of Poly (MAA)—Starch. *Jourbal Appl. Polym. Sci.* **1997**, *68*, 1709–1715. [CrossRef]

145. Beliakova, M.K.; Aly, A.A.; Abdel-Mohdy, F.A. Grafting of Poly(Methacrylic Acid) on Starch and Poly(Vinyl Alcohol). *Starch/Stärke* **2004**, *56*, 407–412. [CrossRef]

146. Ikhuoria, E.U.; Folayan, A.S.; Okieimen, F.E. Studies in the Graft Copolymerization of Acrylonitrile onto Cassava Starch by Ceric Ion Induced Initiation. *Int. J. Biotechnol. Mol. Biol. Res.* **2010**, *1*, 10–14.

147. Burr, R.C.; Fanta, G.F.; Russell, C.R.; Rist, C.E. Influence of Swelling and Disruption of The Starch Granule on The Composition of The Starch-Polyacrylonitrile Copolymer. *J. Macromol. Sci. Part A Chem.* **1967**, *7*, 1381–1385. [CrossRef]

148. Fanta, G.F.; Burr, R.C.; Russell, C.R.; Rist, C.E. Copolymers of Starch and Polyacrylonitrile. Influence of Granule Swelling on Copolymer Composition under Various Reaction Conditions. *J. Macromol. Sci. Part A Chem.* **1970**, *4*, 331–339. [CrossRef]

149. Han, T.L.; Kumar, R.N.; Rozman, H.D.; Noor, M.A.M. GMA Grafted Sago Starch as a Reactive Component in Ultra Violet Radiation Curable Coatings. *Carbohydr. Polym.* **2003**, *54*, 509–516. [CrossRef]

150. Zhu, Z.; Zhuo, R. Controlled Release of Carboxylic-Containing Herbicides by Starch-g-Poly(Butyl Acrylate). *J. Appl. Polym. Sci.* **2001**, *81*, 1535–1543. [CrossRef]

151. Li, M.-C.; Lee, J.K.; Cho, U.R. Synthesis, Characterization, and Enzymatic Degradation of Starch-Grafted Poly(Methyl Methacrylate) Copolymer Films. *Polym. Polym. Compos.* **2013**, *21*, 449–456. [CrossRef]

152. Sangramsingh, N.M.; Patra, B.N.; Singh, B.C.; Patra, C.M. Graft Copolymerization of Methyl Methacrylate onto Starch Using a Ce (IV)—Glucose Initiator System. *J. Appl. Ploymer Sci.* **2003**, *91*, 981–990. [CrossRef]

153. Okieimen, F.E.; Said, O.B. Studies on The Graft Copolymerization of Methyl Mathacrylate onto Starch. *Acta Polym.* **1989**, *40*, 708–710. [CrossRef]

154. Ray-Chaudhuri, D.K. Grafted Starch Acrylates and Their Properties. *Starch/Stärke* **1969**, *21*, 47–52. [CrossRef]

155. Fares, M.M.; El-Faqeeh, A.S.; Osman, M.E. Graft Copolymerization onto Starch-I. Synthesis and Optimization of Starch Grafted with N-Tert-Butylacrylamide Copolymer and Its Hydrogels. *J. Polym. Res.* **2003**, *10*, 119–125. [CrossRef]

156. Zou, W.; Yu, L.; Liu, X.; Chen, L.; Zhang, X.; Qiao, D.; Zhang, R. Effects of Amylose/Amylopectin Ratio on Starch-Based Superabsorbent Polymers. *Carbohydr. Polym.* **2012**, *87*, 1583–1588. [CrossRef]

157. Singh, O.P.; Sandle, N.K.; Varma, I.K. Graft-Copolymerization of Starch With Acrylamide, III. Physical Properties and Enzymatic Degradation. *Die Angew. Makromol. Chemie* **1984**, *122*, 193–201. [CrossRef]

158. Gugliemelli, L.A.; Doane, W.M.; Russell, C.R. Preparation of Soapless Latexes by Sonification of Starch-based Poly(Isoprene–Co–Acrylonitrile) Graft Reaction Mixtures. *J. Appl. Polym. Sci.* **1979**, *23*, 635–644. [CrossRef]

159. Weaver, M.O. Starch-g-Poly(Methyl Acrylate) Latexes for Stabilizing Soil to Water Erosion: Extending the Range of Polymer Add-On. *Starch/Stärke* **1989**, *41*, 106–110. [CrossRef]

160. Gaborieau, M.; de Bruyn, H.; Mange, S.; Castignolles, P.; Brockmeyer, A.; Gilbert, R.G. Synthesis and Characterization of Synthetic Polymer Colloids Colloidally Stabilized by Cationized Starch Oligomers. *J. Polym. Sci. Part A* **2009**, *47*, 1836–1852. [CrossRef]

161. Gugliemelli, L.A.; Swanson, C.L.; Baker, F.L.; Doane, W.M.; Russell, C.R. Cationic Starch-Polyacrylonitrile Graft. *J. Polym. Sci. Polym. Chem. Ed.* **1974**, *12*, 2683–2692. [CrossRef]

162. Patel, N.; Patel, K.C.; Patel, R.D. The Effect of Acrylonitrile on Starch Gelatinization. Morphological Study. *Starch/Stärke* **1985**, *37*, 201–205. [CrossRef]

163. Ghosh, P.; Paul, S.K. Graft Copolymerization of Methyl Methacrylate on Potato Starch Using Potassium Trioxalatomanganate, K3[Mn(C2O4)3], as Initiator. *J. Macromol. Sci.* **2006**, *20*, 179–188. [CrossRef]

164. Carlsohn, H.; Hartmann, M. Mangan(1V)-Initiierte Pfropfcopolymerisation von Methylmethacrylat Und Anderen Acrylmonomeren Auf Starke Und Dextran. *Acta Polym.* **1982**, *33*, 640–643. [CrossRef]

165. Mehrotra, R.; Ranby, B. Graft Copolymerization onto Starch. I. Complexes of Mn3 as Initiators. *J. Appl. Polym. Sci.* **1977**, *21*, 1647–1654. [CrossRef]

166. Qu, B.; Li, H.; Niu, Y. Graft Copolymerization of Poly(Vinyl Acetate) onto Starch Using KMnO4-H2SO4 Redox System. *J. Polym. Eng.* **2013**, *33*, 521–526. [CrossRef]

167. Xiao, L.; Huigen, Y.; Shunying, T.; Qiyun, Z.; Renyun, P. Study on the Graft Copolymerization of Styrene and Binary Monomers Onto Corn Starch by Manganese Pyrophosphate Initiation. *China Synth. Resin Plast.* **1995**, *3*.

168. Chen, S.; Tang, G.D. Emulsion Graft Copolymerization of St/MMA Onto Starch. *J. Nanjing Univ.* **2001**, *6*.

169. Tian, X.; Huang, K.; Ma, T. Study on Graft Copolymerization of Styrene on Corn Starch with Manganese Pyrophosphate as Initiator. *Chem. Adhes.* **2004**, *3*.

170. Trimnell, D.; Fanta, G.F.; Salch, J.H. Graft Polymerization of Methyl Acrylate onto Granular Starch: Comparison of the $Fe^{+2}/H_2O_2$ and Ceric Initiating Systems. *J. Appl. Polym. Sci.* **1996**, *60*, 285–292. [CrossRef]

171. Aravindakshan, P.; Bhatt, A.; Kumar, V.G. Matrix-Induced Variation in Kinetics and Control of Molecular Weight of Methacrylic Acid Polymers During Graft Copolymerization with Starch. *J. Appl. Polym. Sci.* **1997**, *66*, 397–403. [CrossRef]

172. Fanta, G.F.; Burr, R.C.; Doane, W.M.; Russell, C.R. Graft Copolymers of Starch with Mixtures of Acrylamide and the Nitric Acid Salt of Dimethylaminoethyl Methacrylate. *J. Appl. Polym. Sci.* **1972**, *16*, 2835–2845. [CrossRef]

173. Witono, J.R.; Noordergraaf, I.W.; Heeres, H.J.; Janssen, L.P.B.M. Graft Copolymerization of Acrylic Acid to Cassava Starch—Evaluation of The Influences of Process Parameters by an Experimental Design Method. *Carbohydr. Polym.* **2012**, *90*, 1522–1529. [CrossRef] [PubMed]

174. Herold, B.R.; Fouassier, J. Photochemical Grafting of Vinyl Monomers Onto Starch. *Starch/Stärke* **1981**, *33*, 90–97. [CrossRef]

175. Kaur, I.; Sharma, M. Synthesis, Characterization and Applications of Graft Copolymers of Sago and Acrylamide. *Trends Carbohydr. Res.* **2014**, *6*, 51–66.

176. El-Sayed, M.M.; Sorour, M.H.; Abd El Moneem, N.; Shalaan, H.; El Marsafy, S. Grafting of Acrylamide onto Polysaccharides Blend Using Photo-Initiators. *J. Polym. Environ.* **2017**, *25*, 402–407. [CrossRef]

177. Comer, C.; Jessop, J. Photoinitiated Emulsion Graft Polymerization of Synthetic Monomers to Starch. In Proceedings of the RadTech Conference & Exhibition, The Hague, The Netherlands, 14–15 September 2006.

178. Comer, C.M.; Jessop, J.L.P. Evaluation of Novel Back-Flush Filtration for Removal of Homopolymer From Starch-g-PMMA. *Starch/Stärke* **2008**, *60*, 335–339. [CrossRef]

179. Suwanmala, P.; Hemvichian, K.; Hoshina, H.; Srinuttrakul, W.; Seko, N. Preparation of Metal Adsorbent from Poly(Methyl Acrylate)-Grafted-Cassava Starch via Gamma Irradiation. *Radiat. Phys. Chem.* **2012**, *81*, 982–985. [CrossRef]

180. Kaur, I.; Sharma, M. Synthesis and Characterization of Graft Copolymers of Sago Starch and Acrylic Acid. *Starch/Stärke* **2012**, *64*, 441–451. [CrossRef]

181. Zhang, Y.N.; Xu, S.A. Effects of Amylose/Amylopectin Starch on Starch-Based Superabsorbent Polymers Prepared by γ-Radiation. *Starch/Stärke* **2017**, *69*, 1–9. [CrossRef]

182. Zhang, S.; Wang, W.; Wang, H.; Qi, W.; Yue, L.; Ye, Q. Synthesis and Characterisation of Starch Grafted Superabsorbent via 10 MeV Electron-Beam Irradiation. *Carbohydr. Polym.* **2014**, *101*, 798–803. [CrossRef] [PubMed]

183. Zhang, Y.; Cunningham, M.F.; Smeets, N.M.B.; Dubé, M.A. Starch Nanoparticle Incorporation in Latex-Based Adhesives. *Eur. Polym. J.* **2018**, *106*, 128–138. [CrossRef]

184. Nikolic, V.; Velickovic, S.; Antonovic, D.; Popovic, A. Biodegradation of Starch-Graft-Polystyrene and Starch-Graft- Poly(Methacrylic Acid) Copolymers in Model River Water. *J. Serbian Chem. Soc.* **2013**, *78*, 1425–1441. [CrossRef]

185. Samaha, S.H.; Nasr, H.E.; Hebeish, A. Synthesis and Characterization of Starch-Poly(Vinyl Acetate) Graft Copolymers and Their Saponified for m. *J. Polym. Res.* **2005**, *12*, 343–353. [CrossRef]

186. De Graaf, R.A.; Janssen, L.P.B.M. Production of a New Partially Biodegradable Starch Plastic by Reactive Extrusion. *Polym. Eng. Sci.* **2000**, *40*, 2086–2094. [CrossRef]

187. McDowall, D.J.; Gupta, B.S.; Stannett, V.T. Grafting of Vinyl Monomers to Cellulose by Ceric Ion Initiation. *Prog. Polym. Sci.* **1984**, *10*, 1–50. [CrossRef]

188. Kalia, S.; Sabaa, M.W. *Polysaccharide Based Graft Copolymers*, 1st ed.; Kalia, S., Sabaa, M.W., Eds.; Springer: Berlin/Heidelberg, Germany, 2013.

189. De Bruyn, H.; Sprong, E.; Gaborieau, M.; David, G.; Roper, J.A., III; Gilbert, R.G. Starch-Graft-Copolymer Latexes Initiated and Stabilized by Ozonolyzed Amylopectin. *J. Polym. Sci. Part A Polym. Chem.* **2006**, *44*, 5832–5845. [CrossRef]

190. de Bruyn, H.; Sprong, E.; Gaborieau, M.; Roper III, J.A.; Gilbert, R.G. Starch-graft-(synthetic copolymer) Latexes Initiated with $Ce^{4+}$ and Stabilized by Amylopectin. *J. Polym. Sci. Part A Polym. Chem.* **2007**, *45*, 4185–4192. [CrossRef]

191. Nikolic, V.; Velickovic, S.; Popovic, A. Influence of Amine Activators and Reaction Parameters on Grafting Reaction Between Polystyrene and Starch. *J. Polym. Res.* **2014**, *21*, 363. [CrossRef]

192. BelHaaj, S.; Ben Mabrouk, A.; Thielemans, W.; Boufi, S. A One-Step Miniemulsion Polymerization Route Towards The Synthesis of Nanocrystal Reinforced Acrylic Nanocomposites. *Soft Matter* **2013**, *9*, 1975–1984. [CrossRef]

193. Lu, D.; Xiao, C.; Sun, F. Controlled Grafting of Poly(Vinyl Acetate) onto Starch via RAFT Polymerization. *J. Appl. Polym. Sci.* **2012**, *124*, 3450–3455. [CrossRef]

194. Rotureau, E.; Raynaud, J.; Choquenet, B.; Marie, E.; Nouvel, C.; Six, J.L.; Dellacherie, E.; Durand, A. Application of Amphiphilic Polysaccharides as Stabilizers in Direct and Inverse Free-Radical Miniemulsion Polymerization. *Colloids Surf. A Physicochem. Eng. Asp.* **2008**, *331*, 84–90. [CrossRef]

195. Wang, S.; Xu, J.; Wang, Q.; Fan, X.; Yu, Y.; Wang, P.; Zhang, Y.; Yuan, J.; Cavaco-Paulo, A. Preparation and Rheological Properties of Starch-g-Poly(Butyl Acrylate) Catalyzed by Horseradish Peroxidase. *Process Biochem.* **2017**, *59*, 104–110. [CrossRef]

196. Wang, S.; Wang, Q.; Fan, X.; Xu, J.; Zhang, Y.; Yuan, J.; Jin, H.; Cavaco-Paulo, A. Synthesis and Characterization of Starch-Poly(Methyl Acrylate) Graft Copolymers Using Horseradish Peroxidase. *Carbohydr. Polym.* **2016**, *136*, 1010–1016. [CrossRef] [PubMed]

197. Shogren, R.L.; Willett, J.L.; Biswas, A. HRP-Mediated Synthesis of Starch-Polyacrylamide Graft Copolymers. *Carbohydr. Polym.* **2009**, *75*, 189–191. [CrossRef]

198. Lv, S.; Sun, T.; Zhou, Q.; Liu, J.; Ding, H. Synthesis of Starch-g-p(DMDAAC) Using HRP Initiation and the Correlation of Its Structure and Sludge Dewaterability. *Carbohydr. Polym.* **2014**, *103*, 285–293. [CrossRef] [PubMed]

199. Kalra, B.; Gross, R.A. HRP-Mediated Polymerizations of Acrylamide and Sodium Acrylate. *Green Chem.* **2002**, *4*, 174–178. [CrossRef]

200. Pino-Ramos, V.; Meléndez-Ortiz, I.; Ramos-Ballesteros, A.; Bucio, E. *Biopolymer Grafting: Applications*, 1st ed.; Thakur, V.K., Ed.; Elsevier Inc.: London, UK, 2018; pp. 205–250.

201. Dao, V.H.; Cameron, N.R.; Saito, K. Synthesis, Properties and Performance of Organic Polymers Employed in Flocculation Applications. *Polym. Chem.* **2016**, *7*, 11–25. [CrossRef]

202. Mishra, S.; Mukul, A.; Sen, G.; Jha, U. Microwave Assisted Synthesis of Polyacrylamide Grafted Starch (St-g-PAM) and Its Applicability as Flocculant for Water Treatment. *Int. J. Biol. Macromol.* **2011**, *48*, 106–111. [CrossRef] [PubMed]

203. Singh, A.V.; Nath, L.K.; Guha, M. Microwave Assisted Synthesis and Characterization of Sago Starch-g-Acrylamide. *Starch/Stärke* **2011**, *63*, 740–745. [CrossRef]

204. Singh, V.; Tiwari, A.; Pandey, S.; Singh, S.K. Peroxydisulfate Initiated Synthesis of Potato Starch-Graft-Poly(Acrylonitrile) Under Microwave Irradiation. *Express Polym. Lett.* **2007**, *1*, 51–58. [CrossRef]

205. Singh, V.; Tiwari, A.; Pandey, S.; Singh, S.K. Microwave Accelerated Synthesis and Characterization of Potato Starch-g-Poly(Acrylamide). *Starch/Stärke* **2006**, *58*, 536–543. [CrossRef]

206. Cummings, S.; Cunningham, M.; Dubé, M.A. The Use of Amylose-Rich Starch Nanoparticles In Emulsion Polymerization. *J. Appl. Polym. Sci.* **2018**, *135*, 46485. [CrossRef]

207. Cummings, S.; Trevino, E.; Zhang, Y.; Cunningham, M.; Dubé, M.A. Incorporation of Modified Regenerated Starch Nanoparticles in Emulsion Polymer Latexes. *Starch/Stärke* **2018**. [CrossRef]

208. Pei, X.; Zhai, K.; Liang, X.; Deng, Y.; Xu, K.; Tan, Y.; Yao, X.; Wang, P. Fabrication of Shape-Tunable Macroparticles by Seeded Polymerization of Styrene Using Non-Cross-Linked Starch-Based Seed. *J. Colloid Interface Sci.* **2018**, *512*, 600–608. [CrossRef] [PubMed]

209. Rotureau, E.; Marie, E.; Dellacherie, E.; Durand, A. From Polymeric Surfactants to Colloidal Systems (3): Neutral and Anionic Polymeric Surfactants Derived from Dextran. *Colloids Surf. A Physicochem. Eng. Asp.* **2007**, *301*, 229–238. [CrossRef]

210. Agama-Acevedo, E.; Bello-Perez, L.A. Starch as an Emulsions Stability: The Case of Octenyl Succinic Anhydride (OSA) Starch. *Curr. Opin. Food Sci.* **2017**, *13*, 78–83. [CrossRef]

211. Fu, X.; Zhang, G.; Wang, P.; Guo, L.; Qu, Q.; Li, J.; Wu, G. Preparation and Characterization of Starch Graft Soap-Free Styrene-Acrylate Copolymer Emulsion. *Trans. China Pulp Pap.* **2013**, *28*, 32–37.

212. Ortega-Ojeda, F.E.; Larsson, H.; Eliasson, A.C. Gel for mation in Mixtures of Hydrophobically Modified Potato and High Amylopectin Potato Starch. *Carbohydr. Polym.* **2005**, *59*, 313–327. [CrossRef]

213. Jonhed, A.; Järnström, L. Influence of Polymer Charge, Temperature, and Surfactants on Surface Sizing of Liner and Greaseproof Paper with Hydrophobically Modified Starch. *Tappi J.* **2009**, 33–38.

214. Cheng, L.; Guo, H.; Gu, Z.; Li, Z.; Hong, Y. Effects of Compound Emulsifiers on Properties of Wood Adhesive With High Starch Content. *Int. J. Adhes. Adhes.* **2017**, *72*, 92–97. [CrossRef]

215. Jovanović, R.; Dubé, M.A. Emulsion-Based Pressure-Sensitive Adhesives: A Review. *J. Macromol. Sci. Part C Polym. Rev.* **2004**, *44*, 1–51. [CrossRef]

216. Ye, F.; Miao, M.; Jiang, B.; Hamaker, B.R.; Jin, Z.; Zhang, T. Characterizations of Oil-in-Water Emulsion Stabilized by Different Hydrophobic Maize Starches. *Carbohydr. Polym.* **2017**, *166*, 195–201. [CrossRef] [PubMed]

217. Lu, X.; Xiao, J.; Huang, Q. Pickering Emulsions Stabilized by Media-Milled Starch Particles. *Food Res. Int.* **2018**, *105*, 140–149. [CrossRef] [PubMed]

218. Ettelaie, R.; Holmes, M.; Chen, J.; Farshchi, A. Steric Stabilising Properties of Hydrophobically Modified Starch: Amylose vs. Amylopectin. *Food Hydrocoll.* **2016**, *58*, 364–377. [CrossRef]

219. Kraak, A. Industrial Applications of Potato Starch Products. *Ind. Crops Prod.* **1993**, *1*, 107–112. [CrossRef]

220. Athawale, V.D.; Lele, V. Recent Trends in Hydrogels Based on Starch-Graft-Acrylic Acid: A Review. *Starch/Stärke* **2001**, *53*, 7–13. [CrossRef]

221. Guo, C.; Zhou, L.; Lv, J. Effects of Expandable Graphite and Modified Ammonium Polyphosphate on the Flame-Retardant and Mechanical Properties of Wood Flour-Polypropylene Composites. *Polym. Polym. Compos.* **2013**, *21*, 449–456. [CrossRef]

222. Labet, M.; Thielemans, W.; Dufresne, A. Polymer Grafting onto Starch Nanocrystals. *Biomacromolecules* **2007**, *8*, 2916–2927. [CrossRef] [PubMed]

223. Yu, J.; Ai, F.; Dufresne, A.; Gao, S.; Huang, J.; Chang, P.R. Structure and Mechanical Properties of Poly(Lactic Acid) Filled with (Starch Nanocrystal)-Graft-Poly($\varepsilon$-Caprolactone). *Macromol. Mater. Eng.* **2008**, *293*, 763–770. [CrossRef]

224. Brune, D.; Eben-Worllée, R. Von. Starch-Based Graft Polymer, Process for Its Preparation, and Use Thereof In Printing Inks and Operprint Varnishes. U.S. Patent 6,423,775, 23 July 2002.

225. Baruch-Teblum, E.; Mastai, Y.; Landfester, K. Miniemulsion Polymerization of Cyclodextrin Nanospheres for Water Purification from Organic Pollutants. *Eur. Polym. J.* **2010**, *46*, 1671–1678. [CrossRef]

226. Satheesh Kumar, M.N.; Siddaramaiah. Studies on Corn Starch Filled Poly (Styrene-Co-Butyl Acrylate) Latex Reinforced Polyester Nonwoven Fabric Composites. *Autex Res. J.* **2005**, *5*, 227–234.

227. Liu, G.; Gu, Z.; Hong, Y.; Cheng, L.; Li, C. Trends in Food Science & Technology Structure, Functionality and Applications of Debranched Starch: A Review. *Trends Food Sci. Technol.* **2017**, *63*, 70–79.

228. Alcázar-alay, S.C.; Angela, M.; Meireles, A. Physicochemical Properties, Modifications and Applications of Starches from Different Botanical Sources. *Food Sci. Technol (Campinas)* **2015**, *35*, 215–236. [CrossRef]

229. Versino, F.; Lopez, O.V.; Garcia, M.A.; Zaritzky, N.E. Starch-Based Films and Food Coatings: An Overview. *Starch/Stärke* **2016**, 1026–1037. [CrossRef]

230. Li, G.; Zhu, F. Quinoa Starch: Structure, Properties, and Applications. *Carbohydr. Polym.* **2018**, *181*, 851–861. [CrossRef] [PubMed]
231. Dufresne, A. Processing of Polymer Nanocomposites Reinforced With Polysaccharide Nanocrystals. *Molecules* **2010**, *15*, 4111–4128. [CrossRef] [PubMed]
232. Samsudin, H.; Hani, N.M. Use of Starch in Food Packaging. *Starch-Based Mater. Food Packag. Process. Charact. Appl.* **2017**, *2019*, 229–256.
233. Shevkani, K.; Singh, N.; Bajaj, R.; Kaur, A. Wheat Starch Production, Structure, Functionality and Applications-A Review. *Int. J. Food Sci. Technol.* **2017**, *52*, 38–58. [CrossRef]
234. Masina, N.; Choonara, Y.E.; Kumar, P.; du Toit, L.C.; Govender, M.; Indermun, S.; Pillay, V. A Review of the Chemical Modification Techniques of Starch. *Carbohydr. Polym.* **2017**, *157*, 1226–1236. [CrossRef] [PubMed]
235. Bloembergen, S.; McLennan, I.J.; Lee, D.I. Use of Biobased Sugar Monomers in Vinyl Copolymers as Latex Binders and Compositions Based Thereon. U.S. Patent 9,080,290, 14 July 2015.

 *processes*

*Review*

# Trends in Advanced Functional Material Applications of Nanocellulose

Prachiben Panchal, Emmanuel Ogunsona and Tizazu Mekonnen *

Department of Chemical Engineering, University of Waterloo, Waterloo, ON N2L 3G1, Canada;
pjpanchal@edu.uwaterloo.ca (P.P.); e2ogunso@uwaterloo.ca (E.O.)
* Correspondence: tmekonnen@uwaterloo.ca; Tel.: +1-519-888-4567 (ext. 38914)

Received: 24 November 2018; Accepted: 25 December 2018; Published: 30 December 2018

**Abstract:** The need to transition to more sustainable and renewable technology has resulted in a focus on cellulose nanofibrils (CNFs) and nanocrystals (CNCs) as one of the materials of the future with potential for replacing currently used synthetic materials. Its abundance and bio-derived source make it attractive and sought after as well. CNFs and CNCs are naturally hydrophilic due to the abundance of -OH group on their surface which makes them an excellent recipient for applications in the medical industry. However, the hydrophilicity is a deterrent to many other industries, subsequently limiting their application scope. In either light, the increased rate of progress using CNCs in advanced materials applications are well underway and is becoming applicable on an industrial scale. Therefore, this review explores the current modification platforms and processes of nanocellulose directly as functional materials and as carriers/substrates of other functional materials for advanced materials applications. Niche functional attributes such as superhydrophobicity, barrier, electrical, and antimicrobial properties are reviewed due to the focus and significance of such attributes in industrial applications.

**Keywords:** cellulose nanocrystals; functional materials; superhydrophobicity; antimicrobicity; barrier properties

---

## 1. Introduction

Cellulose, lignin, starch, chitosan, protein, triglycerides, natural gums, and polyphenols constitute an interesting nature-derived and renewable feedstock for advanced material applications. Among these feedstocks, cellulose is the most abundant biopolymer on earth with an approximate annual production of $10^{10}$ tons [1,2]. As such, it has been an immense source for the paper and textile industries for thousands of years. Cellulose is a homopolymer with a linear chain of six carbon ring, anhydro-D-glucose unit (AGU) monomers. Each AGU monomer in a chair conformation is linked with a $\beta$ $(1 \rightarrow 4)$ glycosidic bonds [3,4].

Overall, cellulose is a rigid and stiff polysaccharide that has tensile strength comparable to other commercial fibers such as carbon fiber [5]. Reinforcement of high melting temperature polymers with modified biomass has extensively been reported [6–11]. Likewise, it has also been extensively reported that the addition of cellulose fibrils in various forms to produce polymeric composites can greatly enhance mechanical properties of the base polymeric materials at significantly lower loadings compared to other biomasses [12–17]. Forms of cellulose fibrils can range from cellulose powders to microcrystalline cellulose and nanocrystalline cellulose. In the last two decades, a substantial and continuing interest in the utilization of cellulose nanoparticles in material applications is observed. This is accrued from a breakthrough in the largescale production of nanocellulose, as well as their multifaceted benefits in traditional polymer composites and functional materials. Nanocellulose is a material extracted from cellulose which has one or more dimensions in the nanometer range of 100 or less [2,18], and considered as the next generation renewable reinforcing agents for the production of biocomposites, and especially for advanced functional materials applications.

## 1.1. Nanocellulose

Nanocellulose is a generic name referring to different cellulose nanoparticles (CNPs) such as bacterial cellulose, microbial cellulose, cellulose nanocrystals, cellulose nanofibrils, cellulose nanofibers, and cellulose nanowhiskers [19]. Amongst the variety of CNPs, the two major structures are cellulose nanocrystals (CNC) and cellulose nanofibers (CNF) [2,20]. The extraction process of these two structures from the feedstock and their morphology are the two main differences between these CNPs [20,21]. Cellulose nanocrystals are extracted through chemical processes such as hydrolysis and they have a rod -like structure. Contrarily, cellulose nanofibers contain web-like network and are mainly extracted through mechanical and chemo-mechanical processes [5,20,21].

Similar to other bio-based materials used for advanced materials applications [22–24], the renewability, biodegradability, high strength, low density, high crystallinity, high aspect ratio, and unique optical properties make CNCs highly desirable materials in several advanced and functional material applications [20,25]. CNCs can be isolated from bioresources such as wood, cotton, hemp, linen, flax, tunicate (aquatic invertebrate) [5,17,18,20,21,25]. It can also be generated by a bottom-up or biosynthesis from simple molecules. Bacterial nanocellulose produced by bacteria and algae via fermentation of sugar is a typical example of such bottom-up approach [26]. The properties of CNCs can vary depending on bio-sources and the extraction process used. Some of the extensively researched material applications of CNCs include plastic composites, paints and coatings, packaging films, cement, rubber products, biomedical applications (e.g., pharmaceuticals, diagnostic imaging, drug delivery, tissue engineering materials), sensors, and many more [19,27,28].

## 1.2. CNC Characteristics

Cellulose exists in seven allomorphs: Cellulose $I_\alpha$, $I_\beta$, II, $III_I$, $III_{II}$, $IV_I$ and $IV_{II}$, where cellulose I is the most crystalline structure with the highest axial elastic modulus [4,18]. The two polymorph of cellulose I, cellulose $I_\alpha$ and $I_\beta$, coexists where $I_\beta$ is more thermally stable than $I_\alpha$ due to weaker hydrogen bond in $I_\alpha$ [18]. CNC contains 64–98% $I_\beta$ depending on the source [18]. CNCs are found in rod shape. Typically, they can be 3–5 nm wide and 50–500 nm long [18]. However, CNC from tunicate can be as wide as 20 nm [18]. The larger dimensions for CNCs derived from tunicate is a result of highly crystalline tunicate cellulose, which contains less amorphous regions that leave behind larger crystalline region [29]. One of the most attractive properties of CNC is its high aspect ratio. This is because the functional application of CNC is dependent on the aspect ratio in several cases. For instance, the use of CNC as a reinforcing agent relies on its higher aspect ratio. On the other hand, CNCs with uniform crystallinity and lower aspect ratio are beneficial in renewable liquid crystals applications [30,31]. Table 1 shows that the variation in aspect ratio of nanocrystals depending on the source of CNC. Overall, nanocrystals derived from sea plant and animal has significantly higher aspect ratio as compared to those extracted from wood and cotton.

**Table 1.** Aspect Ratio of cellulose nanocrystals (CNC) Derived from Different Sources.

| Source | Length (nm) | Width (nm) | Aspect Ratio * | Reference |
|---|---|---|---|---|
| Bacterial | 640–1070 | 12–22 | 50 | [32] |
| Cotton | 100–300 | 4–10 | 29 | [2] |
| Flax | 100–500 | 10–30 | 15 | [33] |
| Ramie | 150–250 | 6–8 | 29 | [34] |
| Sisal | 150–350 | 3–5 | 62 | [35] |
| Valonia (Sea plant) | 1000–2000 | 20 | 75 | [2] |
| Tunicate (Sea animal) | 500–2000 | 10–20 | 83 | [2] |
| Wood | 100–200 | 3–15 | 17 | [2] |

* Aspect Ratio = Aspect ratio calculated using average length and width.

It is important to highlight that cellulose nanocrystals are a highly crystalline fraction of cellulose. CNCs have degree of crystallinity of 54–88% depending on the amorphous content present in the feedstock, and the production process [18]. Nanocrystals derived from tunicate have been reported to have the highest crystallinity, which ranges from 85–100% [18]. As a result of the higher crystallinity caused by hydrogen bonding, cellulose nanocrystals are rigid and stiff. CNCs are also rich in hydroxyl groups (–OH) since each AGU unit offers one primary and two secondary –OH groups leaving it open to endless modification and functionalization possibilities [18]. Moreover, the high surface area to volume ratio of CNCs further facilities surface functionalization or other –OH chemistry efforts making it suitable for various advanced functional material applications [36].

Due to the sheer numbers of published review articles on CNCs with focuses on their modifications and applications towards polymer composites and nanocomposites, extraction and purification processes and determinations of their properties and characteristics just to mention a few, these subjects will only be highlighted. To the best of our knowledge, only few review articles were published on advances in applications of cellulose nanocrystals covering several topics. For instance, the review article by Grishkewich et al. [37] covered topics ranging from tissue engineering to gene delivery and energy. However, due to growing interests, research being conducted and achieved on CNC continually and especially in niche areas such as wettable surfaces, coatings, several applications of biotechnology, and electronics, there is need to update and disseminate the new developments achieved and knowledge gained. Therefore, this review focuses on the current advanced material functionalization of CNCs geared towards these areas as described in the schematic shown in Figure 1.

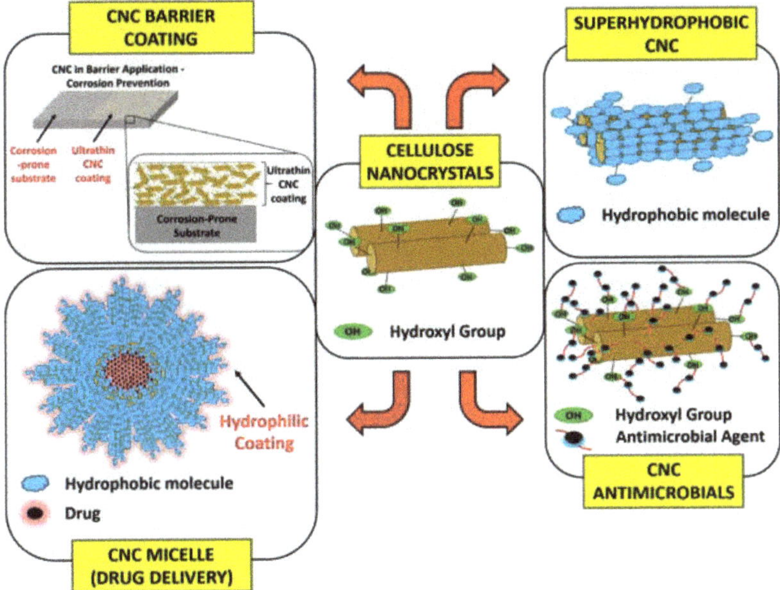


**Figure 1.** Some application of cellulose nanocrystals (CNCs) and nanofibrillated cellulose (NFCs) in advances functional materials applications.

## 2. Cellulose Nanocrystals in Wettable Applications

Wettability of a substrate or surface is the ability of that surface to create intermolecular interactions such as cohesive and adhesive forces when typically, in contact with a liquid. The degree of wettability is determined by the aforementioned forces which also govern the contact angle (CA) between the substrate and liquid. The CA subsequently determines if a material or substrate is hydrophobic, hydrophilic or their extremes. In recent years many researchers have been drawn towards the development of hydrophobic nanomaterials due to the increasing demand in both academic and industrial sectors. Hydrophobicity is an important property to determine the application of nano-composites. Hydrophobic materials with self-cleaning, antifouling, water repellency, and reduced friction properties are highly desirable for industrial applications. CNCs are known for their strength and commonly used as reinforcing agents. However, the application of CNC is currently limited because it cannot be readily incorporated into many polymer matrices that are typically hydrophobic due to its high hydrophilicity. Thus, hydrophobic modification of CNC can lead to better dispersion in non-polar and hydrophobic matrices. Furthermore, hydrophobic CNC surface can also be used as a coating agent for marine vehicles, biomedical devices, windows, textiles, paints, and many other applications. The advancement of hydrophobic CNC surface can open doors to many commercial applications of this abundant biopolymer found in nature.

In nature, butterfly wings, lotus and other plant leaves are greatly studied for their superhydrophobicity [38,39]. Lotus leaves known for its self-cleaning properties has hierarchical rough structure containing two levels of roughness [38]. Superhydrophobic properties of lotus known as lotus-effect have inspired the synthesis of many artificial hydrophobic materials. An important aspect to study while investigating hydrophobicity is the water contact angle (WCA). A hydrophilic material has a WCA of less than 90°, a hydrophobic material has a WCA higher than 90° while materials displaying WCA greater than 150° are superhydrophobic [39,40]. Various relations including Young's equation, Wenzel equation and Cassie equation have been developed to determine WCA [39]. Along with water contact angle, another important characterization property of hydrophobicity is sliding angle. Sliding angle is a measure of tilt angle between the surface and substrate droplet at which droplet starts to roll off of the surface. Superhydrophobic material has sliding angle less than 10°, which is often used to describe the anti-water repellency property [39,41].

Hydrophobicity can be controlled through chemical modification or surface roughness. Reduction of surface energy through chemical modifications and enhanced roughness, both the factors must be simultaneously controlled to achieve superhydrophobicity [42]. Preparation of superhydrophobic materials involves harsh chemical and physical treatments. Chemical modification to obtain hydrophobicity include attachments of low surface energy molecules such as fluorinated agents [43], silanes [42], organic hydrophobic chains [44] and etc. Table 2 lists various chemical treatments and attachments to cellulose -based materials such as cotton, paper, and cellulose nanocrystals.

**Table 2.** Hydrophobic Treatment of Cellulose-based Materials (PFSC: perfluorooctylated quaternary ammonium silane coupling agent; TEOS: Tetraethyl orthosilicate).

| Surface | Modification | WCA | Reference |
|---------|--------------|-----|-----------|
| CNC | Pentafluorobenzoyl chloride | 112° | [43] |
| CNC | Castor Oil | 97° | [44] |
| CNC | Stearyltrimethylammonium chloride | 71° | [45] |
| Cotton | Silica sol treated with PFSC | 145° | [46] |
| Paper | TEOS and tridecafluorooctyl triethoxysilane | 170° | [41] |

Although chemical modification has been the traditional approach to functionalize the surface of a material for a number of applications, a recent trend in enhancing the surface roughness is providing some interesting hydrophobic characteristics in several materials. Increasing surface roughness plays a key role in enhancing water repellency [46]. In this, air, which is very hydrophobic (water contact angle 180°) gets trapped between the grooves of the roughness [47]. When a water droplet rests on the surface it comes in contact with the entrapped hydrophobic air leading to hydrophobicity enhancement [47]. Surface roughness strategies such as etching, laser, electrospinning and etc. are also commonly used [39,48].

Salajkova et al. [45] employed quaternary ammonium salts modification to bring about hydrophobic modification of CNCs. In this study, four different quaternary ammonium salts were used for the CNC modification. Figure 2. shows the attachment of (1) stearyltrimethylammonium chloride as well as the structure of other three quaternary ammonium salts, (2) phenyltrimethyl-ammonium chloride, (3) glycidyl trimethylammonium chloride, (4) diallyldimethylammonium chloride [45]. The highest WCA for stearyltrimethylammonium chloride modified CNC was 71 °C. Although great improvement was observed in WCA of the CNC surface, a higher WCA (>90°) is typically desirable for employing the hydrophobicity of CNC for advanced material applications.

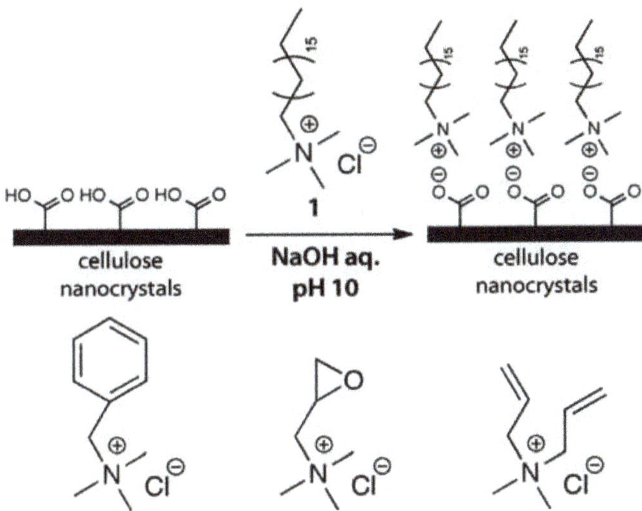

**Figure 2.** CNC modified by Quaternary Ammonium Salts. Reproduced with permission from Salajková et al. [45]. RSC copyright © 2012.

Silica nanoparticles are widely used for hydrophobic modification of cellulosic materials. Tetraethyl orthosilicate (TEOS) prepared through sol-gel process was used for the hydrophobic treatment of cotton fabric [49]. Silica is not hydrophobic therefore fluorinated silane and alkylanated silane, or other alternative water repellent agents are commonly used to achieve water repellency [49]. Silica, TEOS, sol treated with perfluorooctylated quaternary ammonium silane coupling agent (PFSC) showed water contact angle of 145° proving PFSC to be very effective [46].

Paper, a major product from cellulose, was coated with TEOS sol and tridecafluorooctyl triethoxysilane to produce a hydrophobic surface [41]. The modified paper had a WCA greater than 170° and sliding angle less than 7° displaying superhydrophobic properties [41]. Instead of only using silica nanoparticles, the treatment of silica with fluorinated compounds often offers better water repellency. Salam et al. [43] reported fluorine-based modification of nanocellulose to obtain hydrophobic and oleophobic properties. Figure 3 shows the modification of CNC surface through fluorinating agents. The modified CNC displayed a water contact angle of 112° and maintained 80° even after 1200 s [43]. However, due to the surface modification, the crystallinity of cellulose nanocrystals was slightly modified. The modified CNC had a crystallinity index of 82% while the unmodified CNC displayed a higher crystallinity index, 91% [43].

Shang et al. [44] has reported another approach of producing hydrophobic CNC, by attaching castor oil on the surface of the CNC. Castor oil containing hydroxylated fatty ricinoleic acid triglyceride (RTA), was attached onto to CNC surface via grafting-onto modification approach while using diisocyanate as a coupling agent [44]. Only one hydroxyl group was maintained active while other two were terminated in order to link one end of diisocyanate with castor oil and another with the CNC. Successful grafting led to reduction in the surface energy. The increase in water contact angle of grafted material to 97° also suggested hydrophobicity of the nanocellulose [44]. The resulting material was dispersible in non-polar solvents such as toluene and ethyl ether [44].

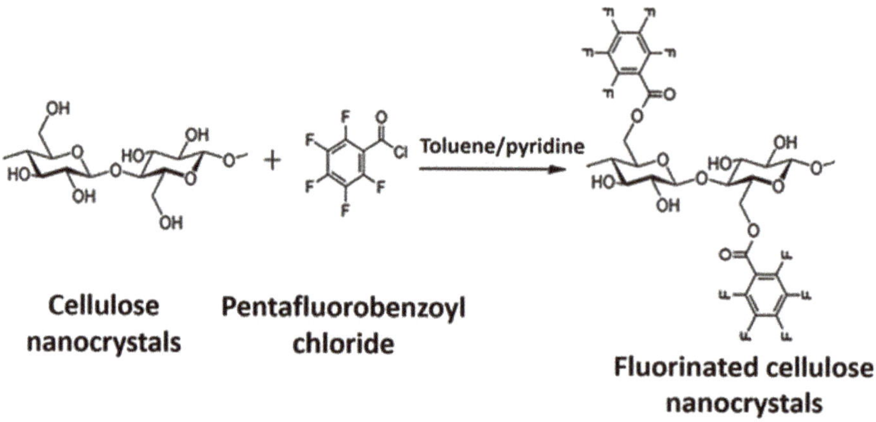

**Figure 3.** Hydrophobic modification of CNC via fluorine-based agents. Reproduced with permission from Salam et al. [43]. Springer copyright © 2015.

Hydrophobic surfaces provide the advantage of efficient and enhanced performance of materials when used in targeted applications. Hydrophobic and superhydrophobic surfaces are highly desirable for marine equipment coating. Watercrafts, marine platforms, offshore rigs and jetties are highly vulnerable to fouling. As a result, operational and maintenance cost for sustaining water-based structure is relatively high. Thus, hydrophobic surfaces with the correct combination of surface roughness and chemical modification can reduce the wetting behavior that result in minimizing attacks from marine organisms and corrosions. CNCs with great reinforcing ability can be modified to achieve both superhydrophobicity and antimicrobial properties. Dual function CNC can greatly enhance the performance of underwater objects and marine equipment.

Surface containing hydrophobic coatings is desirable for self-cleaning windows, satellite dishes, solar energy panels, photovoltaics, exterior architectural glass [39]. Furthermore, water repellent paper can have practical application for valuable printable paper, filter paper, packaging and photographs [47]. Recently, textiles with water repellency, anti-stain and self-cleaning properties are of interest in outdoor sporting textile applications [49]. Due to the increasing demand of hydrophobic materials, researchers are developing various approaches for hydrophobic modification of CNCs. Despite numerous research being conducted in this field, commercial applications of hydrophobic and superhydrophobic materials is limited due to susceptibility of CNCs to environmental degradation over time [41,50].

Zhou et al. recently developed a superhydrophobic micro fibrillated cellulose aerogel (HMFCA) to efficiently separate oil from water [51]. In this study, silanization in an ethanol solution containing methyltriethoxysilane (MTES) was used to modify the aerogel by introducing polysiloxane groups on the surface of the HMFCAs. Prior to silanization, the HMFCA was oxidized to introduce hydroxyl groups, which was used as anchor points for the silanization process. The polysiloxane groups are highly water repulsive (hydrophobic) while allowing the absorption of oil (lipophilic) into the porous structure of the aerogel. The resulting modified aerogel when immersed in oil-contaminated water exhibited oil selective capability of up to 159 g/g. Experimental runs after 30 cycles revealed a reusability capacity of up to 92 g/g which was approximately 58%. Figure 4a shows the contact angle of the unmodified HMFCA and modified HMFCAs with different MTES concentrations of 1, 2 and 3 mL which correspond to HMFCA-1, HMFCA-2 and HMFCA-3, respectively. It was observed that the contact angle increased with increasing MTES concentration up to 151.8 degrees, indicating its superhydrophobic nature. Figure 4b–d show the before and after modification of the aerogels and flotation of the modified aerogel in water due to the reduced surface energy created by the modification process. Figure 3e shows the possible reaction scheme for the modification of the hydrophilic aerogel into lipophilic aerogels. This study shows the great potential of micro fibrillated celluloses and possibly CNCs to be used on a large scale for oil spill clean ups from offshore rigs and leaking underwater pipelines.

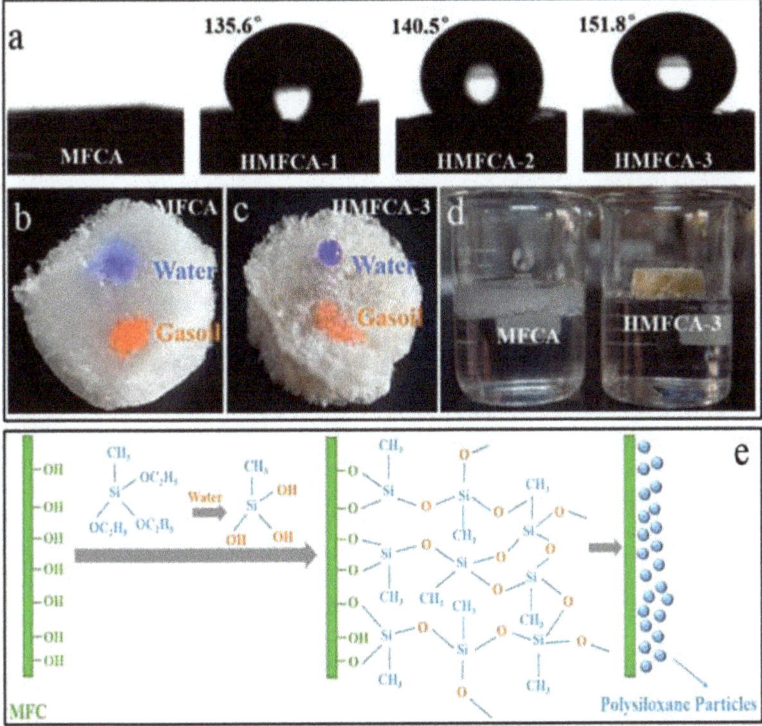


**Figure 4.** (**a**) Contact angle measurement of unmodified and modified micro fibrillated cellulose aerogel (MFCAs) with 1, 2 and 3 mL methyltriethoxysilane (MTES) concentrations. (**b,c**) Water/oil selectivity of unmodified MFCA and modified MFCA with 3 mL of MTES with both water and oil absorbed in the unmodified MFCA while only oil is absorbed in the modified MFCA. (**d**) Modified MFCA floating on water while unmodified MFCA sinking into water. (**e**) Schematic representation of silinization grafting of methyltriethoxysilane (MTES) onto micro fibrillated cellulose (MFC). Reproduced with permission from Zhou et al. [51]. ACS publications copyright © 2016.

Water/oil selectivity of unmodified micro fibrillated cellulose aerogel (MFCA) and modified MFCA with 3 mL of MTES with both water and oil absorbed in the unmodified MFCA while only oil is absorbed in the modified MFCA. Oil-water separation tests were also performed using a stainless-steel mesh treated with the superhydrophobic CNCs. Different oils such as toluene, N-haxane, xylene and cyclohexane were used to determine the efficiency of the treated mesh in separating the oil from the water while reusing the mesh for 40 cycles. The result from this test revealed that the mesh was able to efficiently remove all types of oils with a separation rate of at least 97.37%. Analogous to the above work, oil/water separation using superhydrophobic cotton fabric coated with CNC has been published and shown to have excellent results with a 98% separation efficiency with the coated cotton fabric capable of being reused without detriment to the separation efficiency [52]. The reusability, durability and efficiency of oil water separation most likely stems from the strong covalent bonds on the cellulose surface as well as the excellent mechanical properties of the cellulose nanocrystals such as the strength, stiffness and wear resistance. Similar to the study by Zhou et al. [51], these works demonstrate the capability of superhydrophobic CNCs when geared towards applications such as water contaminant separation.

In a similar study by Huang et al., CNCs were treated with 1H, 1H, 2H, 2H-perfluorodecyltrichlorosilane in the liquid phase through water-ethanol-toluene exchanges to produce superhydrophobic CNCs [53]. In this reaction, the trichlorosilane groups react with the hydroxyl groups of the CNCs to form a covalent bond while leaving the highly fluorinated tail dangling. The group reduces the surface energy and therefore makes it superhydrophobic. The superhydrophobic CNCs were then applied to different substrates such as wood, glass and stainless-steel mesh. Each substrate was first sprayed with a quick-drying adhesive before being sprayed with the superhydrophobic CNCs and allowed to dry. Self-cleaning and water contact angle tests were performed on the treated substrates and are shown in Figure 4. It can be observed from Figure 5a,b that the coated wood and glass substrates exhibited superior hydrophobic characteristics with high contact angles of 158 and 156 degrees, respectively. The self-cleaning efficiency of the uncoated and coated substrates with the superhydrophobic CNCs can be observed in Figure 5c,d, respectively. The high degree of self-cleaning can be applied to marine equipment and vessel surface, thereby allowing for increased buoyance and reduced need for expensive and frequent maintenance. Though, the substrate used by Huang et al. [53] might pose a challenge when it comes to large scale application due to the weight and possible high cost, it is highly durable and can withstand impacts and bumps over several cycles. Those used by Zhou et al. [51] and Cheng et al. [52] are light-weight, readily available and can be mass produced. However, it might not be as durable as the stainless-steel material in the long run. Regardless of the choice of material, the application of modified CNCs for superhydrophobic material applications is promising and on track for industrial scale use.

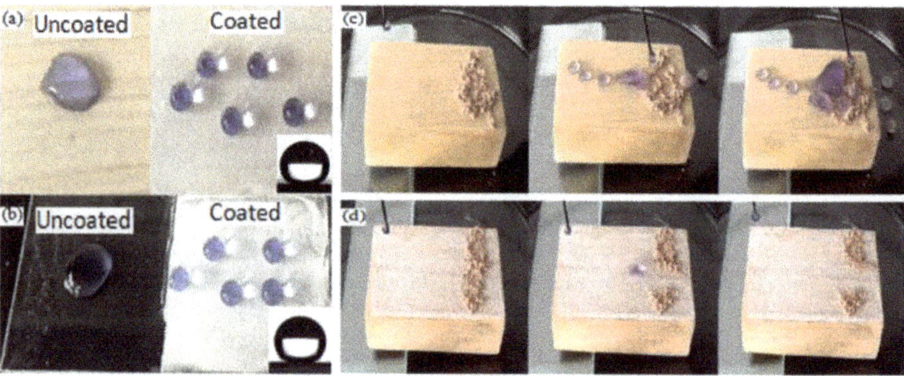


**Figure 5.** Demonstration of superhydrophobic coating from modified CNC with high contact angle on wood (**a**) and glass (**b**). Self-cleaning capability of uncoated (**c**) and coated (**d**) wood. Reproduced with permission from Huang et al. [53]. Elsevier copyright © 2018.

Recently, Gou et al. developed a process to modify the surface of nanofibrillated cellulose (NFC) films to render portions of the films either superhydrophobic or superhydrophilic [54]. In this process, the NFC film was immersed in trichlorovinylsilane (TCVS)-toluene solution to allow for the growth of TCVS silicone nanofilaments on its surface through hydrolysis and polycondensation to form crosslinked porous polymeric silicone nanofilaments on the surface (TCVS-SNFs). This film was further modified to form superhydrophobic surfaces by exposure to ultraviolet light in the presence of hydrophobic thiol compounds such as 20% (*v/v*) 1H, 1H, 2H, 2H-perfluorodecanethiol solution in ethyl acetate or 20% (*v/v*) 1-butanthiol in ethanol. In order to selectively form superhydrophilic surfaces on the superhydrophobized film, it was further exposed to ultraviolet light in the presence of cysteamine or 2-mercaptoethanol as hydrophobic moieties leaving only areas which was intended to be exposed to a UV light. The use of ultraviolet light was to catalyze the reaction between the TCVS-SNFs on the surface of the film and the hydrophobic moieties and between the hydrophobic moieties and the hydrophilic moieties through photo-induced thiol-ene reaction. Likewise, a slippery NFC surface was created by incorporating lubricant fluid into an already superhydrophobic NFC films until complete saturation of the fibrils was acheived. Figure 6a demonstrates the process of superhydrophilic-superhydrophobic dual surface modification of an NFC film using photo-induced thiol-ene reaction by UV light exposure. Figure 6b shows the mapping of dual superhydrophobic-hydrophilic surfaces using secondary ion mass spectrometry coupled with time-of-flight mass analysis. This was used to determine the elemental and molecular identities of the dual surface. Further geometric designs were made and are shown in Figure 6c with different patterns of the hydrophobic-hydrophilic surfaces using dyed water solutions to clearly identify each region. The mechanism of reaction, stepwise functionalization and modification of the NFC films to form both super-hydrophobic and hydrophilic surfaces and slippery NFC films are shown in Figure 6d. Such findings can lead to an extensive utilization in biotechnology and biomedical applications such as specifically designed films, which can be used to repel or attract specific biological moieties in the body due to their polar or nonpolar characteristics.

In the studies above, irreversible superhydrophobic surfaces were developed and shown to be extremely efficient. However, in a very recent study, a reversible superhydrophobic/superoleophilic-superhydrophilic/superoleophobic transition of modified cellulose fabric was developed and designed by Fan et al. [55]. Smart or transition-reversible superhydrophobic-superhydrophilic cellulose fabric was formed by introducing the fabric into an aqueous solution of sodium hydroxide, distilled water and urea for a specific period of time after which it was treated in a zinc chloride aqueous solution to allow zinc cations and chlorine anions be absorbed onto the surface of the cellulose fibers. The fibers are swollen and allowed for the ions to penetrate into the pores. The loaded fabric was steamed to imprint the ions on the fabric fibers. The treated fabric was then washed repeatedly with deionized water and baked thereafter. The baking process was done to shrink the fibers, thereby physically locking the imprinted ions in the pores of the fibers as water is removed and shrinkage occurs. This fabric now loaded with zinc oxide (ZnO-CFs) was further modified to give it a superhydrophobic-superhydrophilic reversible-transition surface by dipping it into a solution of lauric acid and ethanol. Thereafter, soaked in another solution of sodium hydroxide, ethyl alcohol and water. Figure 7a–d, displays the micrographs of the cellulose fibers before and after modification with ZnO, ZnO and lauric acid and ZnO, lauric acid, sodium hydroxide and ethyl alcohol aqueous solution, respectively. It can be observed that there is the presence of ZnO attached to the fiber surface after modification which created a micro-nano rough surface structure which was partially ascribed as a factor in the formation of hydrophobic and hydrophilic surfaces thereafter. Likewise, in Figure 7e, the schematic demonstrates the step taken in achieving this process. Readers should refer to the published work for a well detailed experimental procedure and for more information as needed [55]. It was found that the reversable wettability of the modified cellulose fibric maintained its properties such as separation efficiency and wettability phases even after 20 reversal cycles between hydrophobicity and hydrophilicity. The mechanism behind the hydrophobic to hydrophilic transition of the fabric was ascribed to the formation of sodium carboxylate after the scissioning of chelation between ZnO and carboxylate anion formed during the modification process. The newly formed sodium carboxylate would move to the liquid phase from the solid-liquid interface, thereby increasing the solid-surface free energy as a resulting in the local reduction of water surface tension from the loss of low free energy alkyl chains. This subsequently results to wettability reversal or transition. This study not only advanced the science of oil/water separation but also took it a step further by broadening the application scope to include the removal of water from oil and not just oil from water. In cases where the contaminant is water, it can be removed by using the smart cellulose fabric in its superhydrophilic/superoleophobic form, on the contrary, while it is in its reversed form of superhydrophobicity/superoleophilicity, it can remove oil from water simply by immersing it in a solution to stimulant the transition back to the former as described above. These applications can be geared towards the oil industry where purity of oil is critical to its final use or in industries where oil spill cleanup into large bodies of water is required.

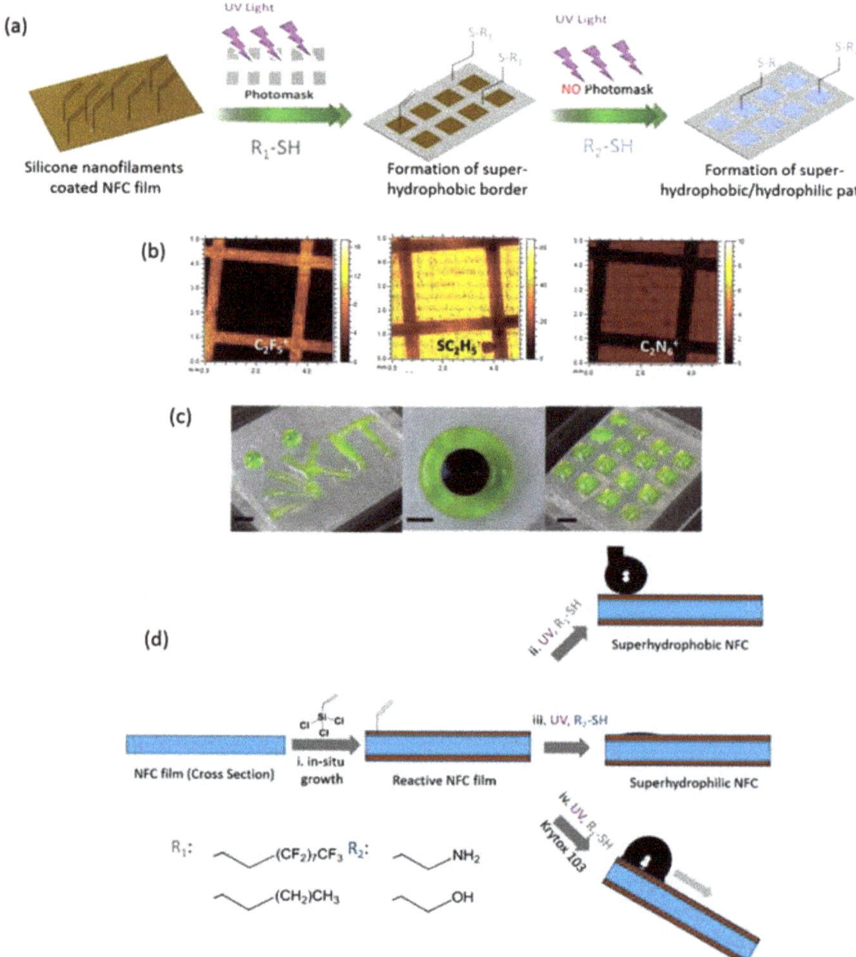



**Figure 6.** (**a**) modification of NFC films to produce superhydrophobic/superhydrophilic patterns. (R1 and R2–superhydrophobic and superhydrophilic moieties grafted on NFC film surfaces), (**b**) secondary ion mass spectroscopy with time of flight 2D mapping of superhydrophobic/superhydrophilic modified films with both positive (left and right) and negative (middle), (**c**) films designed with different patterns of superhydrophobic/superhydrophilic surfaces and highlighted with dyed water solutions. Reproduced with permission from Gou et al. [54]. ACS publications copyright © 2016 and (**d**) schematic of plausible chemical reactions to achieve superhydrophobic/superhydrophilic surfaces. Reproduced with permission from Fan et al. [55]. Elsevier copyright © 2018.

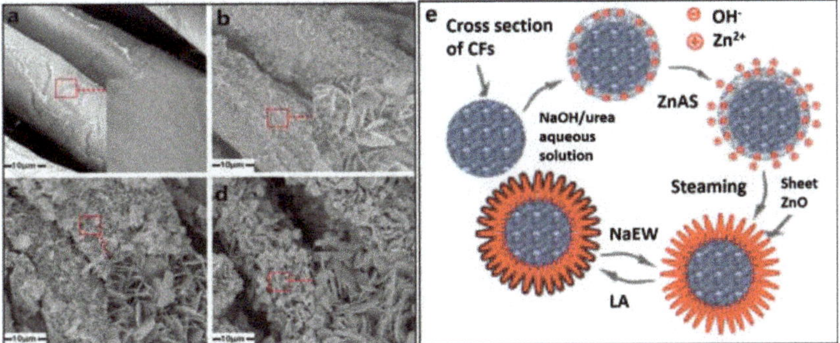



**Figure 7.** (**a**) unmodified, (**b**) ZnO modified, (**c**) ZnO-lauric acid modified, (**d**) ZnO-lauric acid-sodium hydroxide and ethyl alcohol aqueous solution modified cellulose fabric fibers and (**e**) diagrammatic representation of functionalization of reversible superhydrophobic/superhydrophilic cellulose fabric fiber. (NaEW—sodium hydroxide ethyl alcohol aqueous solution, LA—lauric acid). Reproduced with permission from Fan et al. [55]. Elsevier copyright © 2016.

The development of superhydrophobic surfaces has resulted in potential applications in water purification, medicine, biotechnological and materials interface. Modification of cellulose and its derivatives such as MCCs and CNCs have classically been modified using fluorinated moieties which typically exhibit hydrophobic characteristics especially to moisture. Though the technology of CNC is mostly at its infancy stage, the advancement in the use of CNCs in functional materials is a positive step towards moving away from dependence on petroleum-based materials. However, more research is needed in the modification techniques of CNCs to develop the use of fewer chemicals and steps. Reduction in modification steps will result to a lower cost of the end product in addition to the environmental benefits by cutting down on man hours and material cost of the added modificstion step(s).

## 3. CNCs as Antimicrobial Agent Carriers

The majority of the antimicrobial agents used are small in size thus there is always a probability for these active agents to leach out from the material (e.g., fabric, plastics, etc.) containing it [56]. Loss of antimicrobial agents to the fabric, skin and the environment contaminates the material and poses threat to health and environment [56]. The direct addition of antimicrobial agent also reduces the activity and efficiency as a result of leaching out and side reaction with food contents such as proteins and lipids [57]. Some antimicrobial chemical agents pose risk to human health and safety that led to the ban of a number of such agents (e.g., triclosan, iodine complex, phenol, triclocarban, etc.) that were commonly found in over the counter consumer products such as soap [58]. In other cases, the prevalence of antibiotic resistance as a result of microbial mutation and a decreasing effectiveness of antibiotics in treating common microbial infections resulted is another motivation for innovative ways of antimicrobial carrier development [59]. Thus, interest has grown in attaching or incorporating active antimicrobial agents to long chain polysaccharides and polymers especially CNCs.

Commonly used antimicrobial agents are halogens, phenols, silver nanoparticles and quaternary ammonium salts [60]. Transition metal oxides such as Au, Ag, Ti, Mg, NO, ZnO, CuO and $Fe_3O_4$ have antimicrobial properties as well [61]. Antimicrobial enzymes such as lactoperoxidase and lactoferrin and antimicrobial peptides magainins, cecropins, defensins are also often used [62].

Quaternary ammonium compounds (QACs) are positively charged, thus they absorb onto negatively charged microorganisms. They attack the surface cytoplasmic membrane once diffused through the cell wall. Due to the loss of essential constituents of the cell cytoplasm, it becomes difficult for the microorganisms to survive and they eventually die [60,63]. Thus, QACs are one of the most effective antimicrobial agents used. However, they are toxic to bacteria, pathogens (fungi and protozoa) as well as mammals which limit their applications.

Silver nanoparticles are preferred as antimicrobial agents due to the high activity against a wide range of microorganisms. Silver nanoparticles are active against yeast, fungi, viruses as well as both gram negative and gram positive viruses [64]. As the silver particles permeate through microorganisms they disrupt and restrict the respiratory function of the microorganisms, thereby resulting in death [65]. As the size of the silver particles decreases their efficacy increases due to increase in surface area, thus nanosized silver particles are preferred [66]. Silver particles inhibits the growth of *Escherichia coli* (*E. Coli*) and *Staphylococcus aureus* (*S. aureus*) [65,67]. Due to various uncertain health effect of silver nanoparticles, their use for food packaging in certain countries has not been accepted [64]. This has resulted in limited application and research especially in specific industries such as the food industry.

Triclosan is a synthetic biophenol (2,4,4′-trichloro-2′-hydroxydiphenyl ether) with excellent biocide activity. It is effective against both gram-positive and -negative bacteria, yeast and mold [68]. Hence, triclosan has a wide range of applications such as in soap, mouthwash, toys, packaging, textiles and kitchen utensils [56,62,69]. Use of triclosan is regulated by Environmental Protection Agency (EPA) due to its perceived toxicity and environmental risks, and as a result is its use in consumer products is limited [68].

Amongst naturally occurring antimicrobial agents, chitosan is derived from chitin and has gained a lot of popularity in commercial applications. Chitin is extracted from the exoskeleton of insects, as well as algae and fungi [56]. Chitosan is a polycationic polysaccharide with antimicrobial and antifungal activity [70,71]. Chitosan alters or forms a polymeric membrane on the surface of the cell restricting any nutrients from being absorbed, thus resulting in reduced growth and ultimate death of the cells [71,72]. Chitosan has been investigated for many applications such as packaging, drug delivery and other biomedical applications [28,70,71]. However, it is a slow acting antimicrobial agent due to its mechanism as stated previously, therefore, making it not as potent as other antimicrobial agents.

One of the main applications of antibacterial material is in the food packaging industry. Rather than just physical or moisture protection, successful packaging is one that inhibits the growth of microorganism in the food. Thus, active packaging is becoming popular in order to inhibit, reduce and kill microorganisms from the food surface or the surrounding adjacent to the packaging. Recently, consumer demand for low preservative food has increased. Thus, to maintain food quality under low preservation, active packaging can be very attractive. Active food packaging film can increase the product shelf life, maintain the nutritional value of food, provide microbial safety while restricting pathogenic growth. As a result, the need for active antimicrobial agent for food packaging has increased to maintain safe food quality [73].

As the demand for biodegradable and bioactive packaging gains popularity, research has been conducted to make active packaging film with Nisin, showed in Figure 8, as an antimicrobial agent. Nisin, 34-amino acid long bacteriocin, is active against many foodborne gram-positive bacterium such as *Listeria monocytogens, Stephylococcus aureus* and many more [74,75].

Nisin has been used to activate chitin to make an active packaging for pasteurized milk [76]. Additionally, it was reported that microbial growth on oyster and ground beef was delayed when nisin activated film was used for packaging [76]. Since nisin has demonstrated good antimicrobial activity, it was incorporated to polylactic acid-cellulose nanocrystal (PLA-CNC) composite [74]. Overall the nisin active PLA-CNC film contained all three essential packaging properties: antimicrobial activity, strength and biodegradability. Weishaupt et al. [75] reported the self-assembly of nanofibrilled cellulose-nisin biocomposite. TEMPO-oxidized nano-fibrillated cellulose containing carboxylic groups provided a negative surface for nisin to be adsorbed onto the surface [75]. The binding of nisin to nanocellulose was greatly affected by electrostatic interactions. However, at high salt concentration the binding capacity between nisin and nanocellulose was compromised [75]. Nisin bound nanocellulose was investigated against *S. aureus* and reduction in growth was observed [75]. Due to the beneficial properties observed, this biocomposite can be further enhanced to develop a promising application in biomedical industries.

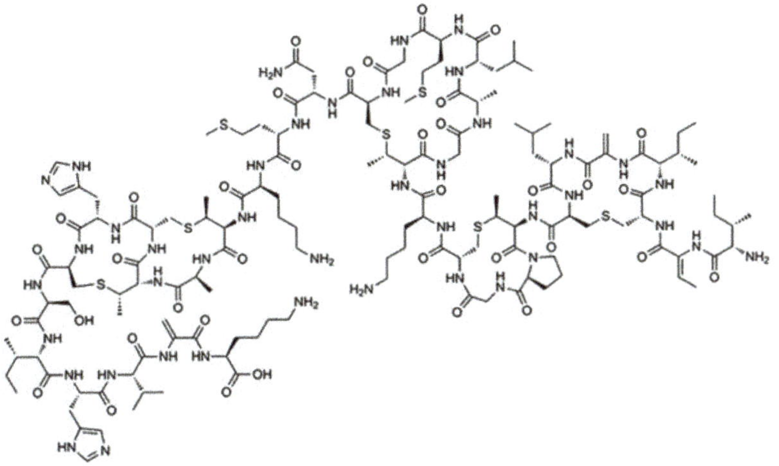

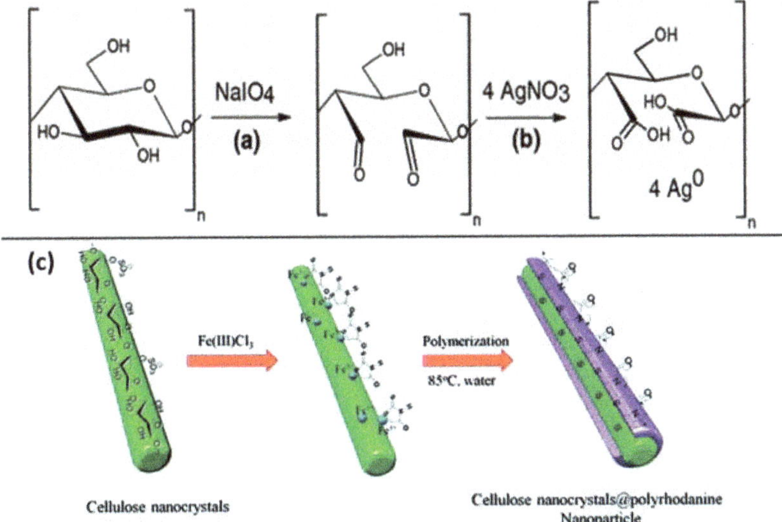

**Figure 8.** Chemical structure of Nisin. Reproduced with permission from Salmieri et al. [74]. Springer Nature copyright © 2014.

Silver nanoparticles are another commonly investigated antimicrobial agents incorporated with cellulose nanocrystals. Drogat at al. [65] reported the activity of silver nanoparticles-CNC composite against *S. aureus* (Gram positive) and *E. coli* (Gram negative). Oxidation of CNC was conducted via periodate (NaIO$_4$) to produce aldehyde, which was then used to reduce silver ions to Ag$^\circ$ to inhibit the growth of both the gram positive and negative bacteria [65]. Figure 9 shows the schematic of reaction of CNC with periodate to produce aldehyde followed by reduction of silver ion.



**Figure 9.** (a) Oxidation of CNC, (b) Reduction of Ag$^+$ to Ag$^\circ$. Reproduced with permission from Drogat et al. [65]. Springer copyright © 2011 and (c) cellulose nanoparticles with polyrhodanine (CNC@PR) Fabrication. Reproduced with permission from Tang et al. [77]. RSC copyright © 2014.

Inhibited growth of bacteria suggests potential application of silver-CNC matrix for wound healing gels, antiseptic solution as well as other biomedical applications. Fortunati et al. also reports antimicrobial activity of PLA-CNC biocomposite containing 1 wt.% Ag nanoparticles against *S. aureus* and *E. coli* [64]. Higher activity was observed against *E. coli* than *S. aureus* due to greater toxicity of silver ions towards *E. coli* while significant amount of activity was also achieved for *S. aureus* [64,78]. PLA biocomposite with excellent antimicrobial activity against both gram-positive and gram-negative bacteria offers more opportunities for the development of active packaging for food, sanitary, and biomedical industries.

Amongst different structures for antimicrobial compounds, core-sheath structure of cellulose nanoparticles with polyrhodanine was reported by Tang et al. [79]. Cellulose nanoparticles with polyrhodanine (CNC@PR) showed antimicrobial activity against *E. coli* and *B. Subtilis* [79]. CNC@PR was synthesized with Fe (III) complex and ferric chloride to negatively charge the CNC surface. Fe (III) was used as an initiator and oxidant for the in situ polymerization of rhodanine on CNC [77,79]. Figure 9c shows the schematic of core-shell nanoparticles of CNC coated with polyrhodanine.

In academia and industry, antimicrobial materials are gaining a lot of interest due to the need to improve on the shelf life of food products. The application of antimicrobial materials is not only limited to food packaging but also a variety of other applications such as textile, coating, military and household equipment. High surface area and shorter diffusion path of nanofibers makes them an excellent choice for attaching antimicrobial molecules while also used as a reinforcing material for polymers as well [80]. Biomedical devices possessing antimicrobial-functionalized polymers can be proven very hygienic and efficient for health care purposes. Similarly, antimicrobial coatings can be used in hospital walls, kitchen counter tops and other pharmaceutical laboratories to maintain microbes-free environment. Household items, sanitary items for bathroom, kitchen utensil such as chopping board, kitchen towels, dish rack, toys for kids made from antimicrobial materials can promote healthy environment.

However, unlike other antimicrobial agent carriers, nanofibers can also provide mechanical strength which is desirable for most of the applications. This is currently an active research area with a focus on acquiring enhanced antimicrobial properties, cost optimization, and scale-up studies to commercialize these products.

## 4. CNCs in Barrier Applications

Limiting the effects of the environment on materials has long been a major concern in numerous industries. In the oil and gas industry where a network of pipping systems is a major vessel for transporting oil and chemicals, corrosion is a major problem and concern. In the marine industry, the exposure of vessels to sea water is constantly corroding and eroding the bottom of the vessels. Ultraviolet light exposure over time can alter the micromolecular properties of materials such as polymers, therefore changing the bulk properties. The use of CNCs in barrier applications are limited except for a few explorations in academia.

### 4.1. UV Protection

It is well known that over time when exposed to UV, the physical appearance of polyurethane (PU) starts to change; that results in yellowing of the material from the photochemical degradation of the surface molecules. In order to prevent or slow this process, research was carried out using CNCs as a UV stabilizer by Zhang et al. [81]. 3-Glycidyloxypropyl trimethoxysilane (GPTMS) was used to modify CNC at different concentrations. The GPTMS was hydrolyzed thereafter the CNC was added and allowed to react over time. Figure 10a shows the proposed reaction scheme during salinization and interaction between the modified CNC and PU. The modified CNC was then incorporated into the PU formulation and homogenized to disperse the modified CNC in the PU as shown in Figure 10b to create the PU/CNC composite. This composite was then exposed to a controlled UV radiation over time. The results from this study showed that the presence of a modified CNC in the PU drastically reduced the yellowing effect of the UV radiation with further reduction observed with increase in the concentration of the modified CNC. With the addition of 1.5% modified CNC, the anti-yellowing effect was increased by approximately 58% which demonstrated the effectiveness of CNC as an antiyellowing agent or a UV barrier agent as observed in Figure 10c. It was postulated that the addition of the modified CNC inhibited the photo degradation of the $CH_2$ group while preventing the scissioning of the urethane group.

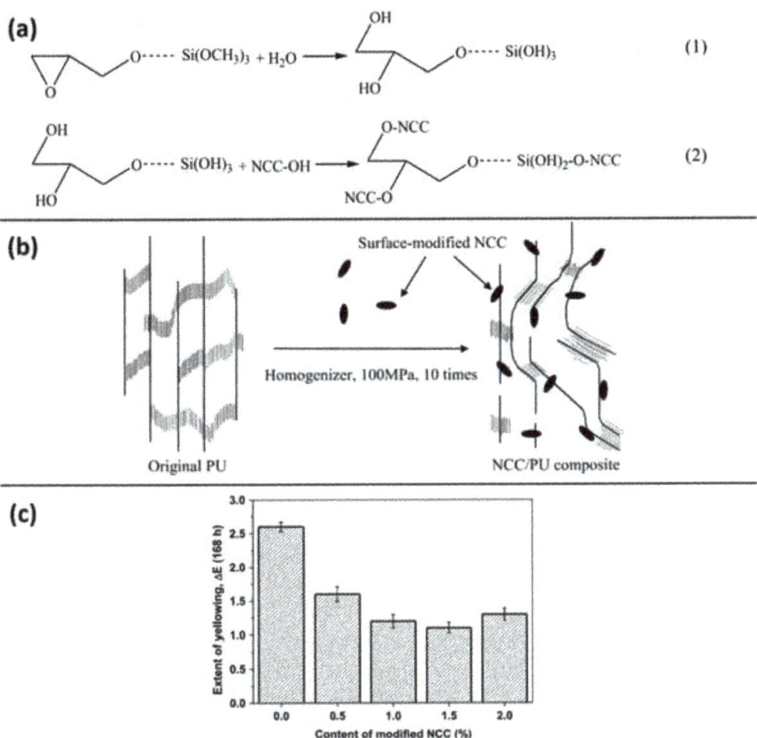

**Figure 10.** (a) reaction scheme for the modification of CNC using 3-Glycidyloxypropyl trimethoxysilane (GPTMS) (reaction scheme 1) and interaction between modified CNC and polyurethane (PU) (reaction scheme 2), (b) diagrammatic representation of the dispersion of modified CNC within the PU and (c) bar graph showing the effect of concentration of the modified CNC on the yellowing of PU upon exposure to UV radiation over a controlled period of time [81].

In another study, CNC was used as a bifunctional filler that provides reinforcement and UV barrier for poly vinyl alcohol polymer [81]. CNC pulp was oxidized using sodium metaperiodate and then reacted with sodium 4-amino-benzoate in an HCl solution to produce modified CNC with photo-active groups attached to it, which was further mechanically disintegrated to form *p*-aminobenzoic acid grafted CNC (PABA-CNC). This modified CNC was added to an aqueous polyvinyl alcohol (PVA) solution at different concentrations. This solution was further degassed and cast to form thin films which were then used for further tests. The results from the addition of PABA-CNC to PVA on the UV transmittance revealed that the presence of PABA-CNC in the PVA in comparison to neat PVA was significantly reduced and was also a function of the concentration of PABA-CNC within the PVA. Increasing the concentrations of PABA-CNC showed further reduction in UV transmittance with PVA films containing 0.5 and 10% PABA-CNC reducing the transmittance to 54 and 12%, respectively in comparison to that of neat PVA film that exhibited 70% transmittance. Likewise, enhancement in the mechanical properties such as the tensile strength and modulus of the film was observed with PABA-CNC incorporation. This increase was also a function of the concentration of PABA-CNC. Similarly, ethyl cellulose nanoparticles (ECNPs) have been studied as a means to confine or protect UV filters which tend to create reactive oxygen species when exposed to UV due to photodegradation [81]. Typically, these filters are used for applications in cosmetics, such as sunscreens and can scavenge carcinogenic reactive oxygen species which can be harmful when they come in contact with the skin.

*4.2. Solvent and Chemical Protection*

Applications of surface coatings to help protect surface from the surrounding environment is a common practice in industry in many products today. Some of the applications include the use of paint on metals to prevent rust, application of clear coat epoxies on plastics and wood to prevent scratching and UV degradation and biodegradation, respectively. However, clear coat epoxies or paints for rust prevention do not always have good strength and chemical resistance. The characteristic properties of CNCs such as excellent strength and reinforcing capabilities at very low loadings due to its crystallinity and nano-size have lent its service in the application as a functional material in epoxy surface coatings of metals in a study by Ma et al. [82]. In this study, epoxy containing 1, 1.5 and 2 wt.% of CNC was thoroughly mixed using a glass rod and then sonicated to ensure proper dispersion and then brush coated onto mild steel in thin layers. The coated steel was allowed to dry and tested for its corrosive resistance using electrochemical impedance spectroscopy (EIS) by immersion in 3.5% sodium chloride over a 30-day period. Similarly, the optical clarity or transmittance was tested using UV vis analysis to determine the effect of CNC in the epoxy coating. The results from the optical transmittance showed that with increasing CNC loading, the transmittance dropped to 20% for the coating containing 2 wt.%. However, the coating containing 1 wt.% CNC was observed to be very clear with a transmittance of 74%. It was further observed that for all composites, the light drop-off transmittance was at approximately 300 to 350 nm, which indicates high light absorption with no light reflections occurring in the UV range of 300 to 400 nm. This indicates that when used in clear coat applications, this coating (especially at 1 wt.% CNC) can act to prevent UV degradation while still having high clarity. The corrosion test on the other hand, revealed that the presence of CNCs in the epoxy significantly increased the corrosion resistance of the epoxy coating. This could be because the CNC acted as a barrier to the ions from sodium hydroxide solution by creating a tortuous path. This resulted in preventing complete penetration of the coating, thereby protecting the mild steel surface. Only the unreinforced epoxy suffered penetration after 1 day of exposure. It was suggested that this was due to the CNC creating a longer path through which the solution had to travel in order to come in contact with the mild steel. Electrochemical analysis of the coatings through bode plots revealed that only the neat epoxy coating showed two-time constants. At day 30, only the unmodified epoxy continued to reveal two-time constants revealing the remarkable ability of the CNC to act as an anticorrosion agent. However, for the 2 wt.% CNC loaded epoxy coating, as the test continued over the 30-day period, the appearance of a two-time constant slowly developed. This was attributed to the possible agglomeration of the CNCs within the epoxy which did not homogeneously disperse like those in the 1 and 1.5 wt.% loaded samples. Therefore, exposing unprotected regions within the epoxy to the solution allowed diffusion of the ions ($Na^+$ and $Cl^-$) towards the mild steel that resulted in corrosion.

## 5. Electrical Applications of CNCs

Research into the development of functional applications of CNCs in electroactive materials such as electrical conductive materials, dielectric materials, microelectronics component etc. is an upcoming research area with some current materials such as starch [24], proteins and peptides [83] already being researched for similar applications. The interest for using CNCs in electrical applications is accrued mainly from its ease of modification, piezoelectric and dielectric properties, and sustainability attributes similar to other bio-derived materials. Csoka et al. asked the question of whether ultrathin films containing well aligned cellulose nanocrystals could exhibit piezoelectric effects stemming from the collective yield of the individual CNCs [84]. Ultrathin films containing different degrees of CNC alignment were produced according to a process well described by the authors in a different publication [85]. The displacement of the film was measured using atomic force microscopy in tapping mode while an electric field was applied to them. It was observed that the higher the degree of alignment, the greater the piezoelectric effect. Therefore, the piezoelectricity of the film was dependent on the orientation of the CNCs. This observation was deduced not only to the alignment of the CNCs but also due to dipolar orientation and the crystallinity of CNCs in the films. Figure 11a shows the schematic of Atomic Force Microscopy (AFM) in tapping mode with a diamond tip over ultrathin films containing aligned CNCs with an electric field applied across it while Figure 11b graphically demonstrates the effect of applied voltage to the thin films in response to the displacement of the CNCs. Given the result from this study, ultrathin films with various degrees of CNC alignment can result to different levels of electro-mechanical actuation, which could potentially be used in applications such as ultra-sensitive micro balances. For examples, having a high degree of CNC alignment and concentration can potentially result in ultra-sensitivity of applied forces.

CNCs are able to change polarization densities due to the high level of crystallinity. Likewise, its dielectric property allows it to be applied as a functional insulating material in different applications. However, moisture plays a huge role in determining the final dielectric property as it acts as a conductor of electricity when present in the CNC. For use in dielectric applications, the moisture levels should typically be ~0.5% [86]. However, due to its hygroscopicity, it usually has a moisture level typically between 4–8%. This range of moisture content stems from the source of the CNCs and highly dependent on the cellulose crystallinity as determined from studies on water sorption of cellulose [87,88]. Overall, the higher the crystallinity of the cellulose, the lower the moisture content will be and vice versa. Bras et al. studied the dielectric property of two nanocelluloses from wood (nanofibrillated cellulose, NFC) and algae (*Cladophora* cellulose) [86]. They found that the moisture sorption capability at low and high humidity was higher for NFC due to its lower crystallinity. However, contrary to the expectations that dielectric properties are highly related to crystallinity, a higher dielectric property was observed for NFC when compared to that of *Cladophora* cellulose. This was due to the high porosity of *Cladophora* cellulose which allowed for air entrapment, subsequently increasing its dielectric loss. It is apparent that CNCs do have excellent dielectric properties which can be harnessed for electrical insulation purpose such as in cable insulation, but the effectiveness of this property not only depends on the source but also on the morphological features [89]. It has also shown that paper from NFCs can be used to make dielectrics and not just as a substrate, with results comparable to wood, polymer and glass dielectric substrates [90].

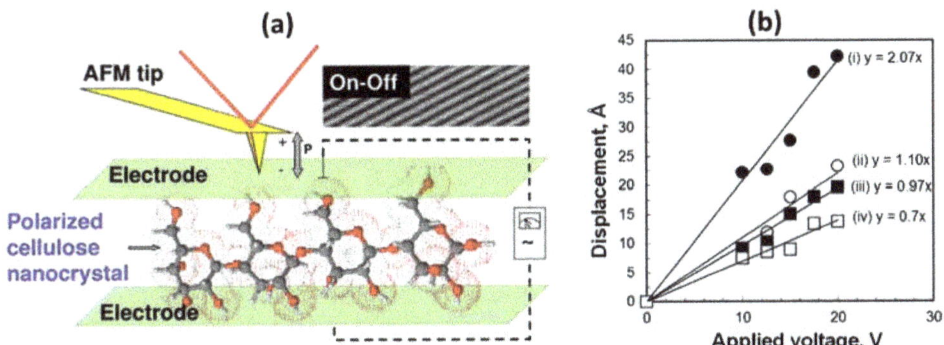


**Figure 11.** (**a**) Characterization of well aligned CNC ultrathin film for its dielectric property using Atomic Force Microscopy (AFM) in tapping mode and (**b**) effect of voltage on CNC displace (piezoelectric effect) [84].

## 6. Other Applications of CNCs in Advanced Functional Materials

Cellulose nanocrystals have gained a lot of interest as a renewable reinforcing filler due its biodegradability, low density, higher aspect ratio and excellent mechanical properties. CNCs are considered for a wide variety of applications since the surface hydroxyl groups of CNC can be functionalized as per requirement. CNCs obtained through sulfuric acid hydrolysis contain negative charge on the surface providing electrostatic repulsion between the CNCs. As a result, CNC can easily be dispersed in polar polymer matrices such as PVA. In the last two decades, applications of CNC have attracted academia and industrial researchers. Other than its use as a reinforcing agent, CNCs have demonstrated interesting potential for biomedical, antimicrobial, personal care and energy applications.

### 6.1. Polymeric Reinforcement

Cellulose nanocrystals are commonly used as a reinforcing filler in polymer matrices to enhance their strength. Traditional polymers alone lack strength required for most structural applications, thus requiring fillers to reinforce them. CNC as a bio-based material provides excellent reinforcement for polymers. Several studies have shown enhancement in both thermal and physical properties of polymers at very low concentrations of CNC incorporation due to its nano-size and ability to efficiently absorb stress from the matrix [21,91,92]. Bras et al. reported the reinforcing effect of CNC in rubber [93]. They found that CNCs were able to enhance the thermomechanical and mechanical properties of rubber. However, due to the hydrophilicity of the CNCs, the water absorption was increased. This study shows that CNCs can be used as polymer property enhancement when specific applications are targeted that do not expose the composites to moisture. Furthermore, reinforcing capabilities of CNC is also evident in cement as investigated by Cao et al. [94]. The water absorbing property of hydrophilic CNCs was an advantage in improving the flexural properties of cement paste. It was postulated that the hydration of the paste was improved when up to 0.2 vol.% of CNC was added, subsequently leading to flexural strength increase of up to 30%. These examples suggest that CNC can be the future of nano-polymer composite materials used for construction, automobile parts, furniture, and other industrial high-strength materials when targeted appropriately for specific materials.

A variety of polymers have been successfully grafted on CNC to modify its surface and obtain modified CNC with desirable properties. For instance, the use of hydrophobic polymer grafted CNC in non-polar polymers usually results in improved dispersion and interaction of the CNC with the polymer and enhances the strength of polymer composites made thereof. In other cases, the grafted polymer provides functionality attribute to the CNC that expands the application range of CNC. Table 3 shows various polymers grafted to CNC and the achieved attribute as a result of the modification.

**Table 3.** Polymer Grafted Nanocellulose.

| POLYMER | REACTION MECHANISM | NOTES | REFERENCES |
|---|---|---|---|
| POLY(4-VINYLPYRIDINE) (PVP) | Ceric ion initiated radical polymerization | Thermal degradation temperature improved by 60 °C | [25] |
| POLYMETHYL METHACRYLATE (PMMA) | Atomic Transfer Radical Polymerizstion (ATRP) | Thermal stability decreased (250 °C from 295 °C) Water contact angle increased by 17 °C | [95] |
| POLYETHYLENE GLYCOL (PEG) | Grafting to | Thermal decomposition of grafted CNC occurs at lower temperature | [96] |
| POLY(ε-CAPROLACTONE) (PCL) | Ring Opening Polymerization (ROP) | Addition of PCL grafted CNC into PCL matrix improved young's modulus from 231 to 582 GPa Elongation at break for same matrix decreased drastically | [97] |
| JEFFAMINES(POLYETHER AMINE) | Grafting-to | Thermo-reversible aggregation Polymer grafted CNC has lower surface charge (zeta potential) | [98] |

**Table 3.** *Cont.*

| POLYMER | REACTION MECHANISM | NOTES | REFERENCES |
|---|---|---|---|
| POLYLACTIC ACID (PLA) | Ring Opening Polymerization (ROP) | Well dispersed in non-polar solvent, chloroform | [99] |
| POLYMETHYL ACRYLATE (PMA) | Nitroxide-mediated Polymaerization (NMP) | CNC grafted PMA was up to 1 wt.% soluble in THF and acetone Grafted CNC was not dispersible in water | [100] |
| POLY TERT-BUTYL ACRYLATE (PTBA) | Grafting-to | Stable suspension in chloroform, toluene and acetone | [101] |
| POLYSTYRENE (PS) | Atomic Transfer Radical Polymerizstion (ATRP) | Can absorb 50% of 1,2,4-Trichlorobenzene (1,2,4-TCB), organic pollutant Water contact angle increased by 50 °C | [102] |

## 6.2. Biomedical Applications

Biomedical applications of CNC include the use of CNC for medical devices, bioimaging, wound healing, scaffolds for tissue engineering and controlled drug delivery [103]. A lot of interest in biomedical application of CNC based materials stems from its biodegradability, biocompatibility, lower toxicity and excellent mechanical properties. Dong and Roman investigated fluorescently labeled CNC for bioimaging applications [104]. In this study, CNC was conjugated with fluorescein-5′-isothiocyanate using epichlorohydrin to study its biodistribution and in vivo interaction using the fluorescence labelling, which can be beneficial in many biomedical applications. This study showed that the modified CNC can be used to study interacion with cells. The authors also acknowledged that the developed functionalized CNC was being further used to study interactions with mammalian cells in another collaborative study. Drug delivery carriers can sometimes cause biological systems such as the human body to have immunological responses. Therefore, it is important that drug delivery systems do not cause adverse effects while performing their functions. A study on the use of CNCs as carriers for cyclodextrin showed that CNC was efficient as a drug delivery agent as well as having no adverse effects; immunological responses were observed to be negligible such as the secretion of pro-inflammatory cytokine, interleukin 1β (IL-1β), by enzyme-linked immunosorbent assay (ELISA) and mitochondria-derived reactive oxygen species (ROS) using fluorescence cell imaging [105].

CNC is extensively studied for tissue engineering where a scaffold device is used for self-healing and regeneration. In order to obtain an enhanced performance appropriate material selection for scaffold is very crucial. Mechanical, physical and biological properties play a vital role in elevated performance and suitable mechanical integration [106]. Thus, properties such as surface roughness, topology, porosity, pore size, surface area to volume ratio must be considered [107]. Moreover, degradation rate of biodegradable material is very important so that the healthy tissue can be restored while scaffold material gets absorbed. In order to achieve both the mechanical and biological properties unfilled polymer materials cannot be used in most cases. Thus, nanofillers are incorporated to form a nanocomposite that provides functional benefits such as mechanical strength, electrical conductivity, adhesiveness and ability to self-assemble [106,107]. A review on CNC-based biomaterial for tissue engineering by Domingues et al. [107] extensively describes the use of CNC-PLA composite material for tissue engineering to fulfill aforementioned requirements for biomedical scaffolding applications. Along with other biomedical, and pharmaceutical products, CNC can open new doors to new personal care products [108,109].

## 6.3. Energy Applications

Energy application of CNC includes the use of cellulose-based composites for energy storage. With growing environmental concerns, interest has grown in making renewable energy source more efficient and feasible. An attempt to produce recyclable solar cell from cellulose nanocrystal by Zhou et al. is a contribution towards the same initiative [110]. While taking advantage of the excellent mechanical properties of CNC and silver, a semitransparent electrode recyclable solar cell was fabricated [110]. However, further improvements were needed to obtain efficient and desirable performance. Kim et al. reported other energy applications of cellulose based materials such as energy harvester, display devices, actuators and paper transistors [111]. Excellent mechanical and biocompatible properties of CNC as well as its ease of functionalization gives it the potential to provide environmentally friendly and sustainable technologies.

*6.4. Smart and Responsive Materials*

The application of smart responsive materials has increased in recent times. Smart responsive materials adapt to external environment and provide response. Changes in stimulus such as exposure to light, heat, chemicals, magnetic fields can be used to generate mechanically adaptive, stimuli-responsive materials. Nanocomposite can respond to exposure to external stimuli in various ways such as swell or shrink, assemble or dissemble or prompts for separation [112]. These changes generated as a response to stimuli variations can be used to develop a smart material. CNC can be used as a stimuli-responsive material for sensors and other applications. CNC displays responsiveness towards pH, light, moisture, heat, chemical and magnetic fields, which adds the stimuli-responsiveness functionality along with reinforcing capabilities. Changes in pH corresponds with changes in rheological properties of CNC composites [113]. Way et al. [113] synthesized carboxylated and amine functionalized CNC to investigate pH responsiveness. By altering the surface chemistry of CNC, the nanocomposites can be reprogrammed to develop various mechanically adaptable materials.

Since modified CNC has the ability to generate a response to changes in external stimuli, they can be readily used in sensing applications. Smart CNC-based sensors can be designed for moisture, ions, organic vapors, and biological species sensing applications. Humidity sensor designed by Kafy et al. [114] is made from CNC-graphene oxide (GO) composite. CNC and graphene both are hydrophilic with higher water uptake capacity. CNC-GO film displayed even increased water uptake capacity which is desirable for moisture sensitivity [114]. The sensing film did not compromise its performance with temperature change, demonstrating practical use of humidity sensor [114]. Similarly, CNC can be functionalized to make gas sensing material which can detect other organic and toxic vapors. Moreover, CNC based sensing material can also be used to detect ionic species. CNC containing pyrene was synthesized to detect ferric ($Fe^{3+}$) ions [115]. This concept can be further explored to design a sensing material for different ions, chemical and biological molecules. Other smart sensors made from CNC include proximity sensor and strain sensor by Sadasivuni et al. [116] and Wang et al. [117] respectively.

Hydrophilicity of CNC can be used to generate water-responsive mechanically adaptive polymer matrix. For instance, Pratheep et al. [118] designed a water sensitive styrene-butadiene-rubber (SBR) and CNC nanocomposite. First of all, the reinforcing capabilities was proven since the modulus of pure SBR (3 MPa) was improved significantly (to 740 MPa) [118]. CNC aids as a hydrophilic channel in a hydrophobic matrix for water uptake. CNC is used to generate water-responsive mechanically adaptive material because upon water swelling, CNC network is disrupted to cause reduction in modulus [118,119]. Thus physiological variation can lead to change in mechanical properties to design a mechanically adaptive materials [118]. Mendez et al. [119] also used the water-responsive ability of CNC to design a smart polyurethane-CNC nanocomposite, which displayed water activated shape memory effect.

Similar to pH and moisture response adaptive materials, thermal and photonic responsive materials were also generated. Cellulose nanocrystals grafted with thermoresponsive brushes of poly (N-isopropylacrylamide) has also been investigated [120–122]. Novel and effective drug therapies have also been investigated to develop a controlled drug delivery method. Controlled drug release provides advantages such as maintaining required therapeutic concentration, localized drug delivery and improve patient compliance, however these novel techniques are yet to be explored to develop drug delivery materials sensitive to physiological changes [123]. Thus, stimuli-responsive polymeric drug delivery processes are typically used for controlled drug delivery. CNCs, being biocompatible and biodegradable polymer, can be functionalized to develop smart material that offers a great potential in targeted, controlled drug delivery systems [111,123].

The study of CNC for smart electronic application has also gained enormous interest. In the Review of Nanocellulose for Sustainable Future Material, Kim et al. [111] has reported various use of CNC based material in energy and electronics applications. Electroactive paper (EAPap), which functions based on two principles i.e., ion migration and piezoelectric effect has application in sensors, actuators, biomimetic robots, and other haptic technology [111,112].

## 7. Challenges, Future Trends and Conclusions

Although CNCs have many desirable properties, their high hygroscopicity, hydrophilicity and tendency to aggregate and form bundles, can drastically restrict their applications [3]. These characteristics lead to the formation of flake-like agglomerates in polymer composites due to surface area and volume effects and the strong intermolecular hydrogen bonding between CNC particles [64]. Aggregation of CNC particles can significantly hinder the performance of the matrices. This suggests that the issue with agglomeration of CNCs must be resolved to achieve its touted reinforcing performance when used as a filler in polymeric and other applications. However, its hydrophilic nature can also play a positive role in enhancing hybrid systems such as cement paste. Therefore, the inherent properties of CNCs are typically deemed as negative can be advantageous and geared towards specific applications. However, the properties can be modified to reduce or completely eliminate its effects as a durable functional material. During CNC modification, it is important to limit the modification only to its surface so as to preserve its structural and morphological integrity [36]. However, the reality is that most CNC modifications involve aggressive oxidization to increase the concentration of hydroxyl groups on its surface to be used as anchor points for further modifications. This usually effects its structural integrity and crystallinity, which could subsequently alter its mechanical properties. Therefore, new and less aggressive modification processed and/or chemicals need to be developed to achieve the same level of modifications as currently used.

Though CNCs have excellent thermal stability in comparison to other bio-based fibers, long term durability of products is of concern when it is incorporated or used as a functional material. Moreover, commercial products intending to apply CNCs in functional applications need multiple and abundant sources. Currently, the majority of CNC production is limited to North America and to some Scandinavian countries (e.g., Sweden, Finland) [124]. Expected rise in CNC production in the future will definitely give raise to the application on an industrial scale [18].

The versality of CNC functionalization can be the future of sustainable, biocompatible, and biodegradable material for a variety of applications, especially with the current environmental concerns such as climate change from greenhouse gases from production and use of petroleum derived materials. Due to its abundance across the globe, bio-renewability and the fast-growing technologies on its production, CNCs are becoming the focus of product development and research as observed from the numerous researches published on CNCs in recent years. Therefore, it is foreseeable that in the nearest future, CNCs will be widely and extensively found in the medical, electronics, oil and gas, biotechnology and food industries just to mention a few. CNCs have also been shown to be a viable option to potential replace synthetic materials for functional applications in various industries. As CNCs continue to gain great attention from researchers all over the world, more understanding and applications with improved technology towards production will rapidly be developed thereby taking CNCs from lab to industrial scale.

**Author Contributions:** Conceptualization, P.P., E.O., T.M.; writing—original draft preparation, P.P. and E.O; writing—review and editing, P.P., E.O., T.M; supervision, T.M.; project administration, T.M.; funding acquisition, T.M.

**Funding:** The authors acknowledge the financial support of InnoTech Alberta through the CNC Challenge grant.

**Acknowledgments:** We thank our colleagues in the same group for the insightful discussions on the subject matter.

**Conflicts of Interest:** The authors declare no conflict of interest.

# References


1. Habibi, Y.; Lucia, L.A.; Rojas, O.J. Cellulose Nanocrystals: Chemistry, Self-Assembly, and Applications. *Chem. Rev.* **2010**, *110*, 3479–3500. [CrossRef] [PubMed]
2. Bettaieb, F.; Khiari, R.; Dufresne, A.; Mhenni, M.F.; Belgacem, M.N. Mechanical and thermal properties of Posidonia oceanica cellulose nanocrystal reinforced polymer. *Carbohydr. Polym.* **2015**, *123*, 99–104. [CrossRef]
3. Lu, P.; Hsieh, Y.-L. Preparation and properties of cellulose nanocrystals: Rods, spheres, and network. *Carbohydr. Polym.* **2010**, *82*, 329–336. [CrossRef]
4. Eyley, S.; Thielemans, W. Surface modification of cellulose nanocrystals. *Nanoscale* **2014**, *6*, 7764–7779. [CrossRef] [PubMed]
5. Wang, B.; Sain, M.; Oksman, K. Study of Structural Morphology of Hemp Fiber from the Micro to the Nanoscale. *Appl. Compos. Mater.* **2007**, *14*, 89–103. [CrossRef]
6. Ogunsona, E.O.; Misra, M.; Mohanty, A.K. Accelerated hydrothermal aging of biocarbon reinforced nylon biocomposites. *Polym. Degrad. Stab.* **2017**, *139*, 76–88. [CrossRef]
7. Ogunsona, E.O.; Misra, M.; Mohanty, A.K. Sustainable biocomposites from biobased polyamide 6,10 and biocarbon from pyrolyzed miscanthus fibers. *J. Appl. Polym. Sci.* **2017**, *134*. [CrossRef]
8. Ogunsona, E.O.; Misra, M.; Mohanty, A.K. Influence of epoxidized natural rubber on the phase structure and toughening behavior of biocarbon reinforced nylon 6 biocomposites. *RSC Adv.* **2017**, *7*, 8727–8739. [CrossRef]
9. Ogunsona, E.O.; Misra, M.; Mohanty, A.K. Effects of Accelerated Aging on the Flammability of Polypropylene Based Biocomposites. In Proceedings of the ANTEC Conference Proceedings, Orlando, FL, USA, 23–25 March 2015.
10. Ogunsona, E.O.; Codou, A.; Misra, M.; Mohanty, A.K. Thermally Stable Pyrolytic Biocarbon as an Effective and Sustainable Reinforcing Filler for Polyamide Bio-composites Fabrication. *J. Polym. Environ.* **2018**. [CrossRef]
11. Ogunsona, E.O.; Misra, M.; Mohanty, A.K. Impact of interfacial adhesion on the microstructure and property variations of biocarbons reinforced nylon 6 biocomposites. *Compos. Part A Appl. Sci. Manuf.* **2017**, *98*, 32–44. [CrossRef]
12. Ilyas, R.A.; Sapuan, S.M.; Sanyang, M.L.; Ishak, M.R.; Zainudin, E.S. Nanocrystalline Cellulose as Reinforcement for Polymeric Matrix Nanocomposites and its Potential Applications: A Review. *Curr. Anal. Chem.* **2018**, *14*, 203–225. [CrossRef]
13. Shojaeiarani, J.; Bajwa, D.S.; Stark, N.M. Green esterification: A new approach to improve thermal and mechanical properties of poly(lactic acid) composites reinforced by cellulose nanocrystals. *J. Appl. Polym. Sci.* **2018**, *135*, 46468. [CrossRef]
14. Sobolčiak, P.; Tanvir, A.; Popelka, A.; Moffat, J.; Mahmoud, K.A.; Krupa, I. The preparation, properties and applications of electrospun co-polyamide 6,12 membranes modified by cellulose nanocrystals. *Mater. Des.* **2017**, *132*, 314–323. [CrossRef]
15. Rahimi, S.K.; Otaigbe, J.U. The effects of the interface on microstructure and rheo-mechanical properties of polyamide 6/cellulose nanocrystal nanocomposites prepared by in-situ ring-opening polymerization and subsequent melt extrusion. *Polymer* **2017**, *127*, 269–285. [CrossRef]
16. Inai, N.H.; Lewandowska, A.E.; Ghita, O.R.; Eichhorn, S.J. Interfaces in polyethylene oxide modified cellulose nanocrystal - polyethylene matrix composites. *Compos. Sci. Technol.* **2018**, *154*, 128–135. [CrossRef]
17. Orts, W.J.; Shey, J.; Imam, S.H.; Glenn, G.M.; Guttman, M.E.; Revol, J.-F. Application of Cellulose Microfibrils in Polymer Nanocomposites. *J. Polym. Environ.* **2005**, *13*, 301–306. [CrossRef]
18. Moon, R.J.; Martini, A.; Nairn, J.; Simonsen, J.; Youngblood, J. Cellulose nanomaterials review: Structure, properties and nanocomposites. *Chem. Soc. Rev.* **2011**, *40*, 3941–3994. [CrossRef] [PubMed]
19. Lee, K.-Y.; Aitomäki, Y.; Berglund, L.A.; Oksman, K.; Bismarck, A. On the use of nanocellulose as reinforcement in polymer matrix composites. *Compos. Sci. Technol.* **2014**, *105*, 15–27. [CrossRef]
20. Jonoobi, M.; Oladi, R.; Davoudpour, Y.; Oksman, K.; Dufresne, A.; Hamzeh, Y.; Davoodi, R. Different preparation methods and properties of nanostructured cellulose from various natural resources and residues: A review. *Cellulose* **2015**, *22*, 935–969. [CrossRef]

21. Xu, X.; Liu, F.; Jiang, L.; Zhu, J.Y.; Haagenson, D.; Wiesenborn, D.P. Cellulose Nanocrystals vs. Cellulose Nanofibrils: A Comparative Study on Their Microstructures and Effects as Polymer Reinforcing Agents. *ACS Appl. Mater. Interfaces* **2013**, *5*, 2999–3009. [CrossRef]

22. Mekonnen, T.H.; Mussone, P.G.; Choi, P.; Bressler, D.C. Development of Proteinaceous Plywood Adhesive and Optimization of Its Lap Shear Strength. *Macromol. Mater. Eng.* **2015**, *300*, 198–209. [CrossRef]

23. Ojogbo, E.; Blanchard, R.; Mekonnen, T. Hydrophobic and Melt Processable Starch-Laurate Esters: Synthesis, Structure-Property Correlations. *J. Polym. Sci. Part A Polym. Chem.* **2018**, *56*, 2611–2622. [CrossRef]

24. Ogunsona, E.; Ojogbo, E.; Mekonnen, T. Advanced material applications of starch and its derivatives. *Eur. Polym. J.* **2018**, *108*, 570–581. [CrossRef]

25. Kan, K.H.M.; Li, J.; Wijesekera, K.; Cranston, E.D. Polymer-Grafted Cellulose Nanocrystals as pH-Responsive Reversible Flocculants. *Biomacromolecules* **2013**, *14*, 3130–3139. [CrossRef] [PubMed]

26. Brown, A.J. XIX.—The chemical action of pure cultivations of bacterium aceti. *J. Chem. Soc. Trans.* **1886**, *49*, 172–187. [CrossRef]

27. Kim, N.; Retsina, T. Innovative nanocellulose process breaks the cost barrier. *TAPPI J.* **2014**, *13*, 19–23.

28. Akhlaghi, S.P.; Berry, R.C.; Tam, K.C. Surface modification of cellulose nanocrystal with chitosan oligosaccharide for drug delivery applications. *Cellulose* **2013**, *20*, 1747–1764. [CrossRef]

29. Peng, B.L.; Dhar, N.; Liu, H.L.; Tam, K.C. Chemistry and applications of nanocrystalline cellulose and its derivatives: A nanotechnology perspective. *Can. J. Chem. Eng.* **2011**, *89*, 1191–1206. [CrossRef]

30. Chen, L.; Wang, Q.; Hirth, K.; Baez, C.; Agarwal, U.P.; Zhu, J.Y. Tailoring the yield and characteristics of wood cellulose nanocrystals (CNC) using concentrated acid hydrolysis. *Cellulose* **2015**, *22*, 1753–1762. [CrossRef]

31. Silvério, H.A.; Flauzino Neto, W.P.; Dantas, N.O.; Pasquini, D. Extraction and characterization of cellulose nanocrystals from corncob for application as reinforcing agent in nanocomposites. *Ind. Crops Prod.* **2013**, *44*, 427–436. [CrossRef]

32. Kalashnikova, I.; Bizot, H.; Bertoncini, P.; Cathala, B.; Capron, I. Cellulosic nanorods of various aspect ratios for oil in water Pickering emulsions. *Soft Matter* **2013**, *9*, 952–959. [CrossRef]

33. Cao, X.; Dong, H.; Li, C.M. New nanocomposite materials reinforced with flax cellulose nanocrystals in waterborne polyurethane. *Biomacromolecules* **2007**, *8*, 899–904. [CrossRef]

34. Habibi, Y.; Foulon, L.; Aguié-Béghin, V.; Molinari, M.; Douillard, R. Langmuir–Blodgett films of cellulose nanocrystals: Preparation and characterization. *J. Colloid Interface Sci.* **2007**, *316*, 388–397. [CrossRef] [PubMed]

35. Garcia de Rodriguez, N.L.; Thielemans, W.; Dufresne, A. Sisal cellulose whiskers reinforced polyvinyl acetate nanocomposites. *Cellulose* **2006**, *13*, 261–270. [CrossRef]

36. Kaboorani, A.; Riedl, B. Surface modification of cellulose nanocrystals (CNC) by a cationic surfactant. *Ind. Crops Prod.* **2015**, *65*, 45–55. [CrossRef]

37. Grishkewich, N.; Mohammed, N.; Tang, J.; Tam, K.C. Recent advances in the application of cellulose nanocrystals. *Curr. Opin. Colloid Interface Sci.* **2017**, *29*, 32–45. [CrossRef]

38. Bravo, J.; Zhai, L.; Wu, Z.; Cohen, R.E.; Rubner, M.F. Transparent superhydrophobic films based on silica nanoparticles. *Langmuir* **2007**, *23*, 7293–7298. [CrossRef] [PubMed]

39. Song, J.; Rojas, O.J. Approaching super-hydrophobicity from cellulosic materials: A Review. *Pap. Chem. Nord. Pulp Pap. Res. J.* **2100**, *28*, 216–238. [CrossRef]

40. Li, S.; Xie, H.; Zhang, S.; Wang, X. Facile transformation of hydrophilic cellulose into superhydrophobic cellulose. *Chem. Commun.* **2007**, 4857–4859. [CrossRef]

41. Teisala, H.; Tuominen, M.; Kuusipalo, J. Superhydrophobic Coatings on Cellulose-Based Materials: Fabrication, Properties, and Applications. *Adv. Mater. Interfaces* **2014**, *1*, 1300026. [CrossRef]

42. Manca, M.; Cannavale, A.; De Marco, L.; Aricò, A.S.; Cingolani, R.; Gigli, G. Durable superhydrophobic and antireflective surfaces by trimethylsilanized silica nanoparticles-based sol-gel processing. *Langmuir* **2009**, *25*, 6357–6362. [CrossRef] [PubMed]

43. Salam, A.; Lucia, L.A.; Jameel, H. Fluorine-based surface decorated cellulose nanocrystals as potential hydrophobic and oleophobic materials. *Cellulose* **2015**, *22*, 397–406. [CrossRef]

44. Shang, W.; Huang, J.; Luo, H.; Chang, P.R.; Feng, J.; Xie, G. Hydrophobic modification of cellulose nanocrystal via covalently grafting of castor oil. *Cellulose* **2013**, *20*, 179–190. [CrossRef]

45. Salajková, M.; Berglund, L.A.; Zhou, Q. Hydrophobic cellulose nanocrystals modified with quaternary ammonium salts. *J. Mater. Chem.* **2012**, *22*, 19798–19805. [CrossRef]

46. Yu, M.; Gu, G.; Meng, W.-D.; Qing, F.-L. Superhydrophobic cotton fabric coating based on a complex layer of silica nanoparticles and perfluorooctylated quaternary ammonium silane coupling agent. *Appl. Surf. Sci.* **2007**, *253*, 3669–3673. [CrossRef]

47. Ogihara, H.; Xie, J.; Okagaki, J.; Saji, T. Simple Method for Preparing Superhydrophobic Paper: Spray-Deposited Hydrophobic Silica Nanoparticle Coatings Exhibit High Water-Repellency and Transparency. *Langmuir* **2012**, *28*, 4605–4608. [CrossRef] [PubMed]

48. Samyn, P. Wetting and hydrophobic modification of cellulose surfaces for paper applications. *J. Mater. Sci.* **2013**, *48*, 6455–6498. [CrossRef]

49. Bae, G.Y.; Min, B.G.; Jeong, Y.G.; Lee, S.C.; Jang, J.H.; Koo, G.H. Superhydrophobicity of cotton fabrics treated with silica nanoparticles and water-repellent agent. *J. Colloid Interface Sci.* **2009**, *337*, 170–175. [CrossRef]

50. Xue, C.-H.; Jia, S.-T.; Zhang, J.; Ma, J.-Z. Large-area fabrication of superhydrophobic surfaces for practical applications: An overview. *Sci. Technol. Adv. Mater.* **2010**, *11*, 033002. [CrossRef]

51. Zhou, S.; Liu, P.; Wang, M.; Zhao, H.; Yang, J.; Xu, F. Sustainable, Reusable, and Superhydrophobic Aerogels from Microfibrillated Cellulose for Highly Effective Oil/Water Separation. *ACS Sustain. Chem. Eng.* **2016**, *4*, 6409–6416. [CrossRef]

52. Cheng, Q.-Y.; Guan, C.-S.; Wang, M.; Li, Y.-D.; Zeng, J.-B. Cellulose nanocrystal coated cotton fabric with superhydrophobicity for efficient oil/water separation. *Carbohydr. Polym.* **2018**, *199*, 390–396. [CrossRef] [PubMed]

53. Huang, J.; Wang, S.; Lyu, S.; Fu, F. Preparation of a robust cellulose nanocrystal superhydrophobic coating for self-cleaning and oil-water separation only by spraying. *Ind. Crops Prod.* **2018**, *122*, 438–447. [CrossRef]

54. Guo, J.; Fang, W.; Welle, A.; Feng, W.; Filpponen, I.; Rojas, O.J.; Levkin, P.A. Superhydrophobic and Slippery Lubricant-Infused Flexible Transparent Nanocellulose Films by Photoinduced Thiol–Ene Functionalization. *ACS Appl. Mater. Interfaces* **2016**, *8*, 34115–34122. [CrossRef] [PubMed]

55. Fan, T.; Qian, Q.; Hou, Z.; Liu, Y.; Lu, M. Preparation of smart and reversible wettability cellulose fabrics for oil/water separation using a facile and economical method. *Carbohydr. Polym.* **2018**, *200*, 63–71. [CrossRef] [PubMed]

56. Varesano, A.; Vineis, C.; Aluigi, A.; Rombaldoni, F. Antimicrobial polymers for textile products. *Sci. Against Microb. Pathog. Commun. Curr. Res. Technol. Adv.* **2011**. Available online: http://www.formatex.info/microbiology3/book/99-110.pdf (accessed on 24 November 2018).

57. Mauriello, G.; De Luca, E.; La Storia, A.; Villani, F.; Ercolini, D. Antimicrobial activity of a nisin-activated plastic film for food packaging. *Lett. Appl. Microbiol.* **2005**, *41*, 464–469. [CrossRef] [PubMed]

58. Kodjak, A. *FDA Bans 19 Chemicals Used in Antibacterial*; Your Health: Salt Lake City, UT, USA, 2016.

59. Laxminarayan, R.; Duse, A.; Wattal, C.; Zaidi, A.K.M.; Wertheim, H.F.L.; Sumpradit, N.; Vlieghe, E.; Hara, G.L.; Gould, I.M.; Goossens, H.; et al. Antibiotic resistance—The need for global solutions. *Lancet Infect. Dis.* **2013**, *13*, 1057–1098. [CrossRef]

60. Roy, D.; Knapp, J.S.; Guthrie, J.T.; Perrier, S. Antibacterial Cellulose Fiber via RAFT Surface Graft Polymerization. *Biomacromolecules* **2008**, *9*, 91–99. [CrossRef]

61. Wang, L.; Hu, C.; Shao, L. The antimicrobial activity of nanoparticles: Present situation and prospects for the future. *Int. J. Nanomed.* **2017**, *12*, 1227–1249. [CrossRef]

62. Appendini, P.; Hotchkiss, J.H. Review of antimicrobial food packaging. *Innov. Food Sci. Emerg. Technol.* **2002**, *3*, 113–126. [CrossRef]

63. Huang, K.-S.; Yang, C.-H.; Huang, S.-L.; Chen, C.-Y.; Lu, Y.-Y.; Lin, Y.-S. Recent Advances in Antimicrobial Polymers: A Mini-Review. *Int. J. Mol. Sci.* **2016**, *17*, 1578. [CrossRef]

64. Fortunati, E.; Rinaldi, S.; Peltzer, M.; Bloise, N.; Visai, L.; Armentano, I.; Jiménez, A.; Latterini, L.; Kenny, J.M. Nano-biocomposite films with modified cellulose nanocrystals and synthesized silver nanoparticles. *Carbohydr. Polym.* **2014**, *101*, 1122–1133. [CrossRef] [PubMed]

65. Drogat, N.; Granet, R.; Sol, V.; Memmi, A.; Saad, N.; Klein Koerkamp, C.; Bressollier, P.; Krausz, P. Antimicrobial silver nanoparticles generated on cellulose nanocrystals. *J. Nanopart. Res.* **2011**, *13*, 1557–1562. [CrossRef]

66. Lala, N.L.; Ramaseshan, R.; Bojun, L.; Sundarrajan, S.; Barhate, R.S.; Ying-jun, L.; Ramakrishna, S. Fabrication of nanofibers with antimicrobial functionality used as filters: Protection against bacterial contaminants. *Biotechnol. Bioeng.* **2007**, *97*, 1357–1365. [CrossRef] [PubMed]

67. Lam, E.; Male, K.B.; Chong, J.H.; Leung, A.C.W.; Luong, J.H.T. Applications of functionalized and nanoparticle-modified nanocrystalline cellulose. *Trends Biotechnol.* **2012**, *30*, 283–290. [CrossRef] [PubMed]

68. Ramos, A.I.; Braga, T.M.; Fernandes, J.A.; Silva, P.; Ribeiro-Claro, P.J.; Almeida Paz, F.A.; de Fátima Silva Lopes, M.; Braga, S.S. Analysis of the microcrystalline inclusion compounds of triclosan with β-cyclodextrin and its tris-O-methylated derivative. *J. Pharm. Biomed. Anal.* **2013**, *80*, 34–43. [CrossRef] [PubMed]

69. Suller, M.T.E.; Russell, A.D. Triclosan and antibiotic resistance in Staphylococcus aureus. *J. Antimicrob. Chemother.* **2000**, *46*, 11–18. [CrossRef] [PubMed]

70. Helander, I.M.; Nurmiaho-Lassila, E.-L.; Ahvenainen, R.; Rhoades, J.; Roller, S. Chitosan disrupts the barrier properties of the outer membrane of Gram-negative bacteria. *Int. J. Food Microbiol.* **2001**, *71*, 235–244. [CrossRef]

71. Rabea, E.I.; Badawy, M.E.T.; Stevens, C.V.; Smagghe, G.; Steurbaut, W. Chitosan as antimicrobial agent: Applications and mode of action. *Biomacromolecules* **2003**, *4*, 1457–1465. [CrossRef]

72. El-tahlawy, K.F.; El-bendary, M.A.; Elhendawy, A.G.; Hudson, S.M. The antimicrobial activity of cotton fabrics treated with different crosslinking agents and chitosan. *Carbohydr. Polym.* **2005**, *60*, 421–430. [CrossRef]

73. Cha, D.S.; Chinnan, M.S. Biopolymer-Based Antimicrobial Packaging: A Review. *Crit. Rev. Food Sci. Nutr.* **2004**, *44*, 223–237. [CrossRef]

74. Salmieri, S.; Islam, F.; Khan, R.A.; Hossain, F.M.; Ibrahim, H.M.M.; Miao, C.; Hamad, W.Y.; Lacroix, M. Antimicrobial nanocomposite films made of poly(lactic acid)-cellulose nanocrystals (PLA-CNC) in food applications: Part A-effect of nisin release on the inactivation of Listeria monocytogenes in ham. *Cellulose* **2014**, *21*, 1837–1850. [CrossRef]

75. Weishaupt, R.; Heuberger, L.; Siqueira, G.; Gutt, B.; Zimmermann, T.; Maniura-Weber, K.; Salentinig, S.; Faccio, G. Enhanced Antimicrobial Activity and Structural Transitions of a Nanofibrillated Cellulose–Nisin Biocomposite Suspension. *ACS Appl. Mater. Interfaces* **2018**, *10*, 20170–20181. [CrossRef] [PubMed]

76. Lee, C.H.; Park, H.J.; Lee, D.S. Influence of antimicrobial packaging on kinetics of spoilage microbial growth in milk and orange juice. *J. Food Eng.* **2004**, *65*, 527–531. [CrossRef]

77. Tang, J.; Song, Y.; Berry, R.M.; Tam, K.C. Polyrhodanine coated cellulose nanocrystals as optical pH indicators. *RSC Adv.* **2014**, *4*, 60249–60252. [CrossRef]

78. Fortunati, E.; Armentano, I.; Zhou, Q.; Iannoni, A.; Saino, E.; Visai, L.; Berglund, L.A.; Kenny, J.M. Multifunctional bionanocomposite films of poly(lactic acid), cellulose nanocrystals and silver nanoparticles. *Carbohydr. Polym.* **2012**, *87*, 1596–1605. [CrossRef]

79. Tang, J.; Song, Y.; Tanvir, S.; Anderson, W.A.; Berry, R.M.; Tam, K.C. Polyrhodanine Coated Cellulose Nanocrystals: A Sustainable Antimicrobial Agent. *ACS Sustain. Chem. Eng.* **2015**, *3*, 1801–1809. [CrossRef]

80. Unnithan, A.R.; Gnanasekaran, G.; Sathishkumar, Y.; Lee, Y.S.; Kim, C.S. Electrospun antibacterial polyurethane–cellulose acetate–zein composite mats for wound dressing. *Carbohydr. Polym.* **2014**, *102*, 884–892. [CrossRef]

81. Zhang, H.; Chen, H.; She, Y.; Zheng, X.; Pu, J. Anti-Yellowing Property of Polyurethane Improved by the Use of Surface-Modified Nanocrystalline Cellulose. *BioResources* **2013**, *9*, 673–684. [CrossRef]

82. Wonnie Ma, I.A.; Shafaamri, A.; Kasi, R.; Zaini, F.N.; Balakrishnan, V.; Subramaniam, R.; Arof, A.K. Anticorrosion Properties of Epoxy/Nanocellulose Nanocomposite Coating. *BioResources* **2017**, *12*, 2912–2929. [CrossRef]

83. Panda, S.S.; Katz, H.E.; Tovar, J.D. Solid-state electrical applications of protein and peptide based nanomaterials. *Chem. Soc. Rev.* **2018**, *47*, 3640–3658. [CrossRef]

84. Csoka, L.; Hoeger, I.C.; Rojas, O.J.; Peszlen, I.; Pawlak, J.J.; Peralta, P.N. Piezoelectric Effect of Cellulose Nanocrystals Thin Films. *ACS Macro Lett.* **2012**, *1*, 867–870. [CrossRef]

85. Csoka, L.; Hoeger, I.C.; Peralta, P.; Peszlen, I.; Rojas, O.J. Dielectrophoresis of cellulose nanocrystals and alignment in ultrathin films by electric field-assisted shear assembly. *J. Colloid Interface Sci.* **2011**, *363*, 206–212. [CrossRef] [PubMed]

86. Le Bras, D.; Strømme, M.; Mihranyan, A. Characterization of Dielectric Properties of Nanocellulose from Wood and Algae for Electrical Insulator Applications. *J. Phys. Chem. B* **2015**, *119*, 5911–5917. [CrossRef] [PubMed]

87. Kocherbitov, V.; Ulvenlund, S.; Kober, M.; Jarring, K.; Arnebrant, T. Hydration of Microcrystalline Cellulose and Milled Cellulose Studied by Sorption Calorimetry. *J. Phys. Chem. B* **2008**, *112*, 3728–3734. [CrossRef] [PubMed]

88. Mihranyan, A.; Llagostera, A.P.; Karmhag, R.; Strømme, M.; Ek, R. Moisture sorption by cellulose powders of varying crystallinity. *Int. J. Pharm.* **2004**, *269*, 433–442. [CrossRef] [PubMed]

89. Gaspar, D.; Fernandes, S.N.; de Oliveira, A.G.; Fernandes, J.G.; Grey, P.; Pontes, R.V.; Pereira, L.; Martins, R.; Godinho, M.H.; Fortunato, E. Nanocrystalline cellulose applied simultaneously as the gate dielectric and the substrate in flexible field effect transistors. *Nanotechnology* **2014**, *25*, 094008. [CrossRef] [PubMed]

90. Pereira, L.; Gaspar, D.; Guerin, D.; Delattre, A.; Fortunato, E.; Martins, R. The influence of fibril composition and dimension on the performance of paper gated oxide transistors. *Nanotechnology* **2014**, *25*, 094007. [CrossRef]

91. Spinella, S.; Lo Re, G.; Liu, B.; Dorgan, J.; Habibi, Y.; Leclère, P.; Raquez, J.-M.; Dubois, P.; Gross, R.A. Polylactide/cellulose nanocrystal nanocomposites: Efficient routes for nanofiber modification and effects of nanofiber chemistry on PLA reinforcement. *Polymer* **2015**, *65*, 9–17. [CrossRef]

92. Xu, S.; Girouard, N.; Schueneman, G.; Shofner, M.L.; Meredith, J.C. Mechanical and thermal properties of waterborne epoxy composites containing cellulose nanocrystals. *Polymer* **2013**, *54*, 6589–6598. [CrossRef]

93. Bras, J.; Hassan, M.L.; Bruzesse, C.; Hassan, E.A.; El-Wakil, N.A.; Dufresne, A. Mechanical, barrier, and biodegradability properties of bagasse cellulose whiskers reinforced natural rubber nanocomposites. *Ind. Crops Prod.* **2010**, *32*, 627–633. [CrossRef]

94. Cao, Y.; Zavaterri, P.; Youngblood, J.; Moon, R.; Weiss, J. The influence of cellulose nanocrystal additions on the performance of cement paste. *Cem. Concr. Compos.* **2015**, *56*, 73–83. [CrossRef]

95. Kedzior, S.A.; Graham, L.; Moorlag, C.; Dooley, B.M.; Cranston, E.D. Poly(methyl methacrylate)-grafted cellulose nanocrystals: One-step synthesis, nanocomposite preparation, and characterization. *Can. J. Chem. Eng.* **2016**, *94*, 811–822. [CrossRef]

96. Araki, J.; Wada, M.; Kuga, S. Steric stabilization of a cellulose microcrystal suspension by poly(ethylene glycol) grafting. *Langmuir* **2001**, *17*, 21–27. [CrossRef]

97. Habibi, Y.; Goffin, A.-L.; Schiltz, N.; Duquesne, E.; Dubois, P.; Dufresne, A. Bionanocomposites based on poly($\varepsilon$-caprolactone)-grafted cellulose nanocrystals by ring-opening polymerization. *J. Mater. Chem.* **2008**, *18*, 5002. [CrossRef]

98. Azzam, F.; Heux, L.; Putaux, J.L.; Jean, B. Preparation by grafting onto, characterization, and properties of thermally responsive polymer-decorated cellulose nanocrystals. *Biomacromolecules* **2010**, *11*, 3652–3659. [CrossRef] [PubMed]

99. Peltzer, M.; Pei, A.; Zhou, Q.; Berglund, L.; Jiménez, A. Surface modification of cellulose nanocrystals by grafting with poly(lactic acid). *Polym. Int.* **2014**, *63*, 1056–1062. [CrossRef]

100. Roeder, R.D.; Garcia-Valdez, O.; Whitney, R.A.; Champagne, P.; Cunningham, M.F. Graft modification of cellulose nanocrystals via nitroxide-mediated polymerisation. *Polym. Chem.* **2016**, *7*, 6383–6390. [CrossRef]

101. Harrisson, S.; Drisko, G.L.; Malmstr€, E.; Hult, A.; Wooley, K.L. Hybrid Rigid/Soft and Biologic/Synthetic Materials: Polymers Grafted onto Cellulose Microcrystals. *Biomacromolecules* **2011**, *12*, 1214–1223. [CrossRef]

102. Morandi, G.; Heath, L.; Thielemans, W. Cellulose Nanocrystals Grafted with Polystyrene Chains through Surface-Initiated Atom Transfer Radical Polymerization (SI-ATRP). *Langmuir* **2009**, *25*, 8280–8286. [CrossRef]

103. Jorfi, M.; Foster, E.J. Recent advances in nanocellulose for biomedical applications. *J. Appl. Polym. Sci.* **2015**, *132*. [CrossRef]

104. Dong, S.; Roman, M. Fluorescently Labeled Cellulose Nanocrystals for Bioimaging Applications. *J. Am. Chem. Soc.* **2007**, *129*, 13810–13811. [CrossRef] [PubMed]

105. Despres, H.; Sunasee, R.; Carson, M.; Pacherille, A.; Nunez, K.; Ckless, K. Cell-based analysis of the immune and antioxidant response of the nanocarrier $\beta$-cyclodextrin conjugated with cellulose nanocrystals. *Free Radic. Biol. Med.* **2018**, *128*, S102. [CrossRef]

106. Armentano, I.; Dottori, M.; Fortunati, E.; Mattioli, S.; Kenny, J.M. Biodegradable polymer matrix nanocomposites for tissue engineering: A review. *Polym. Degrad. Stab.* **2010**, *95*, 2126–2146. [CrossRef]

107. Domingues, R.M.A.; Gomes, M.E.; Reis, R.L. The Potential of Cellulose Nanocrystals in Tissue Engineering Strategies. *Biomacromolecules* **2014**, *15*, 2327–2346. [CrossRef] [PubMed]

108. Yang, X.; Cranston, E.D. Chemically cross-linked cellulose nanocrystal aerogels with shape recovery and superabsorbent properties. *Chem. Mater.* **2014**, *26*, 6016–6025. [CrossRef]

109. Lin, N.; Bruzzese, C.; Dufresne, A. TEMPO-oxidized nanocellulose participating as crosslinking aid for alginate-based sponges. *ACS Appl. Mater. Interfaces* **2012**, *4*, 4948–4959. [CrossRef] [PubMed]

110. Zhou, Y.; Fuentes-Hernandez, C.; Khan, T.M.; Liu, J.-C.; Hsu, J.; Shim, J.W.; Dindar, A.; Youngblood, J.P.; Moon, R.J.; Kippelen, B. Recyclable organic solar cells on cellulose nanocrystal substrates. *Sci. Rep.* **2013**, *3*, 1536. [CrossRef] [PubMed]

111. Kim, J.-H.; Shim, B.S.; Kim, H.S.; Lee, Y.-J.; Min, S.-K.; Jang, D.; Abas, Z.; Kim, J. Review of Nanocellulose for Sustainable Future Materials. *Int. J. Precis. Eng. Manuf. Technol.* **2015**, *2*, 197–213. [CrossRef]

112. Qiu, X.; Hu, S. "Smart" materials based on cellulose: A review of the preparations, properties, and applications. *Materials (Basel)* **2013**, *6*, 738–781. [CrossRef]

113. Way, A.E.; Hsu, L.; Shanmuganathan, K.; Weder, C.; Rowan, S.J. PH-responsive cellulose nanocrystal gels and nanocomposites. *ACS Macro Lett.* **2012**, *1*, 1001–1006. [CrossRef]

114. Kafy, A.; Akther, A.; Shishir, M.I.R.; Kim, H.C.; Yun, Y.; Kim, J. Cellulose nanocrystal/graphene oxide composite film as humidity sensor. *Sens. Actuators A Phys.* **2016**, *247*, 221–226. [CrossRef]

115. Zhang, L.; Li, Q.; Zhou, J.; Zhang, L. Synthesis and Photophysical Behavior of Pyrene-Bearing Cellulose Nanocrystals for $Fe^{3+}$ Sensing. *Macromol. Chem. Phys.* **2012**, *213*, 1612–1617. [CrossRef]

116. Sadasivuni, K.K.; Kafy, A.; Zhai, L.; Ko, H.-U.; Mun, S.; Kim, J. Transparent and Flexible Cellulose Nanocrystal/Reduced Graphene Oxide Film for Proximity Sensing. *Small* **2015**, *11*, 994–1002. [CrossRef] [PubMed]

117. Wang, S.; Zhang, X.; Wu, X.; Lu, C. Tailoring percolating conductive networks of natural rubber composites for flexible strain sensors via a cellulose nanocrystal templated assembly. *Soft Matter* **2016**, *12*, 845–852. [CrossRef] [PubMed]

118. Annamalai, P.K.; Dagnon, K.L.; Monemian, S.; Foster, E.J.; Rowan, S.J.; Weder, C. Water-responsive mechanically adaptive nanocomposites based on styrene-butadiene rubber and cellulose nanocrystals—Processing matters. *ACS Appl. Mater. Interfaces* **2014**, *6*, 967–976. [CrossRef] [PubMed]

119. Mendez, J.; Annamalai, P.K.; Eichhorn, S.J.; Rusli, R.; Rowan, S.J.; Foster, E.J.; Weder, C. Bioinspired mechanically adaptive polymer nanocomposites with water-activated shape-memory effect. *Macromolecules* **2011**, *44*, 6827–6835. [CrossRef]

120. Zoppe, J.O.; Habibi, Y.; Rojas, O.J.; Venditti, R.A.; Johansson, L.-S.; Efimenko, K.; Österberg, M.; Laine, J. Poly(*N* -isopropylacrylamide) Brushes Grafted from Cellulose Nanocrystals via Surface-Initiated Single-Electron Transfer Living Radical Polymerization. *Biomacromolecules* **2010**, *11*, 2683–2691. [CrossRef] [PubMed]

121. Zoppe, J.O.; Österberg, M.; Venditti, R.A.; Laine, J.; Rojas, O.J. Surface Interaction Forces of Cellulose Nanocrystals Grafted with Thermoresponsive Polymer Brushes. *Biomacromolecules* **2011**, *12*, 2788–2796. [CrossRef]

122. Zoppe, J.O.; Venditti, R.A.; Rojas, O.J. Pickering emulsions stabilized by cellulose nanocrystals grafted with thermo-responsive polymer brushes. *J. Colloid Interface Sci.* **2012**, *369*, 202–209. [CrossRef]

123. Bawa, P.; Pillay, V.; Choonara, Y.E.; du Toit, L.C. Stimuli-responsive polymers and their applications in drug delivery. *Biomed. Mater.* **2009**, *4*, 022001. [CrossRef]

124. Lin, N.; Dufresne, A. Nanocellulose in biomedicine: Current status and future prospect. *Eur. Polym. J.* **2014**, *59*, 302–325. [CrossRef]

*Article*

# Development of Environmental Friendly Dust Suppressant Based on the Modification of Soybean Protein Isolate

Hu Jin [1,2], Wen Nie [1,2,3,*], Yansong Zhang [1,2], Hongkun Wang [2], Haihan Zhang [2], Qiu Bao [2] and Jiayi Yan [2]

[1]   Key Laboratory of Ministry of Education for Mine Disaster Prevention and Control, College of Mining and Safety Engineering, Shandong University of Science and Technology, Qingdao 266590, China; 15764250719@163.com (H.J.); hk19941211@163.com (Y.Z.)

[2]   College of Mining and Safety Engineering, Shandong University of Science and Technology, Qingdao 266590, China; SKDWHK@163.com (H.W.); 15065423187@163.com (H.Z.); m15610569565@163.com (Q.B.); yan18739937718@163.com (J.Y.)

[3]   Hebei State Key Laboratory of Mine Disaster Prevention, North China Institute of Science and Technology, Beijing 101601, China

*   Correspondence: niewen@sdust.edu.cn; Tel.: +86-189-5483-5917

Received: 25 February 2019; Accepted: 15 March 2019; Published: 20 March 2019

**Abstract:** Aiming to further improve the dust suppression performance of the dust suppressant, the present study independently develops a new type of biodegradable environmentally-friendly dust suppressant. Specifically, the naturally occurring biodegradable soybean protein isolate (SPI) is selected as the main material, which is subject to an anionic surfactant, i.e., sodium dodecyl sulfonate (SDS) for modification with the presence of additives including carboxymethylcellulose sodium and methanesiliconic acid sodium. As a result, the SDS-SPI cementing dust suppressant is produced. The present study experimentally tests solutions with eight different dust suppressant concentrations under the same experimental condition, so as to evaluate their dust suppression performances. Key metrics considered include water retention capability, cementing power and dust suppression efficiency. The optimal concentration of dust suppressant solution is determined by collectively comparing these metrics. The experiments indicate that the optimal dust suppressant concentration is 3%, at which level the newly developed environmentally-friendly dust suppressant solution exhibits a decent dust suppression characteristic, with the water retention power reaching its peak level, and the corresponding viscosity being 12.96 mPa·s. This performance can generally meet the requirements imposed by coal mines. The peak efficiency of dust suppression can reach 92.13%. Fourier transform infrared spectroscopy (FTIR) and scanning electron microscopy (SEM) were used to analyze the dust suppression mechanism of the developed dust suppressant. It was observed that a dense hardened shell formed on the surface of the pulverized coal particles sprayed with the dust suppressant. There is strong cementation between coal dust particles, and the cementation effect is better. This can effectively inhibit the re-entrainment of coal dust and reduce environmental pollution.

**Keywords:** soybean protein isolate modification; dust suppressant; performance characterization; optimal concentration; analysis of dust suppression mechanism

---

## 1. Introduction

Coal serves as the fundamental source of energy in China. Throughout the foreseeable future, coal will still occupy a significant proportion of primary energy in China [1–5]. During the outdoor storage of coal piles and railroad transportation of coal, a large amount of coal dust may be generated, which not only causes the loss of coal and waste of natural resources, but also leads to severe environmental pollution [6–12]. Besides, as for coal piles, the dust dispersion can adversely impact the normal operation of electric and mechanical equipment as well as the monitoring system, leading to a shortened lifespan. More seriously, the increase of dust concentration can pose a major threat to operational safety, triggering the occurrence of accidents [13–17]. The traditional dust suppression methods include water spray and tarp coverage, which are unfavorable due to high cost and poor long-term dust suppression performance. Therefore, scholars in the international community have started to develop chemical-based dust suppression methods, which have proven to be promising in many dust suppression applications [18–24]. Overall, the chemical dust suppressant is highly favored considering its dust suppression effect, economic viability and environmental-friendliness, and therefore has a vast potential for future development [25–30]. The chemical dust suppressant functions effectively on open dust sources, and therefore this technology has been widely applied to those enterprises with a major generation of coal dust, such as railroad transportation departments, dumping sites and thermal power plants [31–36]. In recent years, the domestic and foreign research on chemical dust suppressant is shifting its focus to enhancing recyclability, environmental-friendliness, and efficiency [37–42]. Bao et al. [43] took corn starch as the main raw material and developed a kind of super absorbent dust suppressant by chemical modification method, which can effectively inhibit the dust diffusion during coal transportation. Grogan [44] conducts research on combining the byproduct of biodiesel production, i.e., glycerol, with surfactant, polyhydroxy esters and acrylic acid compound to produce a dust suppressant, with its performance being characterized. The prepared suppressant has a decent dust suppression effect and is environmentally friendly. Zhang et al. [45] employs glasswort as the raw material, which is blended with sodium dodecylbenzene sulfonate and carboxylmethyl cellulose as the additives to produce an ecologically friendly dust suppressant. Zhang et al. [46] prepared a degradable dust suppressant by chemical modification using a polymer material guar gum as the main raw material. The dust suppressant has good wettability and water retention. Yang et al. [47] utilizes film coalescing aid, fatty alcohol polyoxyethylene ether and polyvinyl alcohol as the raw materials to develop a dust suppressant with a decent ability to withstand rainfall and wind, making it primarily applied to coal transportation. The spray of this suppressant causes a thick hardened layer to form on the particles surface, resulting in a long dust suppression duration. Polat H. et al. [48] uses Polyethylene oxide (PEO)/PO as the raw material to synthesize a dust suppressant, which can effectively wet the coal dust during the mining process while mitigating the equipment corrosion.

The application of cementing dust suppressant is an effective method for preventing and controlling dust dispersion, with cementing power, water retention ability and dust reduction efficiency being the key factors dictating the dust suppression performance of the cementing dust suppressant. Although those aspects have been extensively discussed by domestic and overseas scholars, their studies are mainly focused on certain individual characteristics, without covering the entire suite of metrics [49–53]; also, those cementing dust suppressants are subject to certain disadvantages, including toxicity, lack of biodegradability, and the tendency to cause secondary pollution, etc. [54–60]. To address the aforementioned problems, the present study uses biodegradable and environmentally-friendly soybean protein isolate (SPI) as the main material. SPI is a main ingredient of soybean. It contains a large amount of active functional groups, and is favored for its biodegradability, and available from various resources. It is therefore an environmentally-friendly material for producing environmentally-friendly and pollution-free dust suppressant [61–69]. The present study uses sodium dodecyl sulfonate to modify the SPI while adding carboxymethylcellulose sodium and methanesiliconic acid sodium to the mixture so as to develop a new type of environmentally-friendly naturally-occurring macromolecular dust suppressant, namely SDS-SPI cementing dust suppressant; subsequently, the present study experimentally measures a suite of performance metrics associated with the newly developed dust suppressant, based on which the optimal dust suppressant concentration is determined. At the same time, the research results also provide ideas for the development of other types of high-efficiency environmental dust suppressants.

## 2. Experiments

### 2.1. Main Equipment and Raw Materials

The reagents and equipment used in the present study are listed in Tables 1 and 2.

**Table 1.** Primary raw materials used in the experiment.

| Chemical Name | Chemical Formula | Purity | Manufacturer |
|---|---|---|---|
| Soybean protein isolate | $C_{13}H_{10}N_2$ | BR | Sinopharm Chemical Reagent Co., Ltd., Beijing, China |
| Carboxymethylcellulose sodium | $C_8H_{11}O_7Na$ | CP | Xiya Reagent Co., Ltd., Chengdu, China |
| Methanesiliconic acid sodium | $CH_5SiO_3Na$ | CP | Shandong Yousuo Chemical Technology Co., Ltd., Qingdao, China |
| Sodium dodecyl sulfonate | $C_{12}H_{25}OSO_3Na$ | Tech | Shandong Yousuo Chemical Technology Co., Ltd., Qingdao, China |

Notes: BR indicates that the chemical reagent is a biochemical reagent; CP indicates chemical purity; Tech indicates an industrial reagent.

**Table 2.** Key experimental equipment and specifications.

| Experimental Equipment | Specifications | Manufacturer |
|---|---|---|
| Rotary viscometer | NDI-79 | Shanghai Precision Instrument Co., Ltd. Shanghai, China |
| High-Resolution Scanning Electron Microscope | Nova Nano SEM | Shanghai Casting Gold Analytical Instruments and Equipment Co., Ltd. Shanghai, China |
| Fourier-transform infrared spectroscope | Nicolet (iS10) | Beijing Kaifeng Fengyuan Technology Co., Ltd. Beijing, China |
| Coal mine dust sampler | AKFC-92 | Qingdao Lubo Weiye Environmental Protection Technology Co., Ltd. Qingdao, China |
| Mine energy-saving axialfan | ASZ-11.2 | Zibo Jinhe Fan Co., Ltd. Zibo, China |

### 2.2. Preparation of SDS-SPI Cementing Dust Suppressant

Put 7.50 g of SPI (Sinopharm Chemical Reagent Co., Ltd., Beijing, China), 0.15 g of sodium dodecyl sulfonate (SDS, Shandong Yousuo Chemical Technology Co., Ltd., Qingdao, China), 0.15 g of carboxymethylcellulose sodium (CMC, Xiya Reagent Co., Ltd., Chengdu, China) and 140 mL of water in a three-neck round-bottom flask with sufficient stirring. Raise the temperature of the solution to 60 °C and let the reaction last for one hour at this temperature. Then let the solution cools down. 2.50 g of 30% methanesiliconic acid sodium was added to the solution, followed by stirring it evenly. The experiment finally developed a SDS-SPI cementing dust suppressant with a soy protein isolate concentration of 5%. The stirring speed is controlled to be 60–80 r/min throughout the reaction process to prevent the solution from foaming in a large amount. In the process of application, the dust suppressant solution was diluted by different multiples, resulting in modified SDS-SPI cementing dust suppressant solutions whose SPI concentrations are 1%, 1.5%, 2%, 2.5%, 3%, 3.5%, 4% and 5%, respectively.

### 2.3. Application of SDS-SPI Cementing Dust Suppressant

The coal samples used in the experiment were coking coal from the 20,206 fully mechanized working face of Zaozhuang Corporation's Jiangzhuang, Shandong province. Coal samples were ground into pulverized coal using a planetary motion micro mill, and then fine pulverized coal with particle size less than 3 mm was screened through a sieve with a hole size of 3 mm. The prepared coal sample is dried in a vacuum drying chamber (Shanghai lang gan experimental equipment co., LTD, Shanghai, China) at a temperature of 100 °C. The fine coal particles with a particle size of 0–3 mm was placed in 9 Petri dishes. The extrusion method was used to flatten the surface of the sample and then eight kinds of dust suppressant solutions with different SPI concentrations were evenly sprayed onto the corresponding eight coal samples with a density of 2 L/m$^2$. The ninth coal sample was evenly sprayed with water.

## 3. Characterization of Properties

### 3.1. Experimental Characterization of Water Retention Performance of Dust Suppressant

The water retention of a dust suppressant can be measured through its ability to resist evaporation. Under the same experimental condition, an improved water retention of dust suppressant corresponds to an enhanced anti-evaporation characteristic, leading to an improved dust suppression effect. The anti-evaporation characteristic of a solution can be measured with the evaporation rate, where a low evaporation rate indicates a good anti-evaporation performance. Therefore, the present experiment characterizes the dust suppressant's water retention based on measuring the evaporation rate. The experimental procedure is outlined below: under the same experimental condition, put the desiccated coal sample in a labeled clean Petri dishes; evenly spray the prepared SDS-SPI cementing dust suppressant solutions at different concentrations over the surface of coal sample at 2 L/m$^2$. Weigh the sample after its absorption of solution reaches a sufficient level. Then put the sample in a vacuum drying chamber and keep the temperature at 60 °C (Figure 1). At a set interval, take the sample out of the oven and weigh its mass; the evaporation rate can thereby be calculated with Equation (1). The final results, as shown in Table 3, can be obtained by linear fitting between measurement results and time. The corresponding fitted curves are shown in Figures 2 and 3.

$$\theta = \frac{W_1 - W_2}{A\,T} \tag{1}$$

where $\theta$ denotes the where evaporation rate of dust suppressant (g·m$^{-2}$·s$^{-1}$); $W_1$ denotes the coal sample mass before evaporation (g); $W_2$ denotes the coal sample mass post evaporation (g); $A$ denotes the evaporation area of coal sample (m$^2$); and $T$ denotes the evaporation time (s).

**Figure 1.** Vacuum drying chamber with a preset temperature.

**Table 3.** Evaporation rate of dust suppressant ($g \cdot m^{-2} \cdot s^{-1}$).

| Sample (%) | Time (h) | | | | | | | | | | | |
|---|---|---|---|---|---|---|---|---|---|---|---|---|
| | 0.5 | 1 | 1.5 | 2 | 2.5 | 3 | 4 | 5 | 6 | 8 | 10 | 12 |
| water | 0.062 | 0.055 | 0.051 | 0.042 | 0.038 | 0.023 | 0.015 | 0.012 | 0.008 | 0.008 | 0.005 | 0.005 |
| 1 | 0.058 | 0.052 | 0.047 | 0.039 | 0.035 | 0.021 | 0.013 | 0.013 | 0.008 | 0.007 | 0.003 | 0.003 |
| 1.5 | 0.045 | 0.042 | 0.036 | 0.034 | 0.028 | 0.021 | 0.012 | 0.009 | 0.009 | 0.007 | 0.007 | 0.004 |
| 2 | 0.043 | 0.040 | 0.035 | 0.031 | 0.025 | 0.019 | 0.011 | 0.008 | 0.008 | 0.006 | 0.006 | 0.004 |
| 2.5 | 0.031 | 0.029 | 0.023 | 0.017 | 0.015 | 0.015 | 0.010 | 0.005 | 0.005 | 0.003 | 0.003 | 0.002 |
| 3 | 0.030 | 0.028 | 0.023 | 0.016 | 0.012 | 0.010 | 0.009 | 0.006 | 0.004 | 0.004 | 0.002 | 0.002 |
| 3.5 | 0.034 | 0.030 | 0.025 | 0.018 | 0.013 | 0.013 | 0.011 | 0.008 | 0.007 | 0.004 | 0.004 | 0.003 |
| 4 | 0.042 | 0.038 | 0.034 | 0.029 | 0.024 | 0.014 | 0.010 | 0.009 | 0.008 | 0.008 | 0.005 | 0.005 |
| 5 | 0.048 | 0.043 | 0.035 | 0.026 | 0.022 | 0.018 | 0.014 | 0.011 | 0.010 | 0.009 | 0.006 | 0.006 |

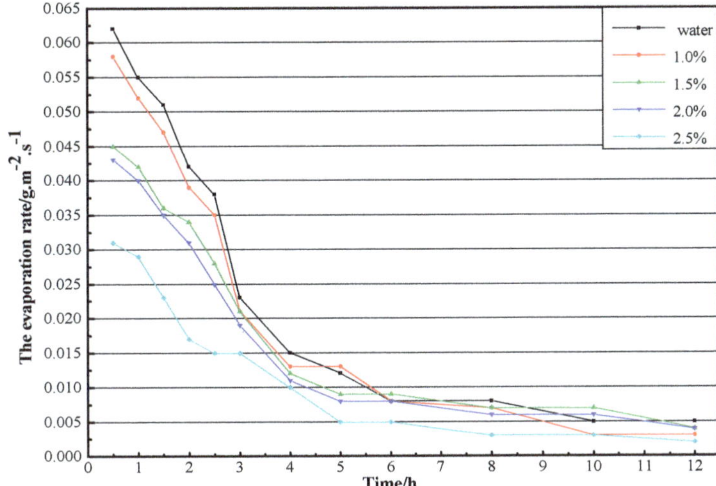

**Figure 2.** Trend chart of sample evaporation rate with time (concentration 0%, 1%, 1.5%, 2%, 2.5%).

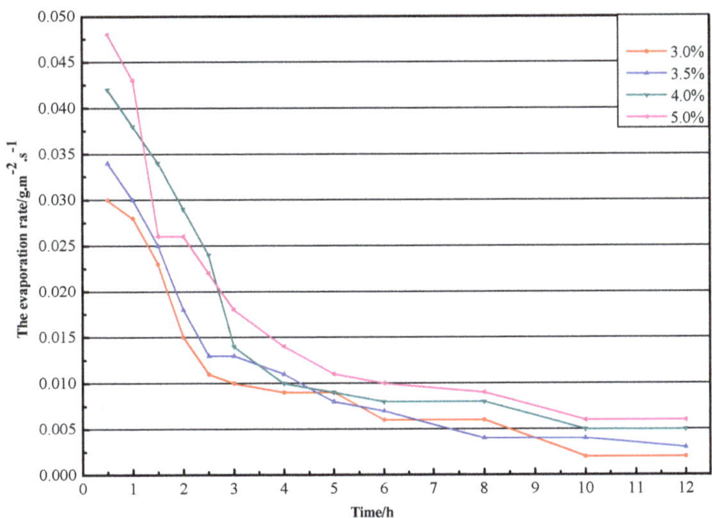

**Figure 3.** Trend chart of sample evaporation rate with time (concentration 3%, 3.5%, 4%, 5%).

It can be seen in Table 3, Figures 2 and 3 that at the same time and under the same experimental condition, the evaporation rate of clean water is the highest, followed by the evaporation rate of 1% dust suppressant solution. The water evaporation rates of dust suppressant solutions with concentrations between 2.5% and 3.5% are relatively small. It can be inferred from Figures 2 and 3 that the 2.5% and 3% dust suppressant solutions exhibit the smallest evaporation rate, and their evaporation rate curves are relatively flat, corresponding to limited variation of the evaporation amount within the same time window; a further comparison indicates that compared to 2.5% dust suppressant, the 3% dust suppressant solution has the lowest evaporation rate, and its curve does not undergo any major change, implying that its anti-evaporation characteristic, water retention ability and dust suppression effect are the best.

*3.2. Viscosity Testing Experiments*

The viscosity of coal dust suppressant is a key technical indicator associated with the applicability of samples. This experiment used an NDJ-79 rotational viscometer (Shanghai Precision Instrument Co., Ltd., Shanghai, China) with its rotating shaft suspended from the equipment. Add a 50 mL fluid sample to the test vessel and insert the spindle into the liquid until the level mark on the top of the spindle is immersed in the water. Then adjust the speed knob to keep the speed at 75 rpm. Based on this condition, the viscosity of the prepared SDS-SPI cement dust suppressant solution diluted to various concentrations was measured under the same conditions. The measurement is repeated for three times for each sample, with the mean value treated as the final viscosity of the sample. The measurement results are shown in Table 4; meanwhile, prepare a solution containing SPI only, and measure its viscosity under the same experimental condition, with the measurement results shown in Table 5; finally, one can compare the measurement results derived from two groups, as illustrated in Figure 4, where (A) denotes the viscosity of SPI solution at various concentrations, and (B) denotes the viscosity of modified SDS-SPI cementing dust suppressant solutions at various concentrations.

**Table 4.** Viscosities of sodium dodecyl sulfonate-soybean protein isolate (SDS-SPI) cementing dust suppressant solutions.

| Number | Concentration | | | | | | | |
|---|---|---|---|---|---|---|---|---|
| | 1% | 1.5% | 2% | 2.5% | 3% | 3.5% | 4% | 5% |
| 1# | 3.68 | 6.20 | 10.20 | 13.28 | 16.20 | 19.40 | 22.48 | 21.84 |
| 2# | 3.48 | 6.76 | 10.72 | 12.84 | 16.76 | 20.84 | 21.80 | 22.72 |
| 3# | 3.04 | 7.20 | 9.56 | 12.76 | 16.60 | 20.72 | 20.64 | 22.88 |
| Average Viscosity (mPa·s) | 3.40 | 6.72 | 10.16 | 12.96 | 16.52 | 20.32 | 21.64 | 22.48 |

**Table 5.** Viscosities of SPI solutions.

| Number | Concentration | | | | | | | |
|---|---|---|---|---|---|---|---|---|
| | 1% | 1.5% | 2% | 2.5% | 3% | 3.5% | 4% | 5% |
| 1# | 1.80 | 3.40 | 5.01 | 6.32 | 8.52 | 10.29 | 11.36 | 12.21 |
| 2# | 1.92 | 3.48 | 5.27 | 6.59 | 8.31 | 10.35 | 11.44 | 12.38 |
| 3# | 1.64 | 3.32 | 5.08 | 7.01 | 8.61 | 10.44 | 11.64 | 12.61 |
| Average Viscosity (mPa·s) | 1.80 | 3.40 | 5.12 | 6.64 | 8.48 | 10.36 | 11.48 | 12.40 |

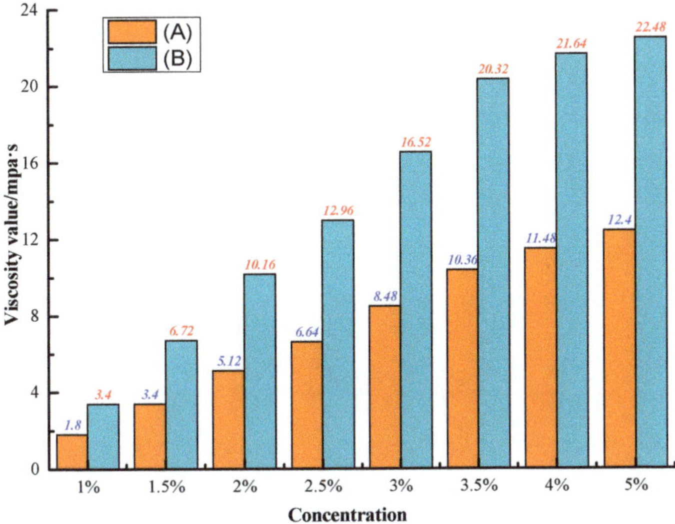


**Figure 4.** Viscosity histogram.

Figure 4 shows that the viscosity of modified SDS-SPI cementing dust suppressant solution increases rapidly with SPI concentration. The viscosity value is increased substantially compared to the viscosity of the solution with SPI only; as the concentration of SPI in the solution reaches 1.0%, the viscosity of dust suppressant solutions 3.40 mPa·s; as the solution concentration reaches 2.5%, the viscosity of the solution with SPI is merely 6.64 mPa·s, whereas the viscosity of the modified dust suppressant solution has reached 12.96 mPa·s, which is more than twice that of pure SPI solution. Therefore, the dust suppression effect of SDS-SPI cementing dust suppressant solution is more pronounced. As for the dust suppressant solution used in coal mines, considering the economic cost and the dust suppression effect, it would be non-ideal to have an excessively high viscosity. Usually the range of viscosity lies between 12.0–20.0 mPa·s. For this reason, the SDS-SPI cementing dust suppressant solution with concentration above 2.5% can meet the requirement posed by coal mines.

*3.3. Measurement of Dust Suppression Efficiency*

The dust suppression efficiency can be employed as a direct metric to indicate the dust suppression effect of a dust suppressant. The present experiment is carried out based on the air tunnel simulation platform located at the Dust Control Laboratory of Shandong University of Science and Technology, as shown in Figure 5. The air tunnel simulation platform consists of aTDI8000-0750G-4T Infinitely Variable Inverter (Yueqing Taida Electrical Technology Co., Ltd., Wenzhou, China) and aSZ-11.2 Axial Fan (Zibo Jinhe Fan Co., Ltd., Zibo, China) (as shown in Figure 6), both of which are made in China. Under the regulation of the inverter, the axial fan can mimic natural wind, whose peak velocity can reach 32 m/s. Place the sample with hardened shell on the platform, and keep the air speed at the sample location at 14~16 m/s. Collect samples, i.e., particles suspended in the air with sizes less than 100 μm, with a coal mine dust sampler (as shown in Figure 7) at the location 3.0 m downstream of the sample. The sampling process is conducted in a continuous mode, with flow rate maintained at 20 L/min. The sampling duration is 30 min. Firstly, one needs to test the air in the experimental chamber for measuring the background concentration; subsequently, the coal sample sprayed with clear water is tested, with the corresponding coal dust concentration measured; as the next step, coal samples treated with dust suppressants of eight different concentrations are tested in a sequential manner, with the corresponding coal dust concentrations in the air recorded. During the post-processing stage, the dust suppression efficiency of each test can be calculated according to Equation (2). Figure 8 illustrates the testing results.

$$\rho = \left(1 - \frac{\theta_x - \theta_0}{\theta_1 - \theta_0}\right) \times 100\%$$

(2)

where $\rho$ denote the dust suppression efficiency (%); $\theta_0$ denote the background dust concentration inside experimental chamber (mg/L); $\theta_1$ denote the dust concentration for coal sample with clear water treatment (mg/L); and $\theta_x$ denote the dust concentrations for coal samples treated with different dust suppressants (mg/L).

**Figure 5.** Air tunnel simulation apparatus.

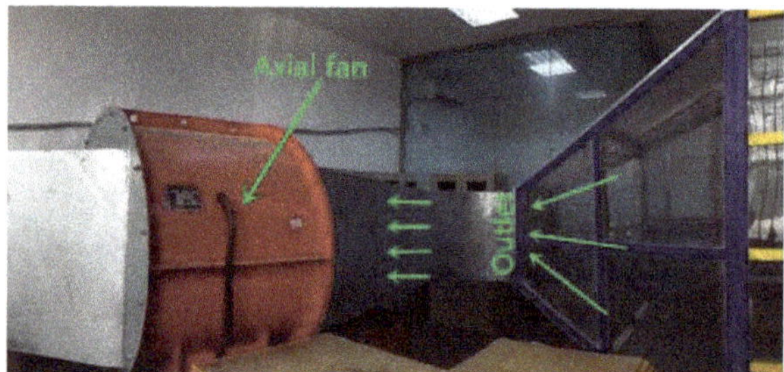


**Figure 6.** SZ-11.2axial fan.

**Figure 7.** Coal mine dust sampler.

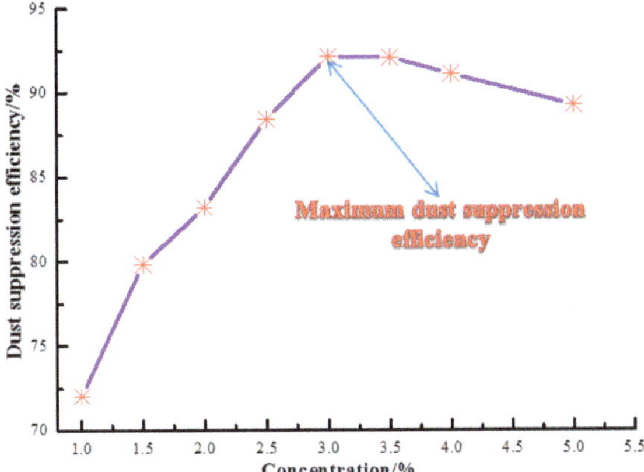


**Figure 8.** Dust suppression efficiency of sodium dodecyl sulfonate-soybean protein isolate (SDS-SPI) cementing dust suppressants.

Figure 8 shows that as the solution concentration of SDS-SPI cementing dust suppressant increases, the dust suppression efficiency climbs rapidly. As the mass concentration reaches 3.0%, the suppressant's dust suppression efficiency on coal dust peaks at 92.13%; subsequently, the dust suppression efficiency starts to drop, which is primarily due to the fact that an increase of viscosity causes the surface tension of SDS-SPI cementing dust suppressant solution on coal dust surface to increase, leading to a subdued wettability with respect to the coal dust. This change compromises the dust suppression efficiency.

In summary, as the mass concentration of SDS-SPI cementing dust suppressant solution stays at 3.0%, its viscosity assumes a value of 16.52 mPa·s, which is considered adequate for coal mine application; meanwhile, the anti-evaporation characteristic and the dust suppression efficiency of the dust suppressant solution both reach optimal level under this concentration. Therefore, based on a comprehensive comparative study covering the factors elaborated above, it is determined that the optimal concentration of the current SDS-SPI cementing dust suppressant solution is 3%.

## 4. Analysis of Dust Suppression Mechanism

### 4.1. Infrared Spectra Tests

The current experiment was tested using a Nicolet iS10 Fourier-Transform infrared spectrometer (Beijing Kaifeng Fengyuan Technology Co., Ltd., Beijing, China). First, the sample was mixed with potassium bromide in an agate mortar at a ratio of 1:200. The ground mixture powder was then made into transparent thin pellets for testing. The test measurement range is between 4000 and 500 cm$^{-1}$, and each sample is scanned 6 times. The infrared spectrum before and after SPI modification is shown in Figure 9, where (A) is the spectrum of SPI, and (B) is the corresponding spectra after SPI modification.

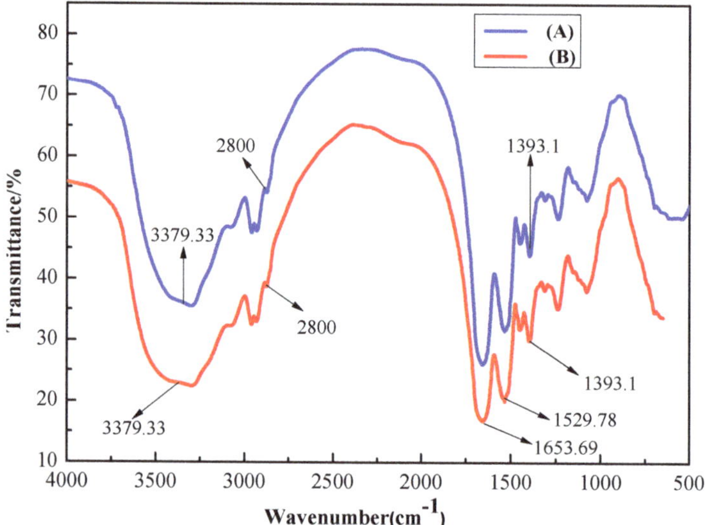

**Figure 9.** The infrared spectra of SPI before and after the SDS modification.

Figure 9 shows the infrared spectra of SPI before and after SDS modification. As can be seen from the figure, the wave number of 3379.33 cm$^{-1}$ mainly corresponds to the spectral peaks of –NH, –OH and sulfonamide bond; The wave number 2800 cm$^{-1}$ corresponds to the absorption peak of CH$_2$. Wave number 1653.69 cm$^{-1}$ corresponds to C=O stretching peak in band I of amide; The wave number of 1529.78 cm$^{-1}$ corresponds to the superposition peaks of N–H bending vibration and C–N stretching vibration in band II, as well as the absorption peaks of sulfa bonds. The wave number of 1393.1 cm$^{-1}$ corresponds to the characteristic peaks of sulfa bond and COO–. The wave numbers of these two peaks are 1653.69 cm$^{-1}$ and 1529.78 cm$^{-1}$ respectively, which correspond to the characteristic spectral peaks and their bending vibration of the benzene ring in the C–H [70–75].

After modification, the spectral peaks at 3379.33 cm$^{-1}$ and 1393.1 cm$^{-1}$ in Figure 9B (representing the sulfonamide bond) moved to the left with a slight increase in amplitude, indicating that the addition of anionic surfactant SDS in the modification of SPI with anionic surfactant SDS would destroy the dense structure existing in SPI and relax SPI. This process may result in the formation of protein-surfactant (SDS-SPI) complex chemicals. In addition, loose connections between domains are compromised, resulting in a decrease in the molecular weight of the protein, which can be enhanced if it occurs within a certain range [76–82]. At the same time, comparing the results before and after the modification, it was found that the modified SDS-SPI moved to the left at the absorption peaks of 3377.33 cm$^{-1}$, 2800 cm$^{-1}$ and 1393.1 cm$^{-1}$ during the strengthening process. This indicates that the amount of newly formed SDS-SPI composite chemicals increases, leading to an increase in the relaxation degree of SPI structure. At the same time, the internal hydrophobic functional group is transformed outward, which improves the water solubility of SPI. The more relaxed the protein structure, the more functional groups, the better the coal dust cementing effect. Although the peaks of 1653.69 cm$^{-1}$ and 1529.78 cm$^{-1}$ were basically stable, the intensity increased slightly, indicating that the benzene ring entered the SPI molecular structure, further confirming the formation of SDS-SPI complex chemistry. At the same time, under the influence of SDS, a group of different hydrophilic functional groups move inward and the hydrophobicity is enhanced. CMC is a commonly used viscosifier that enhances the viscosifying effect of the anionic surfactant SDS. Methanesiliconic acid sodium is a water-resistant enhancer that interacts with CO$_2$ in the air to cause polycondensation between molecules to form a macromolecular structure that is spatially interconnected. The hydrophilic hydroxyl group in the polycondensation process is converted into a hydrophobic Si–O bond, so that the reagent has strong water resistance.

*4.2. Scanning Electron Microscope (SEM) Experiments*

The current experiment employs D-type high-resolution SEM to investigate the surface morphology of the sample. Three sets of samples are selected, and the coal powder used here is coking coal derived from the 20206 mechanized mining face of Shandong Zaozhuang Corporation's Jiangzhuang Coal Mine. A planetary motion micro mill is used to grind the coal chunks into fine powder for the tests. Figure 10 shows the 5000× SEM image of dry coal powder; Figure 11 shows the 5000× SEM photo of coal powder treated with SPI only; Figure 12 shows the 5000× SEM photo of coal powder treated by the developed SDS-SPI cementing dust suppressant solution with optimal spray concentration.

**Figure 10.** 5000× SEM image of dry coal powder.

**Figure 11.** SEM photo of coal powder treated with SPI.

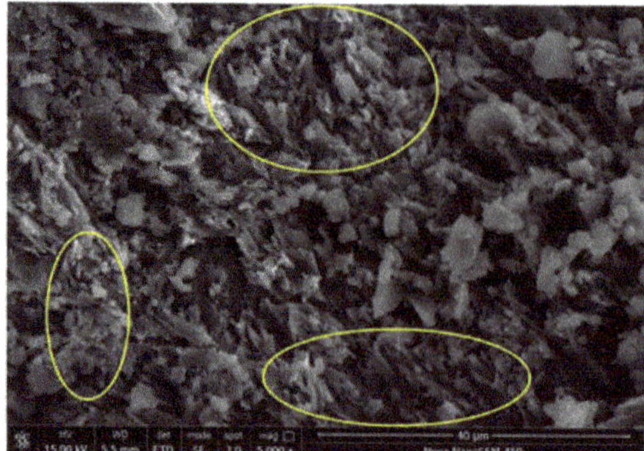

**Figure 12.** 5000× SEM image of coal powder treated by the developed SDS-SPI cementing dust suppressant solution with optimal spray concentration.

In order to study the effect of dust suppressant on the cementation hardening effect of pulverized coal before and after modification, three groups of pulverized coal were scanned by SEM in this study. The surface morphology is shown in Figures 10–12. Figure 10 is a 5000× SEM image of the surface morphology of the dry powder. Under this circumstance, it was found that the shape and size of the coal dust particles changed greatly, the particle spacing was large, and there was a lack of cementation between the particles. In this case, the coal powder is highly susceptible to the dispersion of air, causing dust pollution. Figure 11 shows the SEM pulverized coal 5000× SEM image. The study found that the coal dust particles are tightly bonded and the cementation spacing is small. This is mainly driven by the viscosity of the SPI. Figure 12 is a 5000× SEM image of pulverized coal treated with the optimal concentration of the developed SDS-SPI cementing dust suppression solution. It can be seen from the yellow circle area in the figure that the coal powder particles are tightly cemented, the cementation strength is high, and the cementation spacing between the coal powder particles is small. This indicates that the viscosity of the SDS-SPI compound chemical modification is increased, which is consistent with the viscosity test results of this study; the SDS-SPI cementing dust suppression solution modified by SPI can form a dense layer on the surface of the coal powder. The hardened shell layer cements pulverized coal particles of different particle sizes, and the cementing effect is good, thereby suppressing the re-flying of coal dust.

## 5. Conclusions

(1) The newly developed environmentally-friendly dust suppressant uses SPI as the main cementing ingredient, carboxymethylcellulose sodium as the viscosity enhancer, and methanesiliconic acid sodium as the water resisting additive. Based on modifying SPI with anionic surfactant sodium dodecyl sulfonate, a novel biodegradable environmentally-friendly cementing dust suppressant is prepared.

(2) It is best to use anionic surfactant (SDS) to modify SPI so as to destroy and relax its original compact sphere structure, forming protein-surfactant (SDS-SPI) composite chemical. After the modification, the hydrophobic functional groups inside the protein-surfactant composite chemical move outward, allowing the coal dust to better absorb the hydrophobic functional groups such that the coal dust is more thoroughly cemented for delivering the dust suppression effect.

(3) Compared to SPI solution, the modification further improves the viscosity and water retention of the SDS-SPI cementing dust suppressant solution; based on a comprehensive comparison concerning water retention, viscosity and dust suppression efficiency of dust suppressant, the optimal concentration of dust suppressant solution is finally determined to be 3%. At this level, the

anti-evaporation and water retention characteristics are optimal, with the dust suppression effect being maximized.

(4) The present study conducts SEM high-resolution scanning over the surfaces of native coal powder, coal powder with SPI, and coal powder with dust suppressant at optimal concentration. At 5000×, it is found that the coal powder particles are compactly cemented when the dust suppressant at optimal concentration is applied, exhibiting a strong cementing power. Also, a layer of compact hardened shell forms at the surface, leading to a decent dust suppression effect.

(5) SPI, as the main material used in the present experiment, contains a large number of functional groups, which has an array of favorable properties, including being naturally occurring and being biodegradable. This further demonstrates that the independently developed dust suppressant is biodegradable, environmentally-friendly and clean, and does not pose any threat to the environment.

**Author Contributions:** H.J. and W.N. conceived and designed the experiments; H.J. and Y.Z. performed the experiments; J.Y. and H.Z. analyzed the data; H.W. contributed materials; H.J. and Q.B. wrote the paper; H.J. and W.N. revised the manuscript.

**Funding:** The research was financially supported by the National Natural Science Foundation of China (NO. 51874191 and 51404147), the Focus on Research and Development Plan in Shandong Province (NO.2017GSF20111), the National Key R&D Program of China (2017YFC0805201), the China Postdoctoral Science Foundation (NO. 2015M570601 and 2017T100503), the Open Foundation for Key Laboratory of Mine Disaster Prevention and Control in Hebei Province (KJZH2017K09).

**Acknowledgments:** This work has been funded by the China Postdoctoral Science Foundation (NO. 2015M570601 and 2017T100503), the National Natural Science Foundation of China (NO. 51874191 and 51404147), the Focus on Research and Development Plan in Shandong Province (NO. 2017GSF20111), the National Key R&D Program of China (2017YFC0805201).

**Conflicts of Interest:** The authors declare no conflict of interest.

## References

1. Hattori, T.; Matsuda, M.; Miyake, M. Resource recovery of cupola dust: Study on sorptive property and mechanism for hydrogen sulfide. *J. Mater. Sci.* **2006**, *12*, 3701–3706. [CrossRef]
2. Wang, Y.X.; Guo, P.P.; Dai, F.; Li, X.; Zhao, Y.L.; Liu, Y. Behavior and modeling of fiber-reinforced clay under triaxial compression by combining the superposition method with the energy-based homogenization technique. *Int. J. Geomech.* **2018**, *18*. [CrossRef]
3. Cheng, W.M.; Nie, W.; Zhou, G.; Yu, Y.B.; Ma, Y.Y.; Xue, J. Research and practice on fluctuation water injection technology at low permeability coal seam. *Saf. Sci.* **2012**, *50*, 851–856. [CrossRef]
4. Cai, P.; Nie, W.; Chen, D.W.; Yang, S.B.; Liu, Z.Q. Effect of air flowrate on pollutant dispersion pattern of coal dust particles at fully mechanized mining face based on numerical simulation. *Fuel* **2019**, *239*, 623–635. [CrossRef]
5. Wang, Y.X.; Guo, P.P.; Li, X.; Lin, H.; Liu, Y.; Yuan, H.P. Behavior of Fiber-Reinforced and Lime-Stabilized Clayey Soil in Triaxial Tests. *Appl. Sci.* **2019**, *9*, 900. [CrossRef]
6. Peng, X.L.; Wu, C. New advances in chemical dust suppressants. *J. Saf. Sci. Technol.* **2005**, *5*, 44–47.
7. Cai, J.X.; Dong, B.; Li, Y.Q. New Experimental Research dust suppressants in Bulk coal yard Applicability. *Mine Environ. Prot.* **2011**, *21*, 71–73.
8. Wang, Y.X.; Guo, P.P.; Ren, W.X.; Yuan, B.X.; Yuan, H.P.; Zhao, Y.L.; Shan, S.B.; Cao, P. Laboratory investigation on strength characteristics of expansive soil treated with jute fiber reinforcement. *Int. J. Geomech.* **2017**, *17*. [CrossRef]
9. Yang, S.B.; Nie, W.; Liu, Z.Q.; Peng, H.T.; Cai, P. Effects of spraying pressure and installation angle of nozzles on atomization characteristics of external spraying system at a fully-mechanized mining face. *Powder Technol.* **2019**, *343*, 754–764. [CrossRef]
10. Yuan, B.X.; Sun, M.; Wang, Y.X.; Zhai, L.H.; Luo, Q.; Zhang, X. Full 3D displacement measuring system for 3D displacement field of soil around a laterally loaded pile in transparent soil. *Int. J. Geomech.* **2019**, *19*, 04019028. [CrossRef]

11. Zhang, H.H.; Nie, W.; Liu, Y.H.; Wang, H.K.; Jin, H.; Bao, Q. Synthesis and performance measurement of environment-friendly solidified dust suppressant for open pit coalmine. *J. Appl. Polym. Sci.* **2018**, *135*, 46505. [CrossRef]
12. Nie, W.; Ma, X.; Cheng, W.M.; Liu, Y.H.; Xin, L.; Peng, H.T.; Wei, W.L. A novel spraying/negative-pressure secondary dust suppression device used in fully mechanized mining face: A case study. *Saf. Environ.* **2016**, *103*, 126–135. [CrossRef]
13. Cheng, W.M.; Liu, W.; Nie, W.; Zhou, G.; Cui, X.F.; Sun, X. The Prevention and Control Technology of Dusts in Heading and Winning Faces and Its Development Tendency. *J. Shandong Univ. Sci. Technol. (Nat. Sci.)* **2010**, *4*, 77–82.
14. Cai, P.; Nie, W.; Hua, Y.; Wei, W.L.; Jin, H. Diffusion and pollution of multi-source dusts in a fully mechanized coal face. *Process. Saf. Environ.* **2018**, *118*, 93–105. [CrossRef]
15. Liu, L.; Fang, Z.Y.; Qi, C.C.; Zhang, B.; Guo, L.J.; Song, K.-I. Numerical study on the pipe flow characteristics of the cemented paste backfill slurry considering hydration effects. *Powder Technol.* **2019**, *343*, 454–464. [CrossRef]
16. Wang, M.; Liu, L.; Zhang, X.Y.; Liu, C.; Wang, S.Q.; Jia, Y.H. Experimental and numerical investigations of heat transfer and phase change characteristics of cemented paste backfill with PCM. *Appl. Therm. Eng.* **2019**, *150*, 121–131. [CrossRef]
17. Liu, X.S.; Tan, Y.L.; Ning, J.G.; Lu, Y.W.; Gu, Q.H. Mechanical properties and damage constitutive model of coal in coal-rock combined body. *Int. J. Rock Mech. Min.* **2018**, *110*, 140–150. [CrossRef]
18. Wang, H.T.; Wang, D.M.; Wang, Q.G.; Jia, Z.Q. Novel Approach for Suppressing Cutting Dust Using Foam on a Fully Mechanized Face with Hard Parting. *J. Occup. Environ. Hyg.* **2013**, *3*, 154–164. [CrossRef] [PubMed]
19. Hu, S.Y.; Feng, R.; Ren, X.Y.; Xu, G.; Chang, P.; Wang, Z.; Zhang, Y.T.; Li, Z.; Gao, Q. Numerical study of gas-solid two-phase flow in a coal roadway after blasting. *Adv. Powder Technol.* **2016**, *4*, 1607–1617. [CrossRef]
20. Wang, H.T.; Wang, D.M.; Ren, W.X.; Lu, X.; Han, F.W.; Zhang, Y.K. Application of foam to suppress rock dust in a large cross-section rock roadway driven with roadheader. *Adv. Powder Technol.* **2012**, *1*, 257–262. [CrossRef]
21. Hu, S.Y.; Wang, Z.; Feng, G.R. Temporal and Spatial Distribution of Respirable Dust After Blasting of Coal Roadway Driving Faces: A Case Study. *Minerals* **2015**, *4*, 679–692. [CrossRef]
22. Wang, H.; Nie, W.; Cheng, W.M.; Liu, Q.; Jin, H. Effects of air volume ratio parameters on air curtain dust suppression in a rock tunnel's fully-mechanized working face. *Adv. Powder Technol.* **2018**, *29*, 230–244. [CrossRef]
23. Wang, Y.X.; Guo, P.P.; Lin, H.; Li, X.; Zhao, Y.L.; Yuan, B.X.; Liu, Y.; Cao, P. Numerical Analysis of Fiber-Reinforced Soils based on the Equivalent Additional Stress Concept. *Int. J. Geomech.* **2019**, in press.
24. Liu, Q.; Nie, W.; Hua, Y.; Peng, H.T.; Liu, C.Q.; Wei, C.H. Research on tunnel ventilation systems: Dust diffusion and pollution behaviour by air curtains based on CFD technology and field measurement. *Build. Environ.* **2019**, *147*, 444–460. [CrossRef]
25. Wang, Q.G.; Wang, D.M.; Wang, H.T.; Han, F.W.; Zhu, X.L.; Tang, Y.; Si, W.B. Optimization and implementation of a foam system to suppress dust in coal mine excavation face. *Process Saf. Environ.* **2015**, *96*, 184–190. [CrossRef]
26. Han, F.W.; Liu, J. Flow field characteristics and coal dust removal performance of an arc fan nozzle used for water spray. *PLoS ONE* **2018**, *13*, 0203875. [CrossRef] [PubMed]
27. Wang, Z.; Gao, W.; Zhang, P.; Yan, H.J.; Ren, C.S. Study of the Characteristics of DC and ICP Hybrid Discharge Plasmas. *Plasma Sci. Technol.* **2015**, *3*, 191–195. [CrossRef]
28. Lyu, X.J.; You, X.F.; He, M.; Zhang, W.; Wei, H.B.; Li, L.; He, Q. Adsorption and molecular dynamics simulations of nonionic surfactant on the low rank coal surface. *Fuel* **2018**, *211*, 529–534. [CrossRef]
29. Liu, L.; Fang, Z.Y.; Wu, Y.P.; Lai, X.P.; Wang, P.; Song, K.-I. Experimental investigation of solid-liquid two-phase flow in cemented rock-tailings backfill using Electrical Resistance Tomography. *Constr. Build. Mater.* **2018**, *175*, 267–276. [CrossRef]
30. Nie, W.; Peng, H.T.; Jin, H.; Liu, Y.H.; Wei, W.L. The effect of spray pressure on atomization characteristics of external spray nozzle on coal mining machine. *J. Chin. Univ. Min. Technol.* **2017**, *46*, 41–47.
31. Wang, H.T.; He, S.; Xie, G.R.; Chen, X.Y.; Qin, B.T. Study of the mechanism by which magnetization reduces dust suppressant usage. *Colloids Surf. A* **2018**, *558*, 16–22. [CrossRef]

32. Liu, Y.H.; Nie, W.; Jin, H.; Ma, H.; Hua, Y.; Cai, P.; Wei, W.L. Solidifying dust suppressant based on modified chitosan and experimental study on its dust suppression performance. *Adsorpt. Sci. Technol.* **2018**, *36*, 640–654. [CrossRef]

33. Yuan, Q.H.; Xu, X.J.; Liu, Q.; Zhang, X.J. Medium Consumption Control in Dense Medium Coal Preparation Plant Based on Response Surface Methodology. *J. Shandong Univ. Sci. Technol. (Nat. Sci.)* **2018**, *37*, 11–21. [CrossRef]

34. Kong, B.; Li, Z.H.; Wang, E.Y.; Liu, W.; Chen, L.; Qi, G.S. An experimental study for characterization the process of coal oxidation and spontaneous combustion by electromagnetic radiation technique. *Process Saf. Environ.* **2018**, *119*, 285–294. [CrossRef]

35. Ding, Y.M.; Ezekoye, O.A.; Zhang, J.Q.; Wang, C.J.; Lu, S.X. The effect of chemical reaction kinetic parameters on the bench-scale pyrolysis of lignocellulosic biomass. *Fuel* **2018**, *232*, 147–153. [CrossRef]

36. Cheng, W.M.; Liu, Z.; Yang, H.; Wang, W.Y. Non-linear Seepage Characteristics and Influential Factors of Water Injection in Gassy Seams. *Exp. Therm. Fluid Sci.* **2018**, *91*, 41–53. [CrossRef]

37. Yang, J.; Liu, D.D.; Zhu, X.L.; Fang, X.M. Research progress of chemical dust suppression agent. *Chemistry* **2013**, *4*, 346–353.

38. Sa, Z.Y.; Liu, J.; Liu, J.Z.; Zhang, Y.J. Research on effect of gas pressure in the development process of gassy coal extrusion. *Saf. Sci.* **2019**, *115*, 28–35. [CrossRef]

39. Liu, Y.; Wang, M.Y.; Zhao, S.S.; Liu, Y.Y.; Yang, J. Synthesis and Modification of Hyperbranched Polyester Dust Suppressant. *J. Shandong Univ. Sci. Technol. (Nat. Sci.)* **2018**, *37*, 26–34. [CrossRef]

40. Nie, W.; Peng, H.T.; Liu, Y.H.; Ma, X.; Wei, W.L. Experimental Research on the Coupling and Settlement of Droplets and Dust Particles Influenced by Airflow. *J. Shandong Univ. Sci. Technol. (Nat. Sci.)* **2016**, *35*, 30–36. [CrossRef]

41. Kong, B.; Wang, E.Y.; Li, Z.H. The effect of high temperature environment on rock properties—An example of electromagnetic radiation characterization. *Environ. Sci. Pollut. R.* **2018**, *25*, 29104–29114. [CrossRef] [PubMed]

42. Liu, Z.; Yang, H.; Wang, W.Y.; Cheng, W.M.; Xin, L. Experimental Study on the Pore Structure Fractals and Seepage Characteristics of a Coal Sample Around a Borehole in Coal Seam Water Infusion. *Transp. Porous Med.* **2018**, *125*, 289–309. [CrossRef]

43. Bao, Q.; Nie, W.; Liu, C.Q.; Liu, Y.H.; Zhang, H.H.; Wang, H.K.; Jin, H. Preparation and characterization of a binary-graft-based, water-absorbing dust suppressant for coal transportation. *J. Appl. Polym. Sci.* **2018**, *136*, 47065. [CrossRef]

44. Grogan, R. Mixtures, Compositions, and Methods of Using and Preparing Same. Patent US 20090269499, 29 October 2009.

45. Zhang, L.B.; Jiao, J.; Zhao, X.Y.; Zhang, J.; Zou, D.; Ji, Y.; Shan, C. Study on preparation and properties of eco-friendly dust suppressant. *Trans. Chin. Soc. Agric. Eng.* **2013**, *18*, 218–225.

46. Zhang, H.H.; Nie, W.; Wang, H.K.; Bao, Q.; Jin, H.; Liu, Y.H. Preparation and experimental dust suppression performance characterization of a novel guar gum-modification-based environmentally-friendly degradable dust suppressant. *Powder Technol.* **2018**, *339*, 314–325. [CrossRef]

47. Yang, J.; Wang, K.; Fang, X.M. The Preparation and Performance Test of a New Coal Dust Depressor. *Saf. Coal Mines* **2009**, *5*, 21–24.

48. Polat, H.; Chander, S. Adsorption of PEO/PPO Triblock Co-polymers and Wetting of Coal. *Colloids Surf. A Physicochem. Eng. Asp.* **1999**, *1*, 199–212. [CrossRef]

49. Zhou, Q.; Qin, B.T.; Wang, J.; Wang, H.T.; Wang, F. Experimental investigation on the changes of the wettability and surface characteristics of coal dust with different fractal dimensions. *Colloids Surf. A* **2018**, *551*, 148–157. [CrossRef]

50. Liu, Q.; Nie, W.; Hua, Y.; Peng, H.T.; Liu, Z.Q. The effects of the installation position of a multi-radial swirling air-curtain generator on dust diffusion and pollution rules in a fully-mechanized excavation face: A case study. *Powder Technol.* **2018**, *329*, 371–385. [CrossRef]

51. Liu, Y.H.; Nie, W.; Mu, Y.B.; Zhang, H.H.; Wang, H.K.; Jin, H.; Liu, Z.Q. A synthesis and performance evaluation of a highly efficient ecological dust depressor based on the sodium lignosulfonate–acrylic acid graft copolymer. *RSC Adv.* **2018**, *8*, 11498–11508. [CrossRef]

52. Zhao, Y.L.; Zhang, L.Y.; Wang, W.J.; Wan, W.; Li, S.Q.; Ma, W.H.; Wang, Y.X. Creep behavior of intact and cracked limestone under multi-level loading and unloading cycles. *Rock Mech. Rock Eng.* **2017**, *50*, 1409–1424. [CrossRef]

53. Liu, L.; Fang, Z.Y.; Qi, C.C.; Zhang, B.; Guo, L.J.; Song, K.-I. Experimental investigation on the relationship between pore characteristics and unconfined compressive strength of cemented paste backfill. *Constr. Build. Mater.* **2018**, *179*, 254–264. [CrossRef]

54. Qin, B.T.; Dou, G.L.; Zhong, X.X. Effect of stannous chloride on low-temperature oxidation reaction of coal. *Fuel Process. Technol.* **2018**, *176*, 59–63. [CrossRef]

55. Li, L.; Qin, B.T.; Ma, D.; Zhuo, H.; Liang, H.J.; Gao, A. Unique spatial methane distribution caused by spontaneous coal combustion in coal mine goafs: An experimental study. *Process Saf. Environ.* **2018**, *116*, 199–207. [CrossRef]

56. Zhao, Y.J.; Li, R.Y.; Ji, C.F.; Huan, C.; Zhang, B.; Liu, L. Parametric study and design of an earth air heat exchanger using model experiment for memorial heating and cooling. *Appl. Therm. Eng.* **2019**, *148*, 838–845. [CrossRef]

57. Liu, X.S.; Gu, Q.H.; Tan, Y.L.; Ning, J.G.; Jia, Z.C. Mechanical Characteristics and Failure Prediction of Cement Mortar with a Sandwich Structure. *Minerals* **2019**, *9*, 143. [CrossRef]

58. Nie, W.; Liu, Y.H.; Wang, H.; Wei, W.L.; Peng, H.T.; Cai, P.; Hua, Y.; Jin, H. The development and testing of a novel external-spraying injection dedusting device for the heading machine in a fully-mechanized excavation face. *Process Saf. Environ.* **2017**, *109*, 716–731. [CrossRef]

59. Nie, W.; Wei, W.L.; Ma, X.; Liu, Y.H.; Peng, H.T.; Liu, Q. The effects of ventilation parameters on the migration behaviors of head-on dusts in the heading face. *Tunn. Undergr. Space Technol.* **2017**, *70*, 400–408. [CrossRef]

60. Hua, Y.; Nie, W.; Wei, W.L.; Liu, Q.; Liu, Y.H.; Peng, H.T. Research on multi-radial swirling flow for optimal control of dust dispersion and pollution at a fully mechanized tunnelling face. *Tunn. Undergr. Space Technol.* **2018**, *79*, 293–303. [CrossRef]

61. Dimech, C.C.; Cheeseman, R.; Cook, S.; Simon, J.; Boccaccini, A.R. Production of sintered materials from air pollution control residues from waste incineration. *J. Mater. Sci.* **2008**, *12*, 4143–4151. [CrossRef]

62. Li, T.Y.; Wang, N.; Fang, Q.H. Incorporation of modified soy protein isolate as filier in BR/SBR blends. *J. Mater. Sci.* **2010**, *7*, 1904–1911. [CrossRef]

63. Zhong, X.X.; Qin, B.T.; Dou, G.L.; Xia, C.; Wang, F. A chelated calcium-procyanidine-attapulgite composite inhibitor for the suppression of coal oxidation. *Fuel* **2017**, *217*, 680–688. [CrossRef]

64. Nie, W.; Wei, W.L.; Cai, P.; Liu, Z.Q.; Liu, Q.; Ma, H.; Liu, H.J. Simulation experiments on the controllability of dust diffusion by means of multi-radial vortex airflow. *Adv. Powder Technol.* **2018**, *29*, 835–847. [CrossRef]

65. Wang, Y.X.; Lin, H.; Zhao, Y.L.; Li, X.; Guo, P.P.; Liu, Y. Analysis of fracturing characteristics of unconfined rock plate under edge on impact loading. *Eur. J. Environ Civ. Eng.* **2019**. [CrossRef]

66. Peng, H.T.; Nie, W.; Cai, P.; Liu, Q.; Liu, Z.Q.; Yang, S.B. Development of a novel wind-assisted centralized spraying dedusting device for dust suppression in a fully mechanized mining face. *Environ. Sci. Pollut. Res.* **2018**. [CrossRef] [PubMed]

67. Chen, D.W.; Nie, W.; Cai, P.; Liu, Z.Q. The diffusion of dust in a fully-mechanized mining face with a mining height of 7 m and the application of wet dust-collecting nets. *J. Clean. Prod.* **2018**, *205*, 463–476. [CrossRef]

68. Zhao, Y.L.; Luo, S.L.; Wang, Y.X.; Wang, W.J.; Zhang, L.Y.; Wan, W. Numerical Analysis of Karst Water Inrush and a Criterion for Establishing the Width of Water-resistant Rock Pillars. *Mine Water Environ.* **2017**, *36*, 508–519. [CrossRef]

69. Liu, L.; Zhu, C.; Qi, C.C.; Zhang, B.; Song, K.-I. A microstructural hydration model for cemented paste backfill considering internal sulfate attacks. *Constr. Build. Mater.* **2019**, in press.

70. Kim, J.R.; Netravali, A.N. Parametric study of protein-encapsulated microcapsule formation and effect on self-healing efficiency of 'green' soy protein resin. *J. Mater. Sci.* **2017**, *6*, 3028–3047. [CrossRef]

71. Jin, H.; Nie, W.; Zhang, H.H.; Liu, Y.H.; Bao, Q.; Wang, H.K. The Preparation and Characterization of a Novel Environmentally-Friendly Coal Dust Suppressant. *J. Appl. Polym. Sci.* **2018**, *136*, 47354. [CrossRef]

72. Liu, C.Q.; Nie, W.; Bao, Q.; Liu, Q.; Wei, C.H.; Hua, Y. The effects of the pressure outlet's position on the diffusion and pollution of dust in tunnel using a shield tunneling machine. *Energy Build.* **2018**, *176*, 232–245. [CrossRef]

73. Zhao, Y.L.; Tang, J.Z.; Chen, Y.; Zhang, L.Y.; Wang, W.J.; Liao, J.P. Hydromechanical coupling tests for mechanical and permeability characteristics of fractured limestone in complete stress–strain process. *Environ. Earth Sci.* **2017**, *76*, 1–18. [CrossRef]
74. Zhao, Y.J.; Zhang, Z.H.; Ji, C.F.; Liu, L.; Zhang, B.; Huan, C. Characterization of particulate matter from heating and cooling several edible oils. *Build. Environ.* **2019**, *152*, 204–213. [CrossRef]
75. Ding, Y.M.; Zhou, R.; Wang, C.J.; Lu, K.H.; Lu, S.X. Modeling and analysis of bench-scale pyrolysis of lignocellulosic biomass based on merge thickness. *Bioresour. Technol.* **2018**, *268*, 77–80. [CrossRef] [PubMed]
76. Luo, R.D.; Lin, M.S.; Luo, Y.B.; Dong, J.F. Preparation and properties of a new type of coal dust suppressant. *J. Chin. Coal Soc.* **2016**, *41*, 454. [CrossRef]
77. Wei, Q.H.; Tong, L.; Chen, N.L.; Lin, Q.J. Adhesive properties of a soy-based wood adhesive using sodium dodecyl sulfate. *J. Zhejiang For. Coll.* **2008**, *6*, 772–776.
78. Hua, Y.; Nie, W.; Cai, P.; Liu, Y.H.; Peng, H.T.; Liu, Q. Pattern characterization concerning spatial and temporal evolution of dust pollution associated with two typical ventilation methods at fully mechanized excavation faces in rock tunnels. *Powder Technol.* **2018**, *334*, 117–131. [CrossRef]
79. Zhao, Y.L.; Wang, Y.X.; Wang, W.; Tang, L.; Liu, Q.; Cheng, G. Modeling of rheological fracture behavior of rock cracks subjected to hydraulic pressure and far field stresses. *Theor. Appl. Fract. Mech.* **2019**. [CrossRef]
80. Qi, C.C.; Chen, Q.S.; Fourie, A.; Tang, X.L.; Zhang, Q.L.; Dong, X.J.; Feng, Y. Constitutive modelling of cemented paste backfill: A data-mining approach. *Constr. Build. Mater.* **2019**, *197*, 262–270. [CrossRef]
81. Liu, Z.Q.; Nie, W.; Peng, H.T.; Yang, S.B.; Chen, D.W.; Liu, Q. The effects of the spraying pressure and nozzle orifice diameter on the atomizing rules and dust suppression performances of an external spraying system in a fully-mechanized excavation face. *Powder Technol.* **2019**, in press. [CrossRef]
82. Liu, J.; Zhang, R.; Song, D.Z.; Wang, Z.Q. Experimental investigation on occurrence of gassy coal extrusion in coalmine. *Saf. Sci.* **2019**, *113*, 362–371. [CrossRef]

*Article*

# Pelletization of Torrefied Wood Using a Proteinaceous Binder Developed from Hydrolyzed Specified Risk Materials

Birendra B. Adhikari [1], Michael Chae [1], Chengyong Zhu [1], Ataullah Khan [2], Don Harfield [2], Phillip Choi [3] and David C. Bressler [1,*]

[1] Department of Agricultural, Food and Nutritional Science, Faculty of Agricultural, Life and Environmental Sciences, University of Alberta, Edmonton, AB T6G 2P5, Canada; badhikar@ualberta.ca (B.B.A.); mchae@ualberta.ca (M.C.); chengyon@ualberta.ca (C.Z.)

[2] Thermochemical Processing Group, InnoTech Alberta, Vegreville, AB T9C 1T4, Canada; Ataullah.Mohammed@innotechalberta.ca (A.K.); don.harfield@albertainnovates.ca (D.H.)

[3] Department of Chemical and Materials Engineering, Faculty of Engineering, University of Alberta, Edmonton, AB T6G 1H9, Canada; pchoi@ualberta.ca

* Correspondence: dbressle@ualberta.ca; Tel.: +17-80-492-4986; Fax: +17-80-492-4265

Received: 21 March 2019; Accepted: 16 April 2019; Published: 23 April 2019

**Abstract:** Pressing issues such as a growing energy demand and the need for energy diversification, emission reduction, and environmental protection serve as motivation for the utilization of biomass for production of sustainable fuels. However, use of biomass is currently limited due to its high moisture content, relatively low bulk and energy densities, and variability in shape and size, relative to fossil-based fuels such as coal. In recent years, a combination of thermochemical treatment (torrefaction) of biomass and subsequent pelletization has resulted in a renewable fuel that can potentially substitute for coal. However, production of torrefied wood pellets that satisfy fuel quality standards and other logistical requirements typically requires the use of an external binder. Here, we describe the development of a renewable binder from proteinaceous material recovered from specified risk materials (SRM), a negative-value byproduct from the rendering industry. Our binder was developed by co-reacting peptides recovered from hydrolyzed SRM with a polyamidoamine epichlorohydrin (PAE) resin, and then assessed through pelleting trials with a bench-scale continuous operating pelletizer. Torrefied wood pellets generated using peptides-PAE binder at 3% binder level satisfied ISO requirements for durability, higher heating value, and bulk density for TW2a type thermally-treated wood pellets. This proof-of-concept work demonstrates the potential of using an SRM-derived binder to improve the durability of torrefied wood pellets.

**Keywords:** bioenergy; torrefied wood; pelletization; wood pellets; specified risk materials; binder

---

## 1. Introduction

In recent years, torrefaction of biomass has gained popularity for its ability to improve the fuel properties of the treated material [1–6]. In torrefaction, biomass is heated in the temperature range of 200–300 °C in the absence or very limited supply of oxygen, typically for 1 h. This pre-treatment of biomass removes volatiles and yields a thermally treated product (called torrefied biomass) that typically contains 70% of the mass, but 90% of the energy, of the original crude biomass. The conversion results in a product with an energy density that is roughly 1.3 times higher on a mass basis [1,2]. Additionally, torrefaction of biomass results in a significantly reduced moisture content, and improved grindability and hydrophobicity [1–3,7], which are all desirable attributes for a fuel destined for co-combustion in coal-firing power plants.

Despite the improved qualities achieved through torrefaction, the torrefied biomass has a lower volumetric energy density (typically 4–5 GJ/m$^3$) than that of coal (18–24 GJ/m$^3$) [4], and the process of torrefaction leads to increased dust formation, which poses difficulties for handling and transportation [1,2,4]. The combination of torrefaction and densification (i.e., by pelletization) has been reported to resolve these issues [2,3,7,8]. Torrefied wood pellets have bulk density of up to 850 kg/m$^3$ and a net calorific value in the range of 20–24 MJ/kg, corresponding to a volumetric energy density of 14–18.5 GJ/m$^3$ [4,7], which is comparable to that of sub-bituminous coal (16–17 GJ/m$^3$) [7]. Consequently, torrefied wood pellets, which are also referred to as densified biocoal [1,3,4], are becoming an attractive renewable fuel for major end-use applications such as co-firing and co-gasification in coal fired powerplants [4,8–10].

Even though torrefied wood pellets have several attractive properties and are in many cases superior fuels compared to untreated biomass and biopellets, pelletization of torrefied biomass is more technically challenging and energy intensive compared to pelletization of raw biomass [3,11–16]. Hence, significant research interest has focused on making the pelletization process easier and/or improving the durability of pellets without compromising fuel quality. The use of external binders in the pelletization process has been shown to reduce the energy required for pelletization and/or result in more durable pellets [3,17–19].

Through a detailed analysis of two different process schemes on producing biocoal, Ghiasi et al. concluded that a binder is necessary for pelletization of torrefied wood with reasonable energy consumption [3]. The effect of additives/binders such as lignin, starch, saw dust, caster bean cake, peanut shell, calcium hydroxide, and sodium hydroxide on pelletization of torrefied biomass has been investigated [17–20]. However, in these studies, the binders/additives were used at much higher levels (up to 30%) than allowable by current International Organization for Standardization (ISO) guidelines (≤4%) for premium quality pellets made from thermally treated wood [21]. Additionally, the pellets made with starch had lower bulk density and durability [18]. Two recent reports have thoroughly reviewed the availability of proteinaceous feedstocks and various approaches of developing protein-based adhesives for bonding of cellulosic materials, uncovering the tremendous potential of this renewable resource for development of protein-based wood adhesives/binders [22,23]. In this perspective, proteinaceous material constitutes an attractive feedstock for the development of binders/additives for pelletization of torrefied wood.

Over the years, our group has developed a technology platform for chemical conversion of specified risk materials (SRM), a proteinaceous waste from the rendering industry, into valuable products such as adhesives [24–26], plastics [27], and composites [28]. Recently, we demonstrated that adhesive formulations consisting of 78% peptides and 22% polyamidoamine epichlorohydrin (PAE) resin (dry weight basis) satisfied the performance requirement of the American Standard of Testing and Materials (ASTM) as a plywood adhesive [24]. PAE resin contains azetidinium (four-membered cyclic structure consisting of a quaternary nitrogen atom) functional groups that can easily react with amine, carboxyl, and hydroxyl groups. Hence, the PAE resin acts as a co-reactant for chemical crosslinking of protein/peptides, thus imparting mechanical strength and enhancing adhesive property of protein/peptides-based adhesive formulations for bonding of cellulosic materials [24,25,29,30]. Herein, we report on the potential of an SRM protein-based binder (78% peptides and 22% PAE resin) for pelletization of torrefied wood, as well as the fuel quality of resulting pellets.

## 2. Materials and Methods

### 2.1. Materials and Chemicals

The torrefied wood used in this study was kindly supplied by Airex Energie (Laval, QC, Canada). According to the information provided by the supplier, the feed material was spruce/fir sawdust, and the torrefied material had undergone 30% weight loss on a dry mass basis. This mass loss (30%) upon oven drying is the same value that has been found to be suitable for production of good quality pellets from torrefied SPF (spruce, pine, fir) [13,17]. The SRM was supplied in dried, ground form by a local rendering industry. The received SRM sample was handled following the safe handling and disinfection protocol recommended by the Canadian Food Inspection Agency (CFIA) [31,32]. Kymene™ 557H resin purchased from Solenis (Wilmington, DE, USA) was used as a representative PAE resin. As per the supplier's information, the PAE resins contained 12.5% solids.

The protein/peptide standards for molecular size distribution analysis of hydrolyzed SRM were purchased from Sigma Aldrich (St. Louis, MO, USA) as lyophilized powders. High purity High Performance Liquid Chromatography (HPLC) grade water, acetonitrile (99.9%), and hexane (99.9%) were purchased from Fisher Scientific (Hampton, NH, USA). The filter paper employed for vacuum filtration was Whatman no. 4 filter paper (GE Healthcare, Chicago, IL, USA), and was procured from Fisher Scientific.

### 2.2. Methods

#### 2.2.1. Thermal Hydrolysis of SRM and Recovery and Characterization of Peptides

Regulations set by the Canadian Food Inspection Agency (CFIA) require segregation, removal, and disposal of SRM from slaughtered cattle [31,32]. One of the CFIA approved methods of SRM disposal is thermal hydrolysis at minimum conditions of 180 °C and 1200 kPa for 40 min per cycle [32]. In this study, thermal hydrolysis of SRM and recovery of hydrolyzed protein fragments (referred to as peptides hereafter) from the hydrolysate was achieved following our standard protocol [24,26,28,33]. A high-temperature and high-pressure stainless-steel Parr reactor vessel (5.5 L capacity, Parr 4582, Parr Instrument Company, Moline, IL, USA) equipped with Parr reactor controller (Parr 4848, Parr Instrument Company, Moline, IL, USA) was used for thermal hydrolysis of SRM. In a typical run, 1 kg of SRM was mixed with 1 kg of distilled water in the Parr reactor vessel, and then subjected to hydrolysis at 180 °C and ≥174 psi pressure for 40 min. Recovery of peptides was achieved by diluting the hydrolysate with distilled water (hydrolysate: water ratio = 1:4.5, w/v), agitating the slurry at 200 rpm in a shaker (Innova lab shaker, New Brunswick, Canada) for 10 min, and then allowing the mixture to settle for 10 min. The aqueous fraction was decanted, and then subjected to centrifugation (Avanti J-26 XP high-performance centrifuge, Beckman Coulter, Mississauga, ON, Canada) at 7000× *g* for 40 min. The supernatant was then subjected to vacuum filtration (Whatman no. 4 filter paper, 20–25 µM pore size) to remove any insoluble, if present. The collected filtrate was then washed with hexane (1:1 ratio) three times to remove any residual lipids and/or long-chain fatty acids. Lyophilization of the aqueous fraction afforded a tan colored proteinaceous cake representing 34 ± 2% of the feed material. Prior to hydrolysis, the SRM sample was considered as biohazardous material, and was handled by practicing the safety measures as described in previous reports [24,26,28,33].

The recovered peptides were characterized following the standard characterization methods being practiced in our lab, which comprises analyzing elemental composition, end functional groups (carboxyl and amines), and molecular size distribution by size exclusion high performance liquid chromatography (SEC-HPLC) [24,33]. An Agilent Technologies 1200 series HPLC (Agilent Technologies, Santa Clara, CA, USA) was used for SEC-HPLC analysis of peptides recovered from SRM hydrolysates. The HPLC was equipped with an autosampler and variable wavelength UV detector, and size exclusion was achieved using two size exclusion columns—Superdex™ 200 and Superdex™ Peptide (GE Healthcare Biosciences AB, Uppsala, Sweden)—connected in series, and using 5% acetonitrile in 0.15 M phosphate buffer (pH 9.0) at a flow rate of 0.5 mL/min. Solutions of HPLC standards were prepared using 0.15 M phosphate buffer (pH 9.0). The concentrations of the prepared standards were as follows: Blue dextran, 5 mg/mL; alcohol dehydrogenase, 3 mg/mL; albumin, 5 mg/mL; carbonic anhydrase, 3 mg/mL; cytochrome C, 2 mg/mL; aprotinin, 8 mg/mL; and vitamin B-12, 5 mg/mL. A standards mixture was prepared by combining 1 mL of each of the individual standards, with 5 µL of the mixture injected for each HPLC run. In order to ensure batch-to-batch reproducibility of the starting material (peptides), thermal hydrolysis as well as analysis of recovered peptides were conducted in at least three replicates.

### 2.2.2. Binder Formulation and Characterization

The SRM peptides-based binder was used as a wet formulation, and was prepared as described in a previous report [24]. In a typical experiment, an aqueous solution of peptides prepared by dissolving 35.0 g of peptides in 105 g water was mixed with 80.0 g PAE resin (solid content = 12.5%), and the resulting slurry was stirred for 2 h at room temperature prior to mixing with ground torrefied wood. In the prepared formulation, the ratio of peptides:PAE resin is 78:22 on a dry weight basis.

Lyophilization of the peptides-PAE wet formulation produced a tan colored powder, which was characterized by molecular size distribution and analysis of thermal properties. Molecular size distribution studies were conducted by SEC-HPLC as described in 2.2.1. TA Instruments Q100 Differential Scanning Calorimeter (DSC) and Q50 Thermogravimetric Analyzer (TA Instrument, New Castle, DE, USA) were used for Temperature Modulated Differential Scanning Calorimetry (TMDSC) studies and Thermogravimetric Analysis (TGA) of the binder, respectively. In DSC studies, temperature was modulated in the range of −50–150 °C at a heating rate of 1 °C/min. The thermogravimetric analysis was conducted up to 600 °C at a heating rate of 10 °C/min.

### 2.2.3. Pelletization of Torrefied Wood

Prior to pelletization trials, the torrefied wood feedstock was pre-processed, which involved grinding and sieve-screening to ≤3.18 mm (≤1/8″). The ground torrefied feedstock was moisturized to the desired level (~18.5 wt.%) in a horizontal stand mixer (Hobart N-50 mixer, Hobart Canada, Edmonton, Canada) along with addition of binders at the pre-requisite loadings. Pelletization experiments were conducted using a laboratory scale Kahl flat-die pelletizer (Kahl GmbH, Germany) equipped with a flat die press having 6.35 mm die holes and a length of 21.2 mm. The maximum die temperature noted during pelleting was in the range of 80–85 °C. The collected pellets were air dried at ambient conditions for 20–24 h, and then subjected to durability testing.

### 2.2.4. Analysis of Untreated Wood, Torrefied Wood, and Wood Pellets

The fuel quality of untreated wood (before torrefaction), torrefied wood, and torrefied wood pellets was evaluated through analysis of calorific value and bulk density. An isothermal oxygen bomb calorimeter consisting of four units (1241 Oxygen Bomb Calorimeter, 1710 Calorimeter Controller, 1108 Oxygen Combustion Bomb, and 1541 Water heater (Parr Instruments, Moline, IL, USA)) was employed to determine the calorific value of the torrefied wood pellets as per ASTM D5865-13. The bulk density of different materials was analyzed by evaluating the mass of a given volume of the substance following ASTM 6393. Analysis of chlorine content was performed using Instrumental Neutron Activation Analysis (SLOWPOKE Nuclear Reactor Facility, University of Alberta, Canada).

Durability index (DI) is a measure of the ability of pelletized fuels to resist degradation due to shipping and handling. The durability tests were conducted using a Continental AGRA Equipment Inc (USA) manufactured Dual Tumbler made up of dust tight boxes, with a diagonal baffle to emulate mixing as per ASAE S269.4 methodology. The tumbler simulates the amount of breakage that normally occurs between pellet formation and pellet use. In a typical test, 500 g of pre-screened pellets were tumbled for 10 min at 50 rpm in a dust tight chamber. Durability Index (%) is determined by dividing the post-tumbled weight by the pre-tumbled weight, and multiplying by 100 as shown below:

$$Durability\ Index\ (\%) = \frac{mass\ of\ dust\ free\ pellets\ post\ tumbling}{mass\ of\ dust\ free\ pellets\ pre\ tumbling} \times 100 \tag{1}$$

## 3. Results and Discussion

### 3.1. Characterization of Peptides

The peptides recovered from SRM hydrolysates were characterized by elemental analysis (C, H, N and S), end functional group determination (-NH$_2$ and -COOH), and molecular size distribution analysis (by SEC-HPLC) to ensure batch-to-batch reproducibility, and also to confirm that peptides recovered in these experiments were consistent with those generated in previous studies. The elemental composition and chemical functionality (Table 1) confirmed batch to batch reproducibility and also indicated that the peptides recovered in this study were identical to the peptides recovered in previous studies [24,25,33]. The molecular size of recovered peptides ranged from as small as dipeptides to those having molecular weight of 66 KDa (referenced with albumin in SEC-HPLC). The majority of peptides have molecular weights in the range of 1.36–12.4 KDa (Figure 1), which was again consistent with our previous studies [24,25,33].

**Table 1.** Elemental composition and functional group analysis of peptides recovered from hydrolyzed Specified Risk Materials (SRM).

| Elemental Composition (%) | | | | Amine Groups (mmoles/g) | Carboxyl Groups (mmoles/g) |
| --- | --- | --- | --- | --- | --- |
| C | H | N | S | | |
| 49.3 ± 0.2 | 6.66 ± 0.09 | 14.4 ± 0.1 | 0.54 ± 0.09 | 0.50 ± 0.02 | 1.60 ± 0.05 |

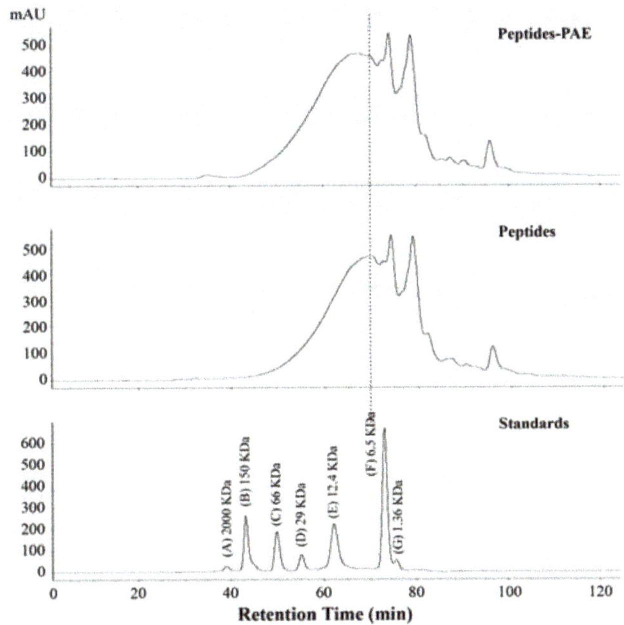


**Figure 1.** SEC-HPLC chromatograms of peptides and peptides-PAE binder formulation. The standards are (from left to right): Blue dextran (A; MW: 2000 KDa), alcohol dehydrogenase (B; 150 KDa), albumin (C; 66 KDa), carbonic anhydrase (D; 29 KDa), cytochrome C (E; 12.4 KDa), aprotinin (F; 6.5 KDa) and vitamin B-12 (G; 1.36 KDa). The molecular weight of the peptides ranged from as small as dipeptides to as large as that of albumin (66 KDa). The majority of peptides have molecular weight in the range of about 1.36–12.4 KDa. The peptides-PAE formulation had limited solubility in water; analysis of the water-soluble fraction showed a general shift towards higher molecular weight species. The dotted line was added as a visual reference.

### 3.2. Thermal Properties of Peptides-PAE Binder

A review of available literature on pelletization of torrefied wood indicated that one of the most important requirements for obtaining good quality pellets is that the die temperature used for pelletization corresponds to that at which softening and/or melting of the binder occurs [4,7,8,13–17]. Hence, identifying the glass transition temperature ($T_g$) of the binder was essential for selecting the pelletization parameters. However, no obvious phase change behavior was observed in differential scanning calorimetry (DSC) of peptides-PAE binder, and $T_g$ of the binder was not discernible in conventional DSC. Accordingly, temperature modulated differential scanning calorimetry (TMDSC) of the binder was conducted, and $T_g$ of peptides-PAE binder was assessed from the plot of reversing heat flow rate as a function of temperature in standard TMDSC of the binder (Figure 2). As evident from the TMDSC plot, onset of glass transition occurred at 45.33 °C and ended at 64.57 °C, the midpoint being measured as 54.56 °C. This suggests that the glass transition (softening) of peptides-PAE binder occurs at about 55 °C, which is a very convenient temperature range for pelletization at industrial scale. Nevertheless, it should also be noted that the glass transition of wood components, which are generally much higher than the glass transition of the binder of this study, also have substantial impact on the mechanical strength/durability of pellets, and must be taken into account in pelleting experiments.

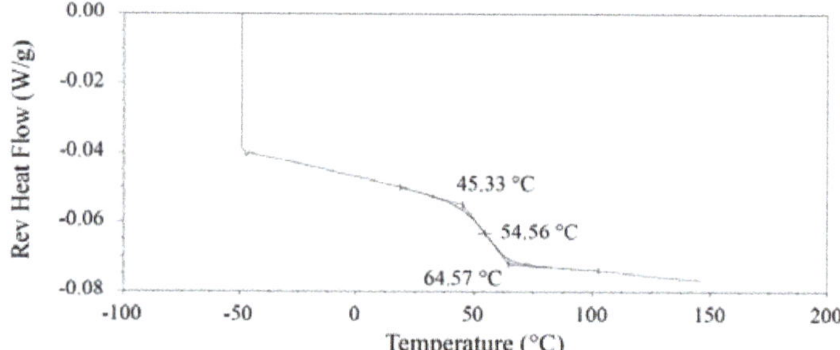


**Figure 2.** Reverse heat flow rate as a function of temperature in temperature modulated differential scanning calorimetry (TMDSC) of the peptides-PAE binder. Parameters of modulation: Equilibrated at −50 °C, heating rate = 1 °C/min, modulation period = 60 s, isothermal for 3 min, amplitude of program temperature = 2 °C/min to 150 °C.

Thermogravimetric analysis was then conducted to determine the degradation temperature of the peptides-PAE binder. The typical plot of weight (%) of a sample of the binder as a function of increasing temperature indicates that the thermal degradation of the binder started at around 190 °C, before which it had undergone small weight loss in two steps (Figure 3). An initial weight loss of about 4% occurring in the temperature range up to 100 °C is attributed to the loss of unbound water (moisture). A shallow weight loss occurring from 100 °C to around 190 °C is likely due to the loss of bound (H-bonded) water and low molecular weight compounds. The TGA plot infers that the onset of thermal degradation of peptides-PAE binder occurs at around 200 °C. From the onset of thermal degradation, one step weight loss occurred up to 500 °C due to the continuous decomposition of the polymer and formation of gaseous reaction products. After the decomposition of the binder, nearly 30–32% mass remained as carbon black.

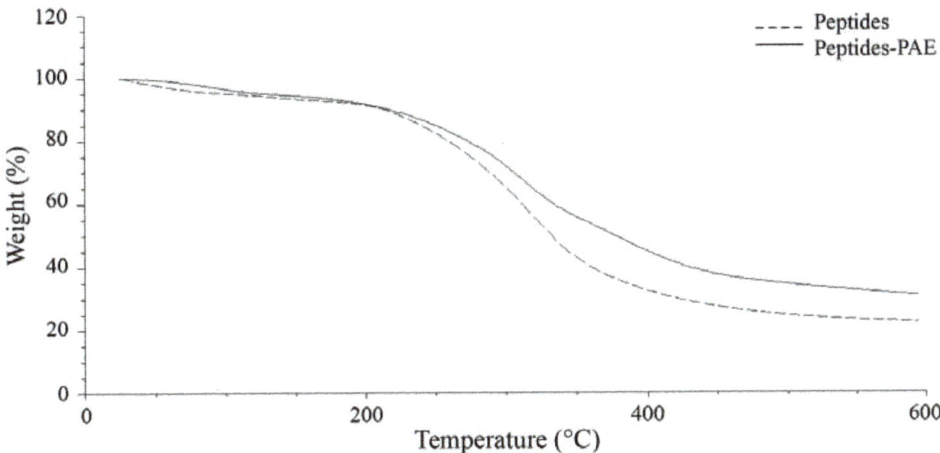

**Figure 3.** Thermogravimetric analysis of peptides and the peptides-PAE formulation. Amplitude of program temperature = 10 °C/min to 600 °C.

### 3.3. Bulk Density and Calorific Value of Raw and Torrefied Wood Chip Feedstocks

On a wet weight basis, the bulk density of torrefied biomass is lower than that of the raw biomass (Table 2). This is attributable to the fact that torrefaction results in removal of volatiles from inside of the particles with insignificant reduction in particle size [13,14,17]. On a dry basis, however, the torrefied wood chips had higher bulk density than the raw wood chips (Table 2), which is due to the removal of moisture and volatiles in the torrefied product. After torrefaction, the high heating value was measured at 22.5 ± 0.1 MJ/Kg, compared to 19.7 ± 0.1 MJ/Kg of raw wood chips (based on analytical duplicates).

**Table 2.** Bulk density and calorific value of raw wood chips and torrefied wood used in this study.

| | Bulk Density (kg/m$^3$) [1] | | Gross Heating Value (MJ/Kg) [3] |
|---|---|---|---|
| | Wet basis [2] | Dry basis [2] | Dry basis |
| **Raw** | 199 ± 1 [a] | 159 ± 1 [b] | 19.7 ± 0.1 |
| **Torrefied** | 182 ± 4 [b] | 177 ± 3 [a] | 22.5 ± 0.1 |

[1] The moisture content of raw wood chips was 21.3 ± 0.7% and that of torrefied wood was 1.8 ± 0.1; [2] Means that do not share the same letter in the same column are significantly different (Tukey test, 95% confidence level); [3] Gross heating values were assessed using analytical duplicates.

### 3.4. Pelletization and Analysis of Torrefied Wood Pellets

#### 3.4.1. Pelletization of Torrefied Wood

Pellet durability is one of the most important parameters that determines the quality of torrefied wood pellets. Dense and strong pellets—characterized by high mechanical strength—have minimum breakage during handling, transportation, and storage, which translates to improved tradability of the pellets. Verhoef et al., suggested that pelletizing torrefied wood requires a proper balance of material quality (e.g., degree of torrefaction), particle size, moisture content, binder, as well as the thickness, hole size and temperature of the pelleting die [2]. For our experiments, the length and bore diameter of the pelleting die were 21.2 mm and 6.35 mm, respectively, and the torrefied wood was ground to ≤3.18 mm (≤1/8″) particle size. The optimal moisture content employed for the pelletization experiments was 18.5%, which was assessed from preliminary experiments evaluating the durability of pellets made from torrefied wood pre-conditioned with moisture content in the range of 16–30%.

It should be noted that the temperature of the pelleting die was 80–85 °C, which is well above the glass transition of the peptides-PAE binder (Figure 2).

According to the International Standard for fuel specifications and classes of solid biofuels (ISO/TS 17225-8:2016), the specifications for the highest quality classes (TW1a and TW2b) of thermally-treated wood pellets require that the additives/binders should be ≤4 wt% on a dry mass basis [21]. On the other hand, specifications from International Maritime Organization (IMO) require that the additives in torrefied wood pellets cannot be more than 3% if they have to be transported through marine vessels. Hence, in our experiments, the effect of binder on the qualities of torrefied wood pellets was evaluated at a 2% and 3% binder level.

The bulk density as well as durability of pellets are important quality parameters for bulk handling, storage, and transportation of pellets as a commodity fuel [34]. Whereas the bulk density refers to the mass of a given volume of pellets, the durability index indicates the mechanical strength and resistance of pellets against breakage and dust formation when handled during storage and transportation. Both of these characteristics were examined to assess the potential of using peptides-PAE as a binder for torrefied wood pellets.

### 3.4.2. Pellet Durability

In the absence of a binder (control experiment), the torrefied wood pellets had an average durability of 91.4 ± 0.3% (Figure 4). The addition of peptides-PAE binder at a 2% level significantly improved the durability of pellets, resulting in an average durability of 95.6 ± 1.4%. On further increasing the binder level to 3%, the average durability of resulting pellets was found to be 96.9 ± 0.2%. The fuel specification standards of ISO/TS 17225-8 for thermally treated wood pellets of TW2a property class requires a minimum durability of 95%. Thus, it is worth mentioning that although there was no statistical difference in the average durability achieved using either 2% or 3% of the peptides-PAE binder, one of the three trials using 2% binder had a durability index of 94.0%; the other two values obtained were both 96.4%. Conversely, the lowest durability index observed using a peptides-PAE binder level of 3% was 96.8%, which was achieved in two of the three trials, with the final value obtained being 97.2%. Thus, the torrefied wood pellets generated using 3% of the peptides-PAE binder satisfied the durability requirement of ISO for TW2a class wood pellets.

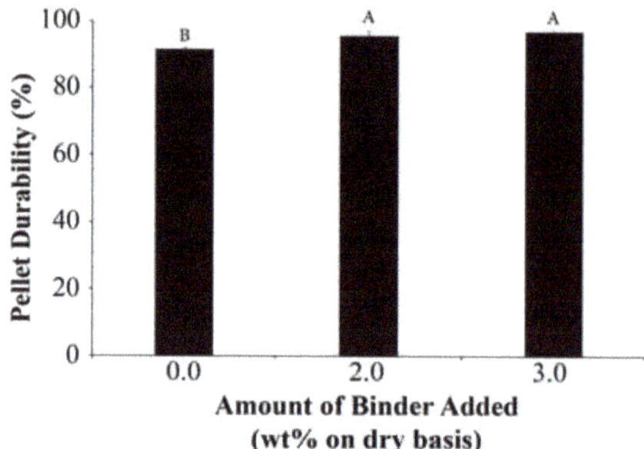


**Figure 4.** Effect of binder level on durability of torrefied wood pellets produced using a bench scale pelletizer. The peptides-PAE binder was generated using 22% PAE resin and 78% SRM peptides on a dry weight basis. The pre-pelletization moisture content was 18.5% and the pelleting die temperature was in the range of 66–85 °C. Triplicate experiments were conducted for pelletization experiments and pellet durability analysis, and the results are presented as average ± standard deviation. Means that do not share a letter are significantly different (Tukey HSD test, 95% confidence level).

The polar functional groups of the peptides-PAE formulation (e.g., hydroxyl, amide, amine, and carboxyl) are expected to enhance attractive intermolecular forces such as polar interactions and hydrogen-bonded connections amongst the wood particles and binder. Additionally, as the peptides-PAE formulation has demonstrated its excellent ability as a binder for wood veneers [24], it is likely that the added binder improves the adhesion forces among the torrefied wood particles and enhances the mechanical strength. Additives such as protein, starch, and lignin have been found to be effective for improving the durability of torrefied and/or biochar pellets through the enhancement of adhesive forces, formation of solid bridges and hydrogen bonded interactions, leading to an improved close-knit structure of wood polymers/particles in pellets [17,19]. It should be noted that since the feedstock was ground to fine powder before pelletization, contributions from mechanical interlocking of particles is minimal. Hence, the forces of adhesion and cohesion are likely the major players in the observed durability of pellets made with peptides-PAE binder. The added binder is expected to have appreciable forces of cohesion due to chemical crosslinking, and it enhances the forces of adhesion by introducing attractive forces such as non-covalent interactions and hydrogen bonding among the particles.

### 3.4.3. Bulk Density and Heating Value of Torrefied Wood Pellets

The bulk density of torrefied wood pellets produced by using peptides-PAE binder (3% binder level) were compared with that of pellets made without binder (control pellets). The torrefied wood pellets made without a binder and with 3% peptides-PAE had bulk densities of 654 ± 33 kg/m$^3$ and 679 ± 26 kg/m$^3$, respectively (Table 3). It should be noted that the two values are not statistically different, which indicates that the addition of our binder did not negatively impact the bulk density of the resulting torrefied wood pellets, while at the same time improving pellet durability (Figure 4). The average bulk density of the torrefied wood feedstock used in this research was 177 ± 3 kg/m$^3$ (Table 2), which indicates that after pelletization, the bulk density was enhanced by a factor of ~3.8.

**Table 3.** Comparison of bulk density and gross heating value of torrefied wood pellets (dry basis) made with or without binder.

|  | **Bulk Density (kg/m$^3$)** [1] | **Gross Heating Value (MJ/Kg)** |
|---|---|---|
| **No Binder** | 654 ± 33 [a] | 22.4 ± 0.0 |
| **Peptides-PAE binder (3%)** | 679 ± 26 [a] | 23.1 ± 0.1 |

[1] Means that do not share the same letter in the same column are significantly different (Tukey test, 95% confidence level).

It is also worth mentioning that the fuel specification standards of ISO/TS 17225-8 for thermally treated wood pellets of TW2a property class requires a minimum bulk density of 650 kg m$^{-3}$. While there was no statistical difference with regards to the bulk densities of torrefied wood pellets obtained, the measured bulk densities for the 3 batches of pellets made without the use of a binder were 692 kg m$^{-3}$, 638 kg m$^{-3}$, and 632 kg m$^{-3}$, indicating that 2 of the 3 batches were below the International Organization for Standardization Technical Specification (ISO/TS) 17225-8 requirement. Conversely, for the batches made with 3% peptides-PAE, a bulk density below 650 kg m$^{-3}$ was not observed (673 kg m$^{-3}$, 707 kg m$^{-3}$, and 656 kg m$^{-3}$).

The calorific values of the torrefied wood pellets generated with and without a binder were measured (Table 3). As the gross heating value reported was obtained from analytical duplicates, a conclusive comparison cannot be made between the torrefied wood pellets made with and without binder. Nevertheless, the data obtained suggest that the torrefied wood pellets made using 3% peptides-PAE binder may possess a heating value that would satisfy the ISO/TS 17225-8 fuel specification standards for TW2a property class pellets. It is also worth mentioning that for the pellets made with peptides-based binder, the final moisture content varied from 2.08 to 6.87, which is less than the 8% limit for TW2a pellets.

### 3.4.4. Chlorine Content

ISO/TS 17225-8 specifies that the chlorine content for thermally treated wood pellets of the fuel quality TW2a class should be ≤0.05%. Analysis of chlorine content indicated that pellets made with the peptides-PAE binder have an average chlorine content of 1200 ± 20 µg/g, which translates to 0.12%. In contrast, pellets made without a binder had an average chlorine content of only 8 ± 1 µg/g. The relatively high chlorine content of wood pellets is likely associated with the risk of corrosion due to emission of HCl gas as effluent while burning the pellets. Thus, the chlorine content of pellets made with peptides-PAE will need to be reduced.

The peptides recovered from SRM hydrolysis contain an average of 1.77 ± 0.09% inorganic chloride. Although the PAE resin is also a source of chlorine in the pellets, since the binder formulation consist of 78% peptides and 22% PAE resin, the major contribution to chlorine in the pellets is most likely from peptides. To address this concern, we examined the possibility of adding an SRM washing step to our thermal hydrolysis protocol. Briefly, 200 g of SRM were washed with Milli-Q water, after which the mass of the wet SRM was brought up to 400 g, then subjected to our standard thermal hydrolysis process. Using this methodology, the chlorine content of the resulting peptides was reduced to 0.513 ± 0.059%. In future pelletization experiments, the SRM pre-washing step will be incorporated to minimize chlorine levels.

## 4. Conclusions

In this research, we explored the potential of using peptides recovered from a negative-value proteinaceous waste from the meat industry as a binder for pelletization of torrefied wood. Our peptides-based binder formulation, which consisted of 78% peptides and 22% PAE resin (dry weight basis), demonstrated remarkable potential as a binder for pelletization of torrefied wood. The torrefied wood pellets obtained using our peptides-PAE binder at a 3% binder level had a net calorific value of ~21 MJ/kg, as well as a durability index and bulk density that satisfied the specifications of ISO/TS 17225-8:2016 for thermally treated wood pellets of the TW2a property class. Although the chlorine content of these pellets was higher than the ISO requirement, prewashing of the SRM prior to thermal hydrolysis was shown to dramatically reduce chlorine levels in the peptides. Future pelletization experiments will employ peptides-PAE derived from prewashed SRM. The research presented in this study provides proof-of-concept that peptides derived from SRM can be utilized for sustainable production of a binder for pelletization of torrefied wood that can potentially serve as a coal replacement.

**Author Contributions:** Conceptualization, B.B.A., M.C., P.C. and D.C.B.; Methodology, B.B.A., M.C., A.K., D.H., P.C., D.C.B.; Software, B.B.A., A.K.; Validation, B.B.A., M.C., A.K., D.H., P.C., D.C.B.; Formal Analysis, B.B.A., C.Z.; Investigation, B.B.A., M.C., C.Z., P.C. and D.C.B.; Resources, B.B.A., M.C., C.Z., A.K., D.H., P.C. and D.C.B.; Data Curation, B.B.A., M.C., C.Z., A.K.; Writing—Original Draft Preparation, B.B.A. and M.C.; Writing—Review & Editing, B.B.A., M.C., A.K., P.C., D.C.B.; Visualization, B.B.A. and M.C.; Supervision, M.C., P.C. and D.C.B.; Project Administration, M.C. and D.C.B.; Funding Acquisition, M.C., P.C. and D.C.B.

**Funding:** This research was funded by Alberta Livestock and Meat Agency (ALMA; 2012R038R), Alberta Prion Research Institute (APRI) and Alberta Agriculture and Forestry (2017A003R), and the Natural Science and Engineering Research Council of Canada (NSERC; 298352-2013).

**Acknowledgments:** The authors gratefully acknowledge the generous financial support received from Alberta Livestock and Meat Agency (ALMA; 2012R038R), Alberta Prion Research Institute (APRI) and Alberta Agriculture and Forestry (2017A003R), and the Natural Science and Engineering Research Council of Canada (NSERC; 298352-2013) for conducting this research. We would also like to recognize Tim Anderson (InnoTech Alberta) who passed away prior to the writing of this manuscript. Anderson was instrumental for this research and performed the pelletization trials and some of the analyses.

**Conflicts of Interest:** The authors declare no conflict of interest.

## References


1. Bergman, P.C.A.; Boersma, A.R.; Zwart, R.W.R.; Kiel, J.H.A. *Torrefaction for Biomass Co-Firing in Existing Coal-Fired Power Stations*; ECN-C-05-013; Energy Research Centre for the Netherlands: Petten, The Netherlands, 2005; pp. 1–71.
2. Verhoeff, H.; Arnuelos, A.A.; Boersma, A.R.; Pels, J.R.; Lensselink, J.; Kiel, J.H.A.; Schukken, H. *Torrefaction Technology for the Production of Solid Bioenergy Carriers from Biomass and Waste*; TorTech Project Report; ECN-E-11-039; Energy Research Centre of the Netherlands (ECN): Petten, The Netherlands, 2011; pp. 1–82.
3. Ghiasi, B.; Kumar, L.; Furubayashi, T.; Lim, C.J.; Bi, X.; Kim, C.S.; Sokhansanj, S. Densified Biocoal from Woodchips: Is it Better to do Torrefaction before Or After Densification? *Appl. Energy* **2014**, *134*, 133–142. [CrossRef]
4. Kumar, L.; Koukoulas, A.A.; Mani, S.; Satyavolu, J. Integrating Torrefaction in the Wood Pellet Industry: A Critical Review. *Energy Fuels* **2017**, *31*, 37–54. [CrossRef]
5. Rousset, P.; Macedo, L.; Commandré, J.; Moreira, A. Biomass Torrefaction under Different Oxygen Concentrations and its Effect on the Cmposition of the Solid by-Product. *J. Anal. Appl. Pyrolysis* **2012**, *96*, 86–91. [CrossRef]
6. Bridgeman, T.G.; Jones, J.M.; Shield, I.; Williams, P.T. Torrefaction of Reed Canary Grass, Wheat Straw and Willow to Enhance Solid Fuel Qualities and Combustion Properties. *Fuel* **2008**, *87*, 844–856. [CrossRef]
7. Bergman, P.C.A. *Combined Torrefaction and Pelletization the TOP Process*; ECN-C-05-073; Energy Research Centre of the Netherlands (ECN): Petten, The Netherlands, 2005; pp. 1–29.
8. Thrän, D.; Witt, J.; Schaubach, K.; Kiel, J.; Carbo, M.; Maier, J.; Ndibe, C.; Koppejan, J.; Alakangas, E.; Majer, S.; et al. Moving Torrefaction Towards Market Introduction—Technical Improvements and Economic-Environmental Assessment Along the overall Torrefaction Supply Chain through the SECTOR Project. *Biomass Bioenergy* **2016**, *89*, 184–200. [CrossRef]
9. Tsalidis, G.; Joshi, Y.; Korevaar, G.; de Jong, W. Life Cycle Assessment of Direct Co-firing of Torrefied and/or Pelletised Woody Biomass with Coal in the Netherlands. *J. Clean. Prod.* **2014**, *81*, 168–177. [CrossRef]
10. Bergman, P.C.A.; Boersma, A.R.; Kiel, J.H.A.; Prins, M.J.; Ptasinski, K.J.; Janssen, F.J.J.G. *Torrefaction for Entrained-Flow Gasification of Biomass*; ECN-C-05-067; Energy Research Centre of the Netherlands (ECN): Petten, The Netherlands, 2005; pp. 1–50.
11. Agar, D.; Gil, J.; Sanchez, D.; Echeverria, I.; Wihersaari, M. Torrefied Versus Conventional Pellet Production—A Comparative Study on Energy and Emission Balance Based on Pilot-Plant Data and EU Sustainability Criteria. *Appl. Energy* **2015**, *138*, 621–630. [CrossRef]
12. Rudolfsson, M.; Stelte, W.; Lestander, T.A. Process Optimization of Combined Biomass Torrefaction and Pelletization for Fuel Pellet Production—A Parametric Study. *Appl. Energy* **2015**, *140*, 378–384. [CrossRef]
13. Peng, J.H.; Ti, H.T.; Lim, C.J.; Sokhansanj, S. Study on Density, Hardness, and Moisture Uptake of Torrefied Wood Pellets. *Energy Fuels* **2013**, *27*, 967–974. [CrossRef]
14. Peng, J.H.; Bi, X.T.; Sokhansanj, S.; Lim, C.J. Torrefaction and Densification of Different Species of Softwood Residues. *Fuel* **2013**, *111*, 411–421. [CrossRef]
15. Shang, L.; Nielsen, N.P.K.; Stelte, W.; Dahl, J.; Ahrenfeldt, J.; Holm, J.K.; Arnavat, M.P.; Bach, L.S.; Henriksen, U.B. Lab and Bench-Scale Pelletization of Torrefied Wood Chips—Process Optimization and Pellet Quality. *Bioenergy Res.* **2014**, *7*, 87–94. [CrossRef]
16. Stelte, W.; Clemons, C.; Holm, J.K.; Sanadid, A.R.; Ahrenfeldt, J.; Shanga, L.; Henriksen, U.B. Pelletizing Properties of Torrefied Spruce. *Biomass Bioenergy* **2011**, *35*, 4690–4698. [CrossRef]
17. Peng, J.; Bi, X.T.; Lim, C.J.; Peng, H.; Kim, C.S.; Jia, D.; Zuo, H. Sawdust as an Effective Binder for Making Torrefied Pellets. *Appl. Energy* **2015**, *157*, 491–498. [CrossRef]
18. Hu, Q.; Shao, J.; Yang, H.; Yao, D.; Wang, X.; Chen, H. Effects of Binders on the Properties of Bio-Char Pellets. *Appl. Energy* **2015**, *157*, 508–516. [CrossRef]
19. Cao, L.; Yuan, X.; Li, H.; Li, C.; Xiao, Z.; Jiang, L.; Huang, B.; Xiao, Z.; Chen, X.; Wang, H.; et al. Complementary Effects of Torrefaction and Co-Pelletization: Energy Consumption and Characteristics of Pellets. *Bioresour. Technol.* **2015**, *185*, 254–262. [CrossRef] [PubMed]
20. Bai, X.; Wang, G.; Gong, C.; Yu, Y.; Liu, W.; Wang, D. Co-Pelletizing Characteristics of Torrefied Wheat Straw with Peanut Shell. *Bioresour. Technol.* **2017**, *233*, 373–381. [CrossRef]

21. ISO. *Solid Biofuels—Fuel Specifications and Classes—Part 8: Graded Thermally Treated and Densified Biomass Fuels*; ISO/TS 17225-8:2016(E); ISO: Geneva, Switzerland, 2016.
22. Adhikari, B.B.; Appadu, P.; Chae, M.; Bressler, D.C. Protein-based wood adhesives: Current trends of preparation and application. In *Bio-Based Wood Adhesives: Preparation, Characterization, and Testing*, 1st ed.; He, Z., Ed.; CRC Press: Boca Raton, FL, USA, 2017; pp. 1–56.
23. Adhikari, B.B.; Chae, M.; Bressler, D.C. Utilization of Slaughterhouse Waste in Value-Added Applications: Recent Advances in the Development of Wood Adhesives Polymers. *Polymers* **2018**, *10*, 176. [CrossRef] [PubMed]
24. Adhikari, B.B.; Appadu, P.; Kislitsin, V.; Chae, M.; Choi, P.; Bressler, D.C. Enhancing the Adhesive Strength of a Plywood Adhesive Developed from Hydrolyzed Specified Risk Materials. *Polymers* **2016**, *8*, 285. [CrossRef] [PubMed]
25. Adhikari, B.B.; Kislitsin, V.; Appadu, P.; Chae, M.; Choi, P.; Bressler, D.C. Development of Hydrolysed Protein-Based Plywood Adhesive from Slaughterhouse Waste: Effect of Chemical Modification of Hydrolysed Protein on Moisture Resistance of Formulated Adhesives. *RSC Adv.* **2018**, *8*, 2996–3008. [CrossRef]
26. Mekonnen, T.H.; Mussone, P.G.; Choi, P.; Bressler, D.C. Adhesives from Waste Protein Biomass for Oriented Strand Board Composites: Development and Performance. *Macromol. Mater. Eng.* **2014**, *299*, 1003–1012. [CrossRef]
27. Mekonnen, T.; Mussone, P.; El-Thaher, N.; Choi, P.Y.K.; Bressler, D.C. Thermosetting Proteinaceous Plastics from Hydrolyzed Specified Risk Materials. *Macromol. Mater. Eng.* **2013**, *298*, 1294–1303. [CrossRef]
28. Mekonnen, T.; Mussone, P.; Alemaskin, K.; Sopkow, L.; Wolodko, J.; Choi, P.; Bressler, D. Biocomposites from Hydrolyzed Waste Proteinaceous Biomass: Mechanical, Thermal and Moisture Absorption Performances. *J. Mater. Chem. A* **2013**, *1*, 13186–13196. [CrossRef]
29. Zhong, Z.; Sun, X.S.; Wang, D. Isoelectric pH of Polyamide–Epichlorohydrin Modified Soy Protein Improved Water Resistance and Adhesion Properties. *J. Appl. Polym. Sci.* **2007**, *103*, 2261–2270. [CrossRef]
30. Li, K.; Peshkova, S.; Geng, X. Investigation of Soy Protein-Kymene® Adhesive Systems for Wood Composites. *J. Am. Oil Chem. Soc.* **2004**, *81*, 487–491. [CrossRef]
31. CFIA. Enhanced Animal Health Protection from BSE—Specified Risk Materials (SRM). SRM. Available online: http://www.inspection.gc.ca/animals/terrestrial-animals/diseases/reportable/bse/srm/eng/1299870250278/1334278201780 (accessed on 24 October 2018).
32. CFIA. Annex D: Specified Risk Materials. Food, 2016. Available online: http://www.inspection.gc.ca/food/meat-and-poultry-products/manual-of-procedures/chapter-17/annex-d/eng/1369768468665/1369768518427 (accessed on 24 October 2018).
33. Mekonnen, T.H.; Mussone, P.G.; Stashko, N.; Choi, P.Y.; Bressler, D.C. Recovery and Characterization of Proteinacious Material Recovered from Thermal and Alkaline Hydrolyzed Specified Risk Materials. *Process Biochem.* **2013**, *48*, 885–892. [CrossRef]
34. Tumuluru, J.S.; Wright, C.T.; Hess, J.R.; Kenney, K.L. A Review of Biomass Densification Systems to Develop Uniform Feedstock Commodities for Bioenergy Application. *Biofuels Bioprod. Biorefin.* **2011**, *5*, 683–707. [CrossRef]

 *processes*

*Article*

# Hydroxymethylation-Modified Lignin and Its Effectiveness as a Filler in Rubber Composites

Nor Anizah Mohamad Aini [1], Nadras Othman [1,*], M. Hazwan Hussin [2], Kannika Sahakaro [3] and Nabil Hayeemasae [3]

[1] School of Materials and Mineral Resources Engineering, Engineering Campus, Universiti Sains Malaysia, Nibong Tebal 14300, Malaysia; noranizah.ma@gmail.com
[2] School of Chemical Science, Universiti Sains Malaysia, Minden 11800, Malaysia; mhh@usm.my
[3] Department of Rubber Technology and Polymer Science, Faculty of Science and Technology, Pattani Campus, Prince of Songkla University, Pattani 94000, Thailand; kannika.sah@psu.ac.th (K.S.); nabil.h@psu.ac.th (N.H.)
* Correspondence: srnadras@usm.my; Tel.: +60-45996177

Received: 26 March 2019; Accepted: 22 May 2019; Published: 25 May 2019

**Abstract:** Kraft lignin was modified by using hydroxymethylation to enhance the compatibility between rubber based on a blend of natural rubber/polybutadiene rubber (NR/BR) and lignin. To confirm this modification, the resultant hydroxymethylated kraft lignin (HMKL) was characterized using Fourier transform infrared (FTIR) and nuclear magnetic resonance (NMR) spectroscopy. It was then incorporated into rubber composites and compared with unmodified rubber. All rubber composites were investigated in terms of rheology, mechanical properties, aging, thermal properties, and morphology. The results show that the HMKL influenced the mechanical properties (tensile properties, hardness, and compression set) of NR/BR composites compared to unmodified lignin. Further evidence also revealed better dispersion and good interaction between the HMKL and the rubber matrix. Based on its performance in NR/BR composites, hydroxymethylated lignin can be used as a filler in the rubber industry.

**Keywords:** lignin; hydroxymethylation; bio-filler; rubber composite

---

## 1. Introduction

Fillers are used in rubber for many reasons. They improve the mechanical and thermal properties, and also reduce the cost and weight of the rubber product. Among commercial fillers, carbon black and silica are the most widely used in the rubber industry. Several studies have reported the substitution of these fillers with other alternative fillers, such as carbon nanotubes, cellulose, protein, starch, and clays [1–8]. Recently, fillers from renewable resources have been widely used in order to replace commercially available fillers due to their sustainability and biodegradability and have been effectively used to improve the mechanical properties of rubber composite. Compared to these renewable materials, lignin is an interesting alternative due to its abundance, low cost, and renewability, which make it potential filler for rubber. Since the 1940s, researchers have reported the preparation of lignin-based rubber composites through patents and academic papers. The reinforcing effect of lignin on rubber composites has been shown by several studies, especially when certain polar materials were used.

Lignin is an aromatic polymer which is mainly comprised of three major phenylpropanoid units, namely *p*-hydroxyphenyl (H), guaiacyl (G), and syringyl (S) units, which form a complex three-dimensional structure. Several linkage types are present in lignin, including the ether linkages (C-O-C), β-β, β-O-4, β-5, etc. [9,10]. Conventionally, lignin is a by-product of the pulp and paper manufacturing industry. A major portion of this lignin is burnt and used as fuel due to its high energy content. Only a small percentage (about 2%) of the lignin produced in the world is converted into value-added products [11–13].

The utilization of lignin as a filler in rubber composites may reduce the dependency on using oil-based materials [14–17] as fillers. However, the disadvantages of using lignin are its poor dispersion and compatibility in non-polar rubber. With the development of new technologies, scientists have been looking to improve manufactured products containing lignin as a filler. Many approaches have been used to enhance the performance of lignin-based rubber composites, including modifying the surface of the lignin [18,19], and using hybrid technologies with carbon black [20], silica [21], montmorillonite [22,23], and layered double hydroxides [24]. Hybridization has been claimed to be an effective method to reduce the viscoelastic loss of rubber compounds [20] and improve the mechanical properties of such compounds with excellent reinforcing effect [22,24]. Subsequent studies involved the modification of lignin particles by acetylation [25,26], alkylation [27,28], etc. Current efforts to develop lignin-based fillers for rubber are focused on the chemical transformation of lignin to increase the crosslinking between lignin macromolecules with formaldehyde as a crosslinking agent. Jiang et al. [29] reported the successful utilization of hydroxymethyl lignin in styrene-butadiene rubber (SBR), improving mechanical properties, such as tensile strength and tear strength. Hydroxymethyl groups were introduced onto the C5 positions of G units in the lignin structure, thus producing further crosslinking condensation reactions between lignin macromolecules. Furthermore, hydroxymethylation modification has also been used to produce lignin nanoparticles [29,30].

To the best of our knowledge, the effect of hydroxymethylated lignin (HMKL)-filled natural rubber/polybutadiene rubber (NR/BR) composites has not been reported. Therefore, the main aim of this work is to focus on the modification of kraft lignin (KL) by using hydroxymethylation and to investigate the effectiveness of HMKL as a reinforcing filler in NR/BR composites. The structure of HMKL was investigated using Fourier transform infrared (FTIR) and nuclear magnetic resonance (NMR) spectroscopy in order to confirm the reaction between formaldehyde and lignin molecules. Additionally, the compounding formulation was also designed to incorporate carbon black (CB). This was done to ensure that hybrid fillers based on modified lignin/carbon black (HMKL/CB) would be able to provide synergistic properties. Modified lignin was used at various concentrations ranging from 5 to 20 parts per hundred rubber (phr) and was hybridized with carbon black with a total filler concentration of 50 phr. The physical, mechanical, and thermal properties of HMKL-filled NR/BR composites were further evaluated.

## 2. Materials and Methods

Natural rubber (NR), SMR 10 was supplied by the Rubber Research Institute of Malaysia (RRIM), and polybutadiene rubber (BR), BR9000 was supplied by Zarm Scientific (M) Sdn. Bhd (Penang, Malaysia). Kraft lignin (weight average molecular weight, $M_W$ = 3526 g/mol, density = 2.64) was extracted in-house from kenaf biomass and sieved to an average particle size of less than 250 μm. Sodium hydroxide (NaOH), sodium sulfide (Na$_2$S), ammonium hydroxide (NH$_4$OH), formaldehyde, acetic anhydride, pyridine, sulphuric acid (H$_2$SO$_4$), and hydrochloric acid (HCl) were purchased from Merck (Petaling Jaya, Malaysia). Deuterated dimethyl sulfoxide (DMSO-$d_6$) was purchased from Sigma-Aldrich (Darmstadt, Germany) and carbon black (N220) was purchased from Cabot Corporation (Alpharetta, GA, USA). Zinc oxide (ZnO), stearic acid (SA), *N-tert*-butyl-2-benzothiazylsulphonamide (TBBS), treated distillate aromatic extracted (TDAE) oil, paraffin wax, *N*-phenyl-*p*-phenylenediamine (6PPD), tetramethylthiuramdisulphide (TMTD), 2,2,4-trimethyl-1,2-dihydroquinoline (TMQ), and sulfur were supplied by Bayer (M) Ltd. (Penang, Malaysia).

The kraft lignin was modified using a hydroxymethylation method outlined by Popa et al. [30]. This modification was performed to substitute hydroxymethyl groups in the lignin structure and was expected to enhance the interaction between the lignin surface and the rubber matrix. A total of 10 g of lignin was mixed with 47 mL of distilled water in a 500 mL round-bottom flask. The mixture was stirred for 2 h at room temperature. Next, 1.29 g of a 50% NaOH solution and 3.14 g of a 25%

NH$_4$OH solution as catalyst were added to the lignin suspension. Then, the mixture was shaken for 2 h. Furthermore, 6.7 g of a 37% formaldehyde solution was introduced into the system. The whole system was placed in a water bath for further reaction at a temperature of 85 °C for 4 h. After that, the desired product was precipitated by adding 1.0 M HCl solution until the pH was 2. Finally, modified lignin was separated by centrifugation and washed three times with excessive distilled water and dried at 50 °C for 24 h. The resultant formaldehyde-modified lignin, i.e., HMKL, had a molecular weight of 4732 g/mol.

Subsequently, NR/BR/HMKL composites were prepared through laboratory-scale open two-roll mills with the formulation listed in Table 1. The NR and BR were first mixed with the lignin, carbon black, and other rubber ingredients (ZnO, SA, TDAE, 6PPD, TMQ, paraffin wax, TBBS, TMTD, and sulfur). The blend was then removed and kept at room temperature for 24 h. Finally, the sample was press-cured into a 2 mm thick sheet at a temperature of 150 °C for the optimum cure time (T$_{90}$), which was determined using an MDR 2000 moving die rheometer (Alpha Technologies, Akron, OH, USA). The rubber composites were named NR/BR/KL or NR/BR/HMKL, where KL indicates unmodified lignin. The NR/BR/KL and NR/BR/CB50 (without lignin) composites were prepared in a similar way to the NR/BR/HMKL composites.

**Table 1.** Formulations of natural rubber/polybutadiene rubber (NR/BR) and modified or unmodified lignin used in this work.

| Ingredients | Amount (phr) | | | | | |
|---|---|---|---|---|---|---|
| | NR/BR/CB50 (without Lignin) | NR/BR/KL10 | NR/BR/ HMKL-5 | NR/BR/ HMKL-10 | NR/BR/ HMKL-15 | NR/BR/ HMKL-20 |
| NR (SMR 10) [1] | 50 | 50 | 50 | 50 | 50 | 50 |
| BR (BR9000) [2] | 50 | 50 | 50 | 50 | 50 | 50 |
| Zinc Oxide (ZnO) | 5 | 5 | 5 | 5 | 5 | 5 |
| Stearic Acid | 2 | 2 | 2 | 2 | 2 | 2 |
| Carbon Black (N220) | 50 | 40 | 45 | 40 | 35 | 30 |
| Unmodified lignin [3] | - | 10 | - | - | - | - |
| HMKL [4] | - | - | 5 | 10 | 15 | 20 |
| TDAE [5] | 5 | 5 | 5 | 5 | 5 | 5 |
| 6PPD [6] | 2 | 2 | 2 | 2 | 2 | 2 |
| TMQ [7] | 1 | 1 | 1 | 1 | 1 | 1 |
| Paraffin Wax | 2.5 | 2.5 | 2.5 | 2.5 | 2.5 | 2.5 |
| TBBS [8] | 1.2 | 1.2 | 1.2 | 1.2 | 1.2 | 1.2 |
| TMTD [9] | 0.35 | 0.35 | 0.35 | 0.35 | 0.35 | 0.35 |
| Sulfur | 1.5 | 1.5 | 1.5 | 1.5 | 1.5 | 1.5 |

[1] Natural rubber, grade standard Malaysian rubber 10. [2] Polybutadiene rubber, grade BR9000 [3] Kraft lignin. [4] Lignin modified by the hydroxymethylation method. [5] Treated distillate aromatic extracted. [6] N-phenyl-p-phenylenediamine. [7] 2,2,4-trimethyl-1,2-dihydroquinoline. [8] N-tert-butyl-2-benzothiazyl sulphonamide. [9] Tetramethylthiuram disulphide.

FTIR spectra were recorded with a Thermo-Nicolet IR 200 Fourier transform infrared (FTIR) with attenuated total reflection (ATR) at a spectral resolution of ±4 cm$^{-1}$ with 32 scans from 600 to 4000 cm$^{-1}$.

$^1$H and $^{13}$C NMR spectra were recorded on a Bruker Avance-500 spectrometer. A mass of approximately 150 mg of acetylated lignin was dissolved in 0.40 mL DMSO-$d_6$. The $^{13}$C NMR spectra were acquired at a temperature of 50 °C in order to reduce the viscosity of the solution. $^1$H and $^{13}$C NMR data were processed offline using the Top Spin processing software (Bruker, Billerica, MA, USA).

The viscosities of the rubber composites were determined using a Mooney viscometer (model AC/684/FD, SPRI) and a single-bore Rosand capillary rheometer according to the American Society for Testing and Materials (ASTM) D1646-04 method. The results are reported in terms of the Mooney viscosity of the large rotor (ML $(1 + 4)_{100}$) at a temperature of 100 °C. The samples were heated for 1 min prior to testing at a temperature of 100 °C.

The filler–filler interaction of the uncured compounds was studied by strain sweep test in the range of 0.6–100% strain at 0.50 Hz and a temperature of 100 °C by a Montech D-RPA 3000, rubber process analyzer (RPA) instrument (Werkstoffprüfmaschinen Gmbh, Buchen, Germany). The differences in the storage shear moduli (G') at low strain (0.6%) and high strain (100%) are reported.

Cure characteristics of the rubber compounds were tested using an MDR 2000 moving die rheometer (Alpha Technologies, Akron, OH, USA) according to the ASTM D2084-01 method. The rubber compounds were tested at a temperature of 150 °C for 30 min under an oscillating rotor at one degree of cure. Cure properties are reported, including scorch time ($t_{S2}$), optimum cure time ($t_{90}$), minimum torque ($M_L$), maximum torque ($M_H$), and torque difference ($M_H$–$M_L$).

The crosslink density was determined by soaking a compound rubber sample with dimensions of 30 × 5 × 2 mm in toluene in a closed bottle for 72 h at room temperature, 23 °C. The sample was removed and quickly wiped and weighed, $W_S$ (swollen sample weight). Then, the swollen samples were dried at 60 °C for 30 min and weighed, $W_d$ (dry sample weight). The crosslink density, $V_C$, was determined by using the Flory–Rehner Equation (1):

$$V_C = \frac{-\left[\ln\left(1 - Q_p\right) + Q_p + \chi Q_{p2}\right]}{V \times \left(Q_{p^{1/3}} - \left(\frac{Q_p}{2}\right)\right)} \tag{1}$$

where $Q_p$ is the volume fraction of rubber in the swollen gel, $\chi$ is the Flory–Huggins interaction parameter between toluene and rubber, $\chi = 0.38$ [31], $V$ is the molecular volume of toluene, and $Q_p$ is the volume fraction of swelling rubber, which was determined by Equation (2):

$$Q_p = \frac{1}{\left[1 + \left(\frac{W_S - W_d}{W_d}\right)\right] \times \frac{\rho_p}{\rho_S}} \tag{2}$$

where $\rho_p$ and $\rho_s$ represent the density of the polymer and solvent, respectively [32].

Tensile tests were performed on the composites according to the ASTM D412 method. The tensile tests were carried out using an H10KS tensometer (Hounsfield Test Equipment Co., Ltd., Croydon, UK) with a die type C dumbbell-shaped specimen and a crosshead speed of 500 mm/min.

Scanning electronic microscopy (SEM) images of lignin-filled rubber compounds were made with a Zeiss SUPRA 35 VP instrument (Carl Zeiss NTS GmbH, Oberkochen, Germany). The fracture surface was coated with a thin gold layer in order to obtain high-quality SEM images.

The hardness of the rubber compounds was measured using a Wallace Shore A durometer (Cambridge, UK), following the ASTM D2240 method.

Compression set tests were performed according to the ASTM D395 method at a temperature of 70 °C over a period of 22 h. The percentage compression set (C%) was calculated using Equation (3):

$$C\% = \frac{t_0 - t_1}{t_1 - t_S} \times 100 \tag{3}$$

where $t_0$, $t_1$, and $t_S$ represent the thickness of the original sample, the thickness of the sample after compression, and the thickness of the spacer, respectively.

The flexing resistance of the rubber composites was tested by subjecting the samples to repeat cycling of flexure using a De Mattia flexing machine. The number of cycles before failure was recorded for this measurement. The tests were performed according to the ASTM D430 method.

Heat build-up tests were carried out using a Goodrich flexometer (Ferry Industry, OH, USA) according to the procedure described in the ASTM D623 method. The frequency of loading was 30 Hz. Measurements of the temperatures of the samples were initially made by means of a thermocouple attached to the surface of the sample, and the temperature was recorded every minute for 25 min.

Aging properties of the rubber compounds were tested under thermal conditions. The influence of thermal aging on the properties of HMKL-filled NR/BR with different loading was investigated and compared to the influence of thermal aging on the same properties of unmodified lignin-filled compound. The specimens were aged at 70 °C for 72 h in a Geer aging oven (Tabai® GPHH-200, Tabai ESPEC

Co., Ltd., Osaka, Japan) according to the ASTM D573 method. After the aging, the specimens were removed and conditioned for 24 h at room temperature before being subjected to mechanical testing for 100% tensile modulus and tensile strength. The effect of aging on the rubber composites was calculated in terms of % strength retention, which was calculated by dividing the aged tensile property (A) by the original tensile property (O) and multiplying by 100.

Thermogravimetric analysis was performed using a Pyris 6 TGA Thermogravimetric Analyzer (PerkinElmer, Inc., Waltham, MA, USA). The measurements were made in a nitrogen atmosphere in the temperature range from 30 to 600 °C at a heating rate of 10 °C/min.

## 3. Results and Discussion

### 3.1. Structural Analysis

The reaction between formaldehyde and lignin in alkaline media results in substitution on the C5 position in the guaiacyl unit and on the side-chain-bearing carbonyl groups. This reaction can continue further with a condensation reaction with the hydroxyl groups, thus reducing the hydroxyl group content [33]. In this study, during the process of hydroxymethylation, formaldehyde was added to the lignin in the alkaline medium. Three possible reactions are proposed, as shown in Figure 1. The main reaction was involved with the attachment of hydroxymethyl groups on the lignin aromatic ring [34] and the substitution of side chains by aliphatic methylol groups, as shown in Figure 1a,b, respectively. Increasing the reaction temperature causes hydroxymethyl groups to react at free positions of other lignin macromolecules to form methylene bonds, as shown in Figure 1c.

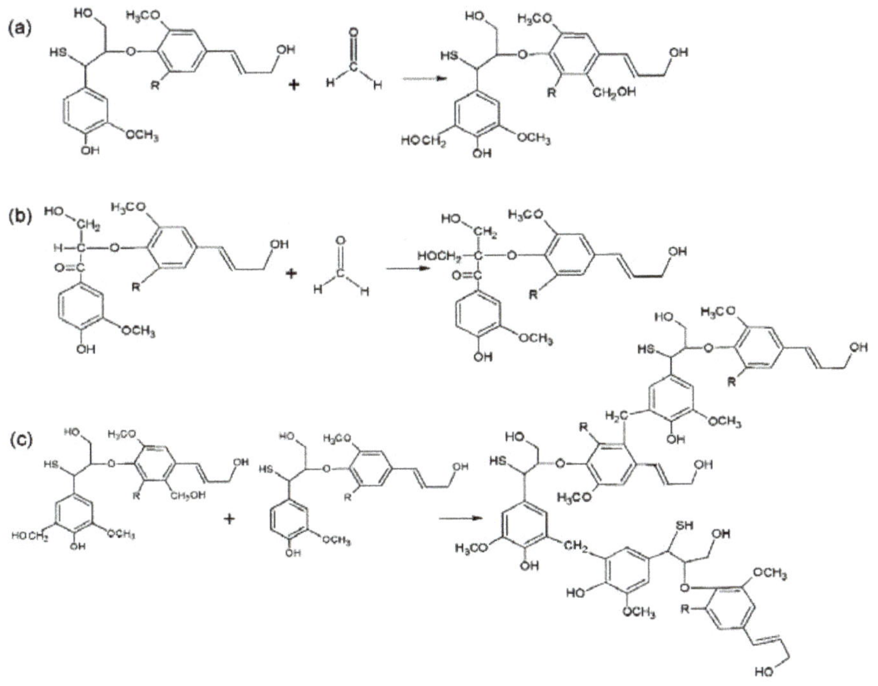

**Figure 1.** Schematic illustration of the proposed reactions of formaldehyde with lignin (a–c) condensation reaction.

### 3.1.1. Fourier Transform Infrared (FTIR) Analysis

Figure 2 displays the FTIR spectra of KL and HMKL. The transmission band at 3400 cm$^{-1}$ is due to O–H stretching vibration, and the bands at 2927 and 2847 cm$^{-1}$ are due to C–H stretching vibration in methyl and methylene groups, respectively. The C–H bending vibration in methyl groups can be assigned to the band at ~1460 cm$^{-1}$. The stretching vibration at 1708 cm$^{-1}$ is due to the carbonyl groups, C=O conjugated with the aromatic ring. The absorption bands at 1601, 1509, 1454, and 1422 cm$^{-1}$ are attributed to skeletal vibrations of aromatic ring macromolecules, C=C and C–H. The band for C–O stretching vibration in the guaiacyl ring is identified at 1270 cm$^{-1}$, and the bands for C–O stretching vibration in the syringyl ring are identified at 1328 and 1111 cm$^{-1}$. The band at 1213 cm$^{-1}$ is due to the phenolic hydroxyl group and ether in syringyl and guaiacyl. The band at 1035 cm$^{-1}$ can be attributed to the C–O deformation in the secondary and primary alcohol or aliphatic esters [35]. The bands at 1328, 1270, 1213, and 1111 cm$^{-1}$ observed in both types of lignin indicate that both lignins contain the same functional groups but in different amounts.

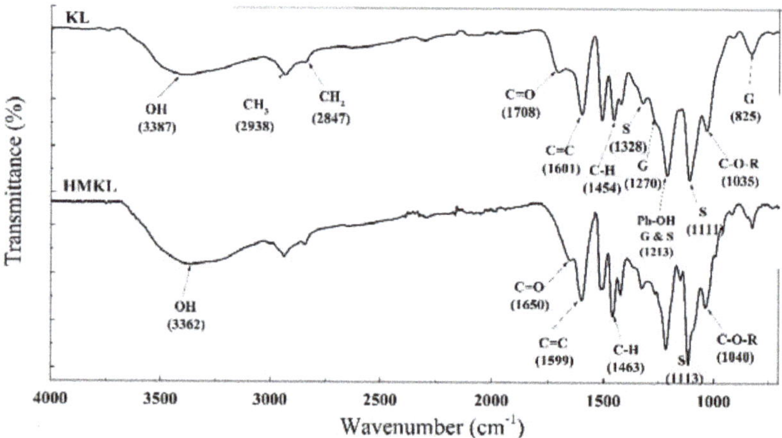


**Figure 2.** The Fourier transform infrared (FTIR) spectra of unmodified kraft lignin (KL) and hydroxymethylated kraft lignin (HMKL).

Lignin mainly contains highly reactive guaiacyl (G) moieties with a free C5 position in the aromatic ring, with small amounts of syringyl (S) and *p*-hydroxyphenyl (H) units. Therefore, the reactive G unit can actively react with formaldehyde rather than the stable S unit, which is linked to methoxy groups at the C3 and C5 positions [36,37]. In the creation of the formaldehyde-modified lignin, i.e., the HMKL, the formaldehyde reacts with the free site of G units and transforms into hydroxymethyl groups.

In the FTIR spectrum of HMKL, the band at ~3400 cm$^{-1}$, which is due to the hydroxyl group (–OH), is broader than the same band in the spectrum of KL. Intense C–H bands at 2938 and 2840 cm$^{-1}$ correspond to methoxyl groups and hydroxymethyl groups. Furthermore, the intensities of the C–H asymmetric vibration band (1463 cm$^{-1}$) and the C–O stretching vibration band for primary and secondary alcohol and ether (1040 cm$^{-1}$) are intense in the FTIR spectrum of HMKL than that of KL. On the other hand, the band at 825 cm$^{-1}$, which corresponds to the aromatic C–H out-of-plane deformation vibration of the G units at carbon positions 2, 5, and 6, is weaker in the FTIR spectrum of HMKL than that of KL. This analysis suggests the presence of hydroxymethyl groups in the lignin structure, as well as condensation resulting in the partial crosslinking of these hydroxymethyl lignin molecules. The introduction of methylol groups in the lignin structure is indicated by the increased intensity of the bands at 3400, 1463, and 1113 cm$^{-1}$ after the hydroxymethylation reaction, proving that the reaction took place within the system.

### 3.1.2. Nuclear Magnetic Resonance (NMR)

Figure 3 shows the $^1$H NMR spectra of both KL and HMKL. The $^1$H NMR spectra proves that the hydroxymethylation reaction took place between lignin and formaldehyde. The peak intensity of the spectrum of the modified lignin is observed in the range of 8–6 ppm. This is due to the substitution of hydroxymethyl groups in the aromatic nuclei of G or H units. Intense peaks in the 4.0–2.7 ppm range are related to protons in the methoxyl group. The peak related to hydroxymethylated lignin was observed to be broader than that of unmodified lignin (i.e., KL). This is due to the introduction of a functional group through the hydroxymethylation reaction. The signal from DMSO protons can be seen at 2.5 ppm. The modification of hydroxymethylated lignin can be observed in the range of 2.49–1.5 ppm, which is where the signals from acetyl groups are observed. The spectra allow the aromatic peaks to be differentiated from the aliphatic acetate peaks. These peaks correspond to the proposed reactions (a) and (b) shown in Figure 1. Proton signals, as exhibited in the range of 1.5–0.8 ppm, can be attributed to the aliphatic signal, which is related to methyl and methylene groups from the lignin macromolecules.

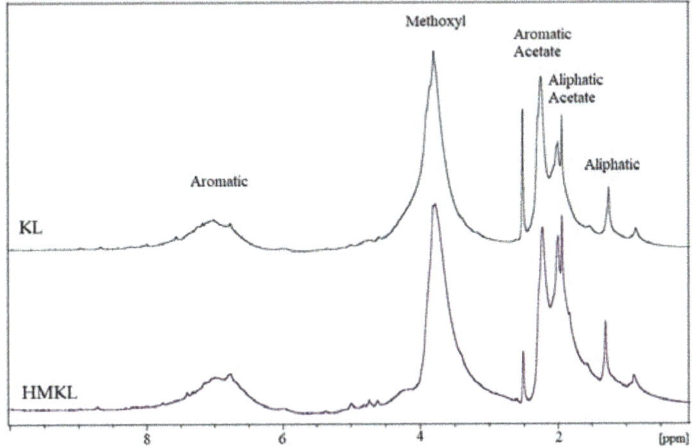


**Figure 3.** $^1$H Nuclear magnetic resonance (NMR) spectra of unmodified KL and HMKL.

Figure 4 displays the $^{13}$C NMR spectra for KL and HMKL, which are similar. The increased intensity of the peak at 154 ppm after the hydroxymethylation modification is attributed to the substitution of the hydroxymethyl functional group, which converts the G unit to a stable S unit. The peak observed at 61.3 ppm for both HMKL and KL is due to methylene or $CH_2$ resonance [38].

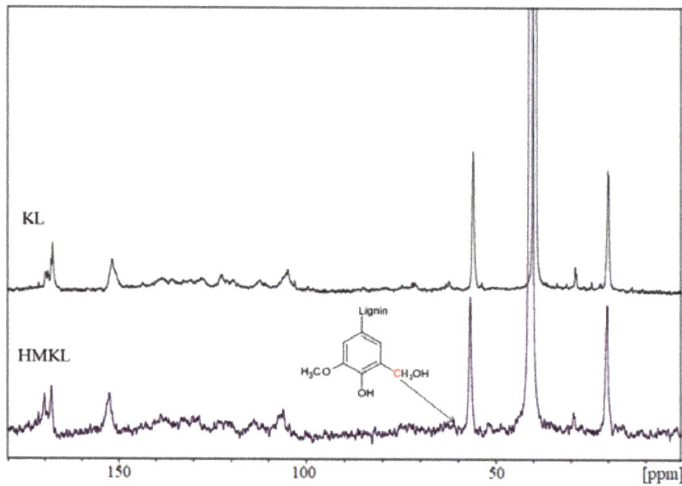


**Figure 4.** $^{13}$C NMR spectra of unmodified KL and HMKL.

### 3.2. Mooney Viscosity and Payne Effect of Compounds

The effect of different HMKL contents on the Mooney viscosity of the NR/BR composites was investigated and compared to the Mooney viscosity of the unmodified lignin (KL10)-filled NR/BR, as displayed in Figure 5a. The highest viscosity was observed at a concentration of 5 phr of NR/BR with HMKL compared to NR/BR/KL and also NR/BR/CB50. This indicates that there is a high restriction of the mobility of the macromolecules due to the greater interaction between lignin and the rubber matrix. However, at the same lignin content, the Mooney viscosity of HMKL-filled rubber composite is higher than NR/BR/KL composite. The Mooney viscosity was found to decrease with increasing HMKL content. This is due to the higher compatibility between the filler and the rubber matrix. The reduction of the Mooney viscosity with increasing HMKL content may be attributed to the plasticizing effect of HMKL. HMKL contains low-molecular-weight lignin, which easily diffuses into the rubber chain and helps adhesion. This has a direct effect on the behavior of the rubber compound [39]. Thus, the reduction of the Mooney viscosity results in good lignin miscibility, which enhances the efficiency of the processing of rubber compounds [40–42]. Therefore, it could increase the chain mobility and improve the flow behavior of rubber compounds.

In the same way, the inclusion of lignin also affected the tensile modulus of rubber compounds, especially when high strain was applied. This phenomenon occurred since the filler networks were destroyed when high strain was applied, which suggests that the inclusion of modified lignin in the rubber matrix weakened the filler network interaction and reduced the Payne effect, as shown in Figure 5b. The Payne effect of lignin-filled rubber compounds is normally used to explain the degree of filler–filler interactions, which are mainly caused by hydrogen bonding, leading to the formation of the filler network in the rubber matrix. By decreasing the filler–filler interaction, the processability of the rubber compounds was enhanced in the present study. The other reason for the enhanced processability of the rubber compounds, i.e., the partial replacement of CB with HMKL, also plays a role in decreasing the Payne effect of the rubber compounds. However, at the same lignin content, there are no significant changes in the Payne effect of the NR/BR/KL and NR/BR/HMKL rubber composites. This is due to disruption of the well-structured filler–filler network of CB in the rubber matrix, which leads to a reduction in the strength of the filler–filler network.

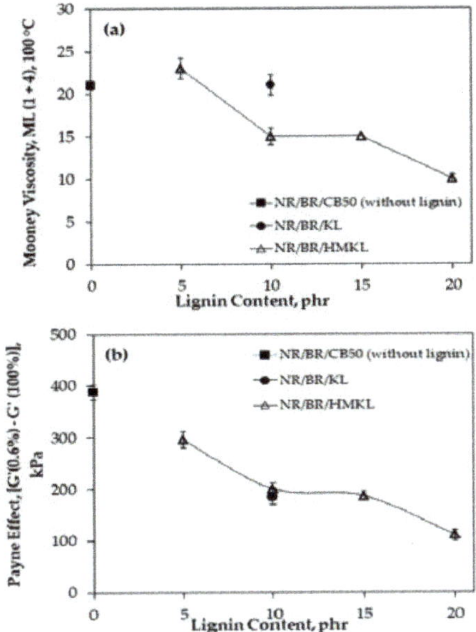

**Figure 5.** The effect of HMKL content on (**a**) the Mooney viscosity and (**b**) the Payne effect of lignin-filled natural rubber/polybutadiene rubber (NR/BR) compounds.

### 3.3. Curing Characteristics and Crosslink Densities

An analysis of the curing of NR/BR composites with KL and HMKL is illustrated in Table 2. It can be observed that the optimum curing time, $t_{90}$, and the scorch time, $t_{S2}$, of composites with HMKL are lower than those of composites with KL. This suggests that the presence of HMKL enhanced the vulcanization of the NR/BR composite due to the inclusion of the hydroxymethyl group in the lignin structure and the possibility of a reaction occurring between the hydroxymethyl functional groups of HMKL and the rubber molecules. However, the optimum curing time and scorch time of HMKL-filled NR/BR composites increased significantly with increasing HMKL content. This indicates that high levels of HMKL affect the efficiency of crosslinks in the vulcanizates and delay the vulcanization of rubber compounds [43].

**Table 2.** Curing characteristics and crosslink density of NR/BR/kraft lignin (KL) compared to NR/BR/hydroxymethylated kraft lignin (HMKL) composites.

| Sample | Optimum Cure Time, $t_{90}$ (min) | Scorch Time, $t_{S2}$ (min) | Minimum Torque, $M_L$ (dN·m) | Maximum Torque, $M_H$ (dN·m) | Torque Difference, $\Delta M$ ($M_H$–$M_L$) (dN·m) | Crosslink Density, $V_C$ ($\times 10^{-4}$ mol/m$^3$) |
|---|---|---|---|---|---|---|
| NR/BR/CB50 (without lignin) | 4.57 | 1.76 | 1.45 | 13.12 | 11.67 | 6.5 ± 0.0 |
| NR/BR/KL10 | 6.23 | 2.27 | 1.09 | 9.41 | 8.32 | 5.2 ± 0.2 |
| NR/BR/HMKL5 | 3.54 | 1.89 | 1.33 | 11.78 | 10.45 | 10.3 ± 0.1 |
| NR/BR/HMKL10 | 3.97 | 1.99 | 1.10 | 10.41 | 9.31 | 10.9 ± 0.4 |
| NR/BR/HMKL15 | 4.82 | 2.22 | 1.08 | 9.88 | 8.80 | 6.8 ± 0.1 |
| NR/BR/HMKL20 | 5.49 | 2.36 | 0.91 | 9.03 | 8.12 | 5.0 ± 0.1 |

With increasing HMKL content, the maximum torque, $M_H$, of NR/BR/HMKL displays a reducing trend, but remains higher than that of NR/BR/KL with an HMKL content of up to 15 phr. This shows that rubber compounds containing HMKL formed more crosslinks between the lignin and the rubber matrix compared to rubber compounds containing KL, which restricted the mobility of the rubber chains. Thus, the presence of HMKL enhanced the interfacial adhesion between the filler and the rubber matrix. A similar trend is observed for minimum torque, $M_L$, which is mainly associated with the physical crosslinking between the lignin and the rubber matrix before vulcanization [44]. At low HMKL content, the rubber composite shows a high value of $M_L$, which suggests stronger physical crosslinking between the filler (CB/HMKL) and the rubber matrix. However, the trend of $M_L$ is reduced with increasing HMKL content as the lignin acts as a plasticizer in the rubber compound. It is well known that a relatively low value of $M_L$ leads to better rubber processability [45]. Furthermore, the observed $\Delta M$ indicates a reduction in crosslink density (see $V_C$ values in Table 2). All HMKL-filled NR/BR compounds exhibited a higher $\Delta M$ compared to KL-filled NR/BR compounds, except for the compound with an HMKL content of 20 phr. This is due to the higher crosslink density in HMKL-filled NR/BR compounds and could also be due to the formation of a filler–rubber network between the HMKL and rubber matrix. This will enhance the restriction of the mobility of rubber chains and significantly reinforce the strength of the rubber composite.

*3.4. Tensile Properties*

Figures 6 and 7 show the tensile properties of NR/BR composites containing KL and HMKL. For the samples with HMKL, a decreasing trend of tensile strength and tensile modulus (M100, M300, and reinforcement index, M300/M100) is observed with increasing HMKL content. However, the opposite trend was observed for elongation at break when the HMKL content was increased.

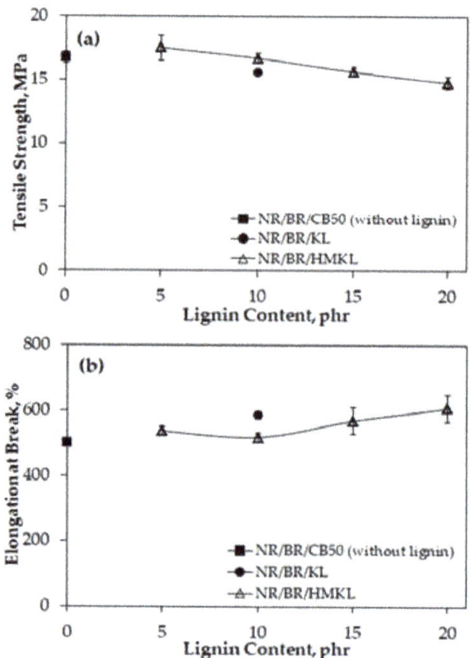


**Figure 6.** The effect of HMKL content on (**a**) tensile strength and (**b**) elongation at break, of lignin-filled NR/BR composites.

At a filler content of 10 phr, the tensile strength of the NR/BR/HMKL composite is slightly higher than that of the NR/BR/KL composite and is comparable to that of the NR/BR composite with CB50 (i.e., without lignin). This indicates that HMKL has a reinforcing effect on NR/BR and slightly enhances the rubber performance. The greater tensile strength of the NR/BR/HMKL composite is also attributed to the compatibility between the lignin and the NR/BR matrix, which can be concluded to have improved the rubber–filler interaction. As a consequence, the filler was able to support the uniform stress transferred from the rubber molecular chains before breakage occurred [46], resulting in improved tensile strength. However, the elongation at break of the HMKL-filled NR/BR composites was lower at low HMKL contents of up to 10 phr. This is due to the rigidity of the network that formed between the HMKL and the rubber matrix, which decreased the flexibility and elasticity of the composite and reduced the deformation resistance of the rubber chains.

The tensile moduli at strains of 100% (M100) and 300% (M300) are shown in Figure 7. For both M100 and M300, the modulus decreases with increasing lignin content. However, at a lignin content of 10 phr, the modulus of the NR/BR/HMKL composite is higher than that of the NR/BR/KL composite. This could be due to the reaction between hydroxymethyl groups and the double bond of NR or BR, which could form a covalent bond. Moreover, the modulus continues to reduce with increasing lignin content at contents above 10 phr. This is due to the reduction of free volume, which leads to a decrease in the flexibility and an increase in the stiffness of the composites. This observation correlates well with the aforementioned high value of ΔM and crosslink density. Additionally, the reinforcement indexes (M300/M100) of the HMKL-filled NR/BR composites were higher than those of the KL-filled NR/BR composites at lignin contents of up to 10 phr (Figure 7c). This could be due to a reduction in the size of the lignin particles, as claimed by Jiang et al. [29] and Popa et al. [30], leading to better dispersion of lignin throughout the rubber matrix. Thus, the reinforcement index of the rubber composite for a lignin content of up to 10 phr is slightly higher than for lignin contents of 15 and 20 phr.

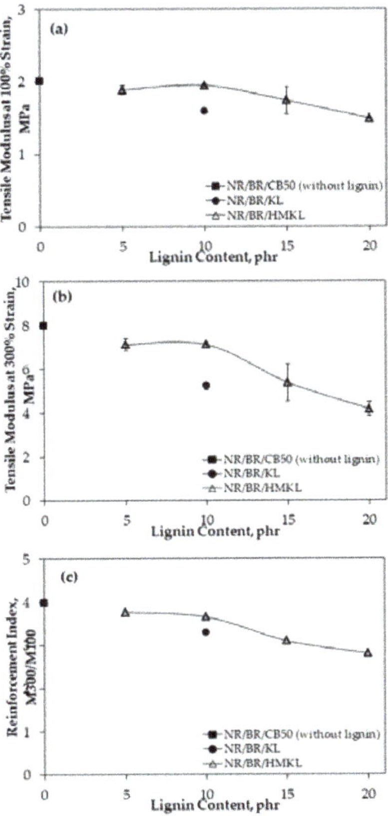

**Figure 7.** The effect of HMKL content on (**a**) the tensile modulus of lignin-filled NR/BR compounds at 100% strain, (**b**) the tensile modulus of lignin-filled NR/BR compounds at 300% strain, and (**c**) the reinforcement index of lignin-filled NR/BR compounds.

A proposed mechanism to elaborate the relationship between experimental results (e.g., tensile strength, crosslink density, and reinforcement index) with hydroxymethylated lignin content is shown in Figure 8. As depicted in the schematic illustration, the gray particles representing the HMKL formed linkages with NR/BR. Thus, the adhesion between the modified lignin and the rubber matrix is improved.

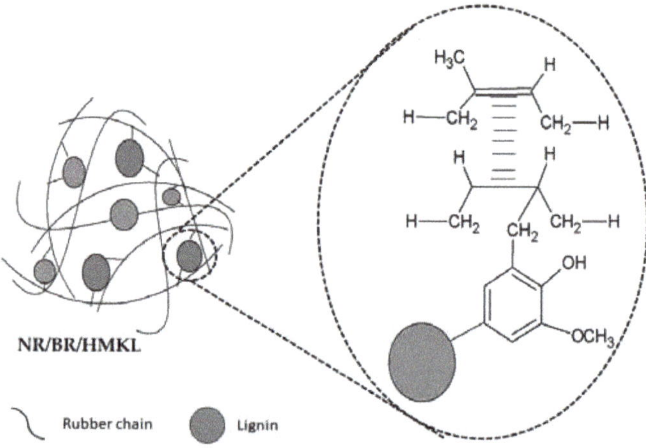


**Figure 8.** Schematic illustration of the proposed reinforcing mechanism of HMKL on NR/BR.

Filler particle size, filler dispersion, and filler–rubber interaction are three important factors that influence the reinforcement of rubber composites. The HMKL particles are uniformly dispersed in the NR/BR matrix. The HMKL particles are expected to react with the double bonds of NR or BR to form a rigid rubber network during the mixing and curing processes [29]. Under stress, the formation of covalent bonds between the NR/BR and HMKL will restrict the slippage of rubber molecules, which leads to the ultimate maximal stress at smaller deformation. However, at higher HMKL content (more than 10 phr), the lignin particles tend to self-aggregate and form agglomerations in the NR/BR matrix due to polarity differences. Ultimately, these agglomerations cannot withstand the external stress, which deteriorates the performance of the rubber compound. Therefore, the increase in the interfacial adhesion between the NR/BR and HMKL at low HMKL contents of up to 10 phr is beneficial for the transfer of external stress from the rubber matrix to the HMKL particles and results in a better reinforcing effect of the NR/BR composite.

*3.5. Compatibility of Lignin-Filled NR/BR*

The SEM images obtained from tension-fractured surfaces of the KL- and HMKL-filled NR/BR compounds are shown in Figure 9. The fractured surfaces of the KL-filled NR/BR composites (see Figure 9a) contained many holes and large amounts of filler agglomerate. This may be due to the incompatibility and poor interfacial adhesion between KL and the rubber matrix. Meanwhile, as shown in Figure 9b, the HMKL-filled NR/BR composites with an HMKL content of 10 phr had a homogeneous filler distribution due to the interaction between hydroxymethyl groups of the HMKL surface and the rubber molecules. Additionally, the morphology of the surface might have generated more shear force, leading to the break-up of the filler agglomerate. The use of more HMKL resulted in a rough surface with large holes due to the detachment of lignin from the fractured surface. This proves that the presence of low levels of HMKL improved interfacial bonding and filler dispersion.

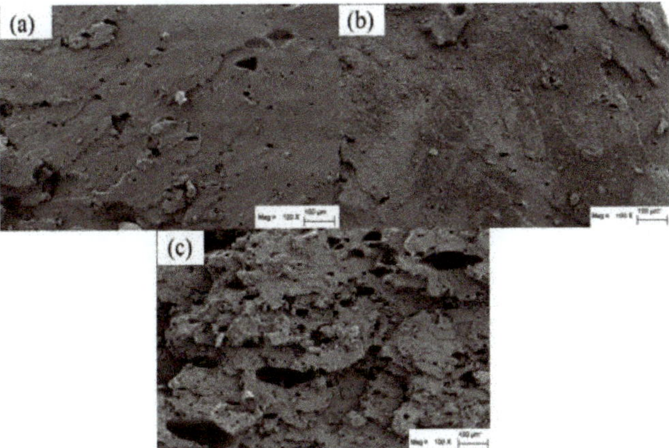

**Figure 9.** Scanning electronic microscopy (SEM) images of (**a**) NR/BR/KL10 at a lignin content of 10 phr, (**b**) NR/BR/HMKL at a lignin content of 10 phr, and (**c**) NR/BR/HMKL at a lignin content of 20 phr.

## 3.6. Hardness

The hardness of a rubber composite refers to the ability of the surface of the rubber to resist the penetration of an indenter and is related to the deformation of the rubber surface [47]. Figure 10 shows the hardness of the NR/BR composites containing HMKL for different HMKL contents. As expected, at low lignin contents of up to 10 phr, the HMKL-filled NR/BR composites have slightly higher hardness values compared to the KL-filled NR/BR composites. At the same lignin content, HMKL-filled NR/BR composites have higher stiffness compared to NR/BR/KL and NR/BR/CB50 (without lignin) composites. This is due to the formation of crosslinking between the HMKL and the rubber chains. However, a reduction in stiffness is observed with increasing lignin content, which is related to the reduction of crosslink density. This phenomenon is due to the adsorption of zinc complexes on the surface of the modified lignin, which disrupts the efficiency of sulfur vulcanization [48].

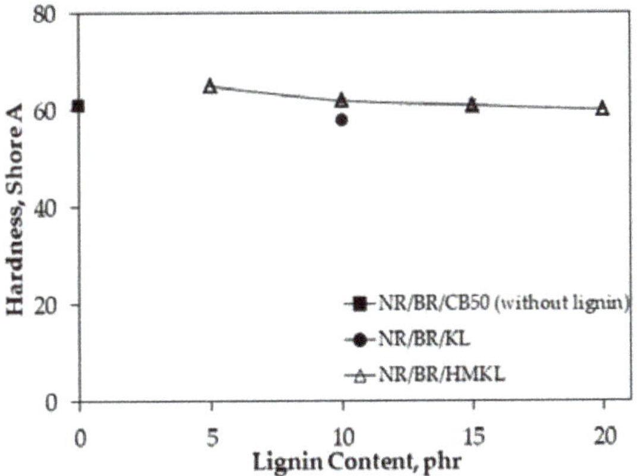

**Figure 10.** The effect of HMKL content on the hardness of lignin-filled NR/BR compounds.

## 3.7. Compression Set

Compression set tests were performed by applying stress to the NR/BR composites at certain times to determine their ability to retain their elastic properties with respect to the lignin content. The lower the compression set, the higher the ability of rubber vulcanizates to retain their elasticity.

Figure 11 displays the value of the compression set (%) for NR/BR composites containing different HMKL contents.

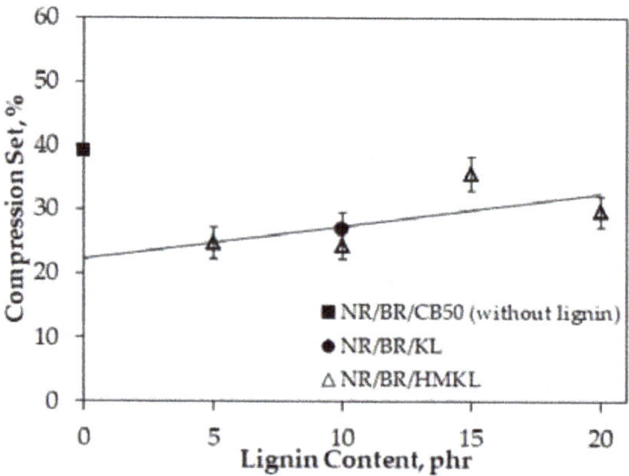

**Figure 11.** The effect of HMKL content on the compression set of lignin-filled NR/BR compounds as obtained from compression tests.

The compression set of NR/BR/HMKL composites slightly decrease at low lignin contents of up to 10 phr. This might be due to the high crosslink density, as discussed earlier [49,50]. The decrease of the compression of NR/BR/HMKL composites is also correlated to the HMKL-filled NR/BR composites, which have high values of torque, crosslink density, tensile modulus, and hardness compared to NR/BR/KL composites. This behavior is also due to the presence of hydroxymethyl functional groups in the HMKL structure, which promotes filler–rubber interaction and the formation of crosslinks, whereby it can prevent the recovery of the rubber molecular chains. However, when the HMKL content is more than 10 phr, the compression value continuously increases with increasing HMKL content. This increase is probably due to post-curing reactions which increase the tensile modulus and reducing the mobility of rubber chains.

### 3.8. Flexing Resistance

The flexing resistance of NR/BR composites for different HMKL contents is shown in Figure 12. The results show a significant decrease in the number of cycles needed for failure as the HMKL content increases. The De Mattia flexing machine is operated at a high strain level rather than at a high stress level [51,52]. Therefore, the outcomes of the flexing resistance tests are related to the tensile modulus of the rubber composites. The larger the modulus, the shorter the total flexure lifetime. Moreover, a high modulus also indicates a better per-cycle energy input. This is due to the fact that the growth of flexure cracks is an energy-dependent process resulting in poor flexure life. Furthermore, the reduction of rubber-chain flexibility increases the stiffness of rubber composites containing HMKL filler. This also results in the reduction of flexing resistance. The other assumption is the coiling of rubber chains onto the surface of the filler particles or the formation of lignin–rubber agglomerates, which result in reduced flexing resistance [53]. Thus, failure begins at the position of a lignin–rubber agglomeration, which breaks, leading to crack formation. This leads to failure when further crack propagation processes occur. However, the flexing process is very complex, since it also involves the mechano-oxidative aging process [21,54,55]. The lignin could act as an antioxidant which could partially avoid the harmful effects of flexing.

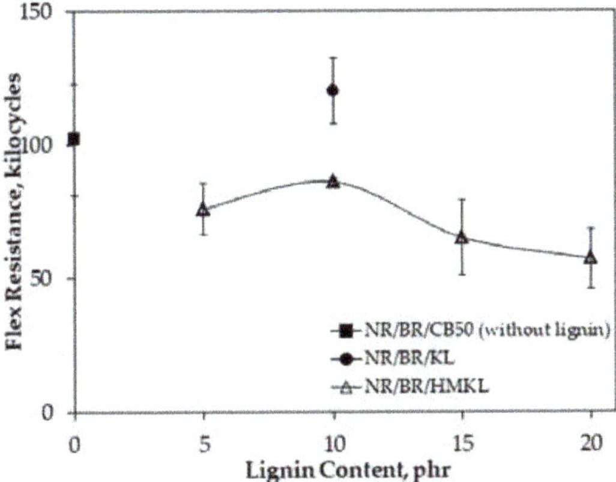

**Figure 12.** The effect of HMKL content on the flexing resistance of lignin-filled NR/BR compounds.

*3.9. Heat Build-Up*

The heat energy dissipated due to the friction between the filler particles and the rubber matrix under repeated cycles of deformation and recovery is known as heat build-up [21]. The hysteresis that generates heat build-up within rubber can lead to failure and reduce the durability of the material. Figure 13 displays the heat build-up of the NR/BR composites containing HMKL and KL.

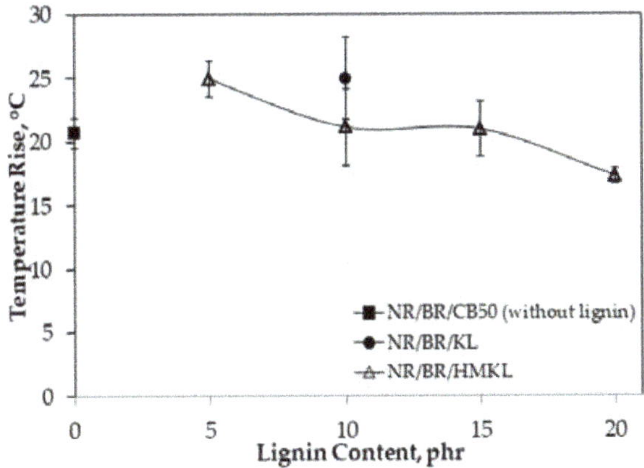

**Figure 13.** The effect of HMKL content on the heat build-up in lignin-filled NR/BR compounds.

The results show that the heat build-up reduces steadily with increasing lignin content. The heat build-up of composites with HMKL is lower than for composites with KL, except for composites with an HMKL content of 5 phr. For HMKL contents of more than 5 phr, the observed decreasing trend of heat build-up (°C) with increasing lignin content reflects good hysteresis properties. This is possibly due to the flexibility of the chain backbones of rubber composites, which correlated well with elongation at break value [48,56]. Gusev [57] reported that interfacial phenomena, rather than the filler network, were responsible for the dissipation of heat energy in rubber. This is due to the high friction between the lignin and the rubber matrix when the lignin content is increased. However, most studies of heat generation report that rubber composites with low heat build-up have low hardness [58]. Thus, the results show that as HMKL content increases, the heat build-up in NR/BR composites reduced.

### 3.10. Thermal Stability

#### 3.10.1. Aging Resistance

The aging of NR occurs due to the impact of oxygen, heat, and stress during processing, storage, and service. The unsaturated backbone of NR is easily broken down due to environmental stresses, such as attack by oxygen molecules, especially at high temperature. This phenomenon leads to the chain scission of long molecular chains and reduces the molecular weight of rubber, which leads to less entanglement. Therefore, the ability of the rubber matrix to transfer stress is reduced, which leads to the deterioration of mechanical properties [59,60].

A variety of antioxidants were used to prevent the oxidative deterioration of the rubber composites, as well as boost their heat stability. Figure 14 displays the percentage of aging retention of NR/BR composites containing KL and HMKL. The composites were exposed to hot air conditions for 72 h, and their retention of tensile properties was then calculated. Lignin that contains many hindered phenolic hydroxyl and methoxy groups can form a special chemical structure which is able to capture free radicals and terminate the chain reaction [60–62]. Thus, it can improve the aging resistance of rubber.

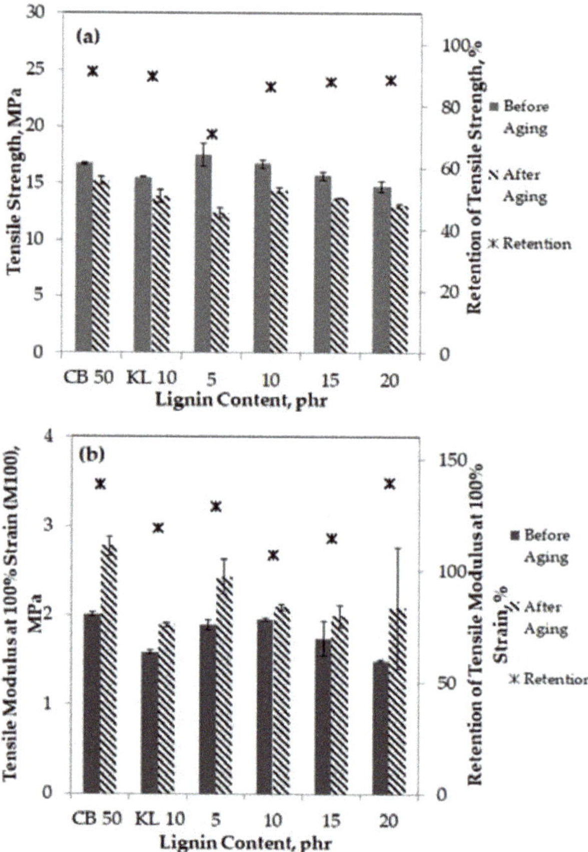

**Figure 14.** The effect of HMKL content on the retention of (**a**) tensile strength and (**b**) tensile modulus at 100% strain in lignin-filled NR/BR compounds.

As shown in Figure 14a, the tensile strengths of NR/BR/KL and HMKL-filled NR/BR with an HMKL content of 5 phr are lower than that of NR/BR with CB50. When more than 5 phr of HMKL is incorporated into the NR/BR matrix, the tensile strength increases slightly, to almost the same value as the KL-filled NR/BR composites. The percentage of tensile strength retention of HMKL-filled NR/BR composites with HMKL contents of 10–20 phr shows almost the same value as that of the KL-filled NR/BR composites. This means that less degradation occurred in HMKL-filled NR/BR composites with higher HMKL contents. It is likely that the unchanged basic structure of lignin after modification with highly branched phenolic groups prevents the rubber matrix from atmospheric degradation [44,63,64].

The increase in the retention values for M100 is shown in Figure 14b. The maximum measured value is more than 110%. This might be due to the effect of chain scission and crosslinking reactions. A macro-radical reaction occurred with the unsaturated double bond (C=C) in the rubber after hot air aging, which increased the stiffness and hardness of the rubber due to further crosslinking [65] during the post-curing process. This phenomenon might contribute to increasing brittleness, and hence, the damping of rubber composites could suffer from increased brittleness. However, as also shown in Figure 14b, the retention value of M100 is lower than that of NR/BR composites with CB50 (without lignin). Therefore, HMKL-filled NR/BR is a more stable compound during the service life under exposure to heat compared to compounds with CB50 (without lignin).

### 3.10.2. Thermogravimetric Analysis (TGA)

The thermal characteristics (thermogram, TG, and derivative weight loss, DTG) measured for KL-filled NR/BR and HMKL-filled NR/BR are shown in Figure 15.

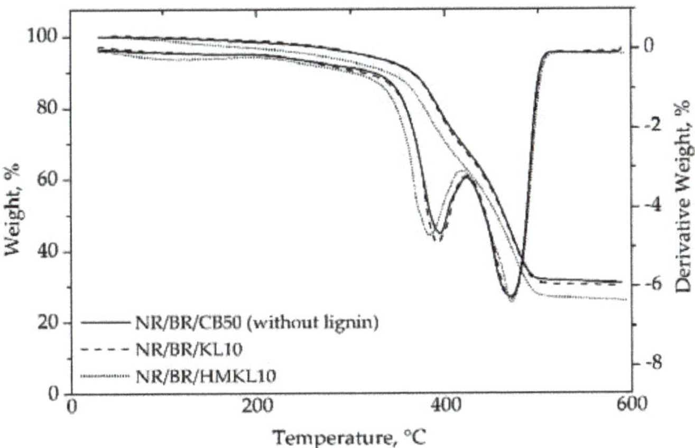


**Figure 15.** The effect of HMKL content on the thermal stability, thermogravimetric analysis (TGA), and derivative weight loss (DTG) of lignin-filled NR/BR compounds.

The thermogram curves in Figure 15 show the degradation of the rubber composites, which involved the degradation of the main chain of the rubber matrix via scission of the C–C bonds. Two steps of degradation can be observed in the thermogram curves. The weight loss of the rubber samples started at temperatures of around 180–200 °C. This is associated with volatile components, i.e., stearic acid, as well as the adsorption of water at around 300 °C. The first degradation started at 350 °C and terminated at approximately 400 °C. In this phase, the natural rubber segment (polyisoprene) was degraded, as indicated by the presence of a small peak in the DTG curve [66].

Natural rubber degradation is sensitive to the presence of an oxidized structure and the depletion of sulphidic crosslinks. However, the thermogram in Figure 15 shows that the DTG peak ($T_{max}$) for NR/BR/HMKL was shifted towards a lower temperature, i.e., from 392 °C to 383 °C for NR/BR/KL and NR/BR/HMKL compounds, respectively. This shows there was a slight reduction in the thermal stability of the NR/BR composite. Calco-Flores et al. [66] reported that unmodified lignin is extremely effective at protecting rubber matrix from oxidation and high temperature due to the stabilizing effect of lignin's hindered phenolic hydroxyl groups towards reactions induced by oxygen and its radical species. However, the modification of lignin slightly minimized the oxidative degradation of the rubber. Additionally, it reduced the radical scavenging efficacy [40] of the attached hydroxymethylated groups at the free reactive site (C5) of the phenolic structure and secondary aliphatic groups.

The second phase of decomposition occurred between 450 and 550 °C. This phase (shown by a major peak in the DTG curve) was due to the cleavage of cross-linked BR and conjugated polyene left after the first phase of degradation [67]. During this phase, a similarly large DTG peak ($T_{max}$) is visible, which indicates that the thermal stability was slightly higher at high temperature. Meanwhile, there was more char residue from the compounds with KL than from the compounds containing HMKL. This shows that the compounds with the unmodified lignin were more stable than HMKL-filled compounds.

## 4. Conclusions

In this study, a detailed characterization of structural kraft lignin, both unmodified and modified by hydroxymethylation treatment, was accomplished. The results show that the hydroxymethylation treatment formed methylene bridges which bound together with the macromolecular lignin. It can be concluded that the inclusion of HMKL in rubber composites weakened the filler–filler interaction and improved the rubber's processability. Furthermore, the compatibility and high interfacial adhesion between the HMKL and the rubber matrix increased the cure rate of the rubber compound, and hence, increased the crosslink density compared to KL-filled NR/BR composites. A consequent improvement of the mechanical properties of the rubber composite at low HMKL contents of up to 10 phr was clearly observed. However, the aging resistance and thermal stability did not alter significantly for HMKL-filled NR/BRs as compared to the KL-filled NR/BRs. As a result, these properties influenced the flexing resistances of NR/BR composites containing HMKL, whose flexing resistances were lower than those of the NR/BR/KL composites. Replacing carbon black with HMKL at a content of up to 10 phr improves the rheological, curing, and mechanical performance of rubber composites. However, it slightly lowers their thermal stability and flexing resistance.

**Author Contributions:** Conceptualization, N.A.M.A. and N.O.; methodology, N.A.M.A., N.O., and M.H.H.; Project administration, N.O. and N.A.M.A.; formal analysis, N.A.M.A., N.O., K.S., and N.H.; investigation, N.A.M.A.; validation, N.O., K.S., N.H., and M.H.H.; resources, N.O., M.H.H., and K.S.; data curation, N.A.M.A., N.O., and K.S.; writing—review and editing, N.A.M.A., N.O., K.S., N.H., and M.H.H.; visualization, N.A.M.A.; supervision, N.O., K.S., and N.H.; funding acquisition, N.O. and M.H.H.

**Funding:** This research was funded by the Fundamental Research Grant Scheme, grant number 203.PBAHAN.6071350 and by USM Bridging Grants, grant number 304.PKIMIA.6316041.

**Acknowledgments:** The authors acknowledge the support of the School of Materials and Mineral Resources Engineering (SMMRE), Universiti Sains Malaysia (USM), and Prince of Songkla University, Pattani Campus (PSU). One of the authors, Nor Aniza Mohamad Aini (N.A.M.A.) is thankful to the Ministry of High Education (MoHE) of Malaysia for the MyPhD scholarship and Postgraduate Research Attachment (PGRA) Fund, Institute of Postgraduate Studies, USM.



## References

1. Ismail, H.; Ramly, F.; Othman, N. Multiwall carbon nanotube-filled natural rubber: The effects of filler loading and mixing method. *Polym. Plast. Technol. Eng.* **2010**, *49*, 260–266. [CrossRef]
2. Xu, S.H.; Gu, J.; Luo, Y.F.; Jia, D.M. Effects of partial replacement of silica with surface modified nanocrystalline cellulose on properties of natural rubber nanocomposites. *Express Polym. Lett.* **2012**, *6*, 14–25. [CrossRef]
3. Cummings, S.; Zhang, Y.; Smeets, N.; Cunningham, M.; Dubé, M.A. On the use of starch in emulsion polymerizations. *Processes* **2019**, *7*, 140. [CrossRef]

4.  Rattanasom, N.; Prasertsri, S. Mechanical properties, gas permeability and cut growth behavior of natural rubber vulcanizates: Influence of clay types and clay/carbon black ratios. *Polym. Test.* **2012**, *31*, 645–653. [CrossRef]

5.  Abraham, E.; Deepa, B.; Pothan, L.A.; John, M.; Narine, S.S.; Thomas, S.; Anandjiwala, R. Physicomechanical properties of nanocomposites based on cellulose nanofibre and natural rubber latex. *Cellullose* **2013**, *20*, 417–427. [CrossRef]

6.  Rajisha, K.R.; Maria, H.J.; Pothan, L.A.; Ahmad, Z.; Thomas, S. Preparation and characterization of potato starch nanocrystal reinforced natural rubber nanocomposites. *Int. J. Biol. Macromol.* **2014**, *67*, 147–153. [CrossRef]

7.  Jong, L. Modulus enhancement of natural rubber through the dispersion size reduction of protein/fiber aggregates. *Ind. Crops Prod.* **2014**, *55*, 25–32. [CrossRef]

8.  Jong, L. Influence of protein hydrolysis on the mechanical properties of natural rubber composites reinforced with soy protein particles. *Ind. Crops Prod.* **2015**, *65*, 102–109. [CrossRef]

9.  Gosselink, R.J.A.; Abächerli, A.; Semke, H.; Malherbe, R.; Käuper, P.; Nadif, A.; van Dam, J.E.G. Analytical protocols for characterization of sulphur-free lignin. *Ind. Crops Prod.* **2004**, *19*, 271–281. [CrossRef]

10. Saito, K.; Kato, T.; Tsuiji, Y.; Fukushima, K. Identifying the characteristic secondary ions of lignin polymer using ToF-SIMS. *Biomacromolecules* **2005**, *6*, 678–683. [CrossRef] [PubMed]

11. Lora, J.H.; Glasser, W.G. Recent industrial applications of lignin: A sustainable alternative to nonrenewable materials. *J. Polym. Environ.* **2002**, *10*, 39–48. [CrossRef]

12. Abejón, R.; Pérez-Acebo, H.; Clavijo, L. Alternatives for chemical and biochemical lignin volarization: Hot topics from a bibliometric analysis of the research published during the 2000–2016 period. *Processes* **2018**, *6*, 98. [CrossRef]

13. Abejón, R.; Rabadán, J.; Lanza, S.; Abejón, A.; Garea, A.; Irabien, A. Supported ionic liquid membranes for separation of lignin aqueous solutions. *Processes* **2018**, *6*, 143. [CrossRef]

14. Deka, H.; Mohanty, A.; Misra, M. Renewable-resource-based green blends from poly(furfuryl alcohol) bioresin and lignin. *Macromol. Mater. Eng.* **2014**, *299*, 552–559. [CrossRef]

15. Sahoo, S.; Misra, M.; Mohanty, A.K. Biocomposites from switchgrass and lignin hybrid and poly(butylene succinate) bioplastic: Studies on reactive compatibilization and performance evaluation. *Macromol. Mater. Eng.* **2014**, *299*, 178–189. [CrossRef]

16. Miao, C.; Hamad, W.Y. Controlling lignin particle size for polymer blend applications. *J. Appl. Polym. Sci.* **2017**, *134*. [CrossRef]

17. Ten, E.; Vermerris, W. Recent developments in polymers derived from industrial lignin. *J. Appl. Polym. Sci.* **2015**, *132*. [CrossRef]

18. Bahl, K.; Jana, S.C. Surface modification of lignosulfonates for reinforcement of styrene-butadiene rubber compounds. *J. Appl. Polym. Sci.* **2014**, *131*. [CrossRef]

19. Frigerio, P.; Zoia, L.; Orlandi, M.; Hanel, T.; Castellani, L. Application of sulphur-free lignins as a filler for elastomers: Effect of hexamethylenetetramine treatment. *BioResources* **2014**, *9*, 1387–1400. [CrossRef]

20. Bahl, K.; Miyoshi, T.; Jana, C. Hybrid fillers of lignin and carbon black for lowering of viscoelastic loss in rubber compounds. *Polymer* **2014**, *55*, 3825–3835. [CrossRef]

21. Yu, P.; He, H.; Jia, Y.; Tian, S.; Chen, J.; Jia, D.; Luo, Y. A comprehensive study on lignin as a green alternative of silica in natural rubber composites. *Polym. Test.* **2016**, *54*, 176–185. [CrossRef]

22. Cao, X.V.; Ismail, H.; Rashid, A.A.; Takeichi, T.; Vo-Huu, T. Maleated natural rubber as a coupling agent for recycled high density polyethylene/natural rubber/kenaf powder biocomposites. *Polym. Plast. Technol. Eng.* **2012**, *51*. [CrossRef]

23. Jiang, C.; He, H.; Yu, P.; Wang, D.K.; Zhou, L.; Jia, D.M. Plane-interface-induced lignin based nanosheets and its reinforcing effect on styrene-butadiene rubber. *Express Polym. Lett.* **2014**, *8*, 619–634. [CrossRef]

24. Xiao, S.; Feng, J.; Zhu, J.; Wang, X.; Yi, C.; Su, S. Preparation and characterization of lignin-layered double hydroxide/styrene-butadiene rubber composites. *J. Appl. Polym. Sci.* **2013**, *130*, 1308–1312. [CrossRef]

25. Rowell, R.M. Acetylation of wood. *For. Prod. J.* **2006**, *56*, 4–12.

26. Rowell, R.M. Chemical modification of wood: A short review. *Wood Mater. Sci. Eng.* **2006**, *1*, 29–33. [CrossRef]

27. Siochi, E.J.; Ward, T.C.; Haney, M.A.; Mahn, B. The absolute molecular weight distribution of hydroxypropylated lignins. *Macromolecules* **1990**, *23*, 1420–1429. [CrossRef]

28. Zhang, T.; Zhou, Y.; Liu, D.; Petrus, L. Qualitative analysis of products formed during the acid catalyzed liquefaction of bagasse in ethylene glycol. *Bioresour. Technol.* **2007**, *98*, 1454–1459. [CrossRef] [PubMed]
29. Jiang, C.; He, H.; Yao, X.; Yu, P.; Zhou, L.; Jia, D. The aggregation structure regulation of lignin by chemical modification and its effect on the property of lignin/styrene butadiene rubber composites. *J. Appl. Polym. Sci.* **2018**, *135*. [CrossRef]
30. Popa, V.I.; Căpraru, A.M.; Grama, S.; Mǎluţan, T. Nanoparticles based on modified lignins with biocide properties. *Cellullose Chem. Technol.* **2011**, *45*, 221–226.
31. Castro, D.F.; Suarez, J.C.M.; Nunes, R.C.R.; Visconte, L.L.Y. Influence of mixing procedures and mica addition on properties of NR/BR vulcanizates. I. *J. Appl. Polym. Sci.* **2004**, *94*, 1574–1585. [CrossRef]
32. Pongdong, W.; Nakason, C.; Kummerlöwe, C.; Vennemann, N. Influence of filler from a renewable resource and silane coupling agent on the properties of epoxidized natural rubber vulcanizates. *J. Chem.* **2015**, *2015*. [CrossRef]
33. Benar, P.; Gonçalves, A.R.; Mandelli, D.; Schuchard, U. Eucalyptus organosolv lignins: Study of the hydroxymethylation and use in resols. *Bioresour. Technol.* **1999**, *68*, 11–16. [CrossRef]
34. Malutan, T.; Raluca, N.; Valentin, I. Contribution to the study of hydroxymetylation reaction of alkali lignin. *BioResources* **2007**, *3*, 13–20.
35. Zhao, J.; Xiuwen, W.; Hu, J.; Liu, Q.; Shen, D.; Xiao, R. Thermal degradation of softwood lignin and hardwood lignin by TG-FTIR and Py-GC/MS. *Polym. Degrad. Stabil.* **2014**, *108*, 133–138. [CrossRef]
36. Boeriu, C.; Bravo, G.; Gosselink, D.; Richard, J.A.; van Dam, E.G. Characterisation of structure-dependent functional properties of lignin with infrared spectroscopy. *Ind. Crops Prod.* **2004**, *20*, 205–218. [CrossRef]
37. Tejado, A.; Pĕna, C.; Labidi, J.; Echeverria, J.M.; Mondragon, I.I. Physico-chemical characterization of lignins from different sources for use in phenol-formaldehyde resin synthesis. *Bioresour. Technol.* **2007**, *98*, 1655–1663. [CrossRef]
38. Constant, S.; Basset, C.; Dumas, C.; Renzo, F.D.; Robitzer, M.; Barakat, A.; Quignard, F. Reactive organosolv lignin extraction from wheat straw: Influence of Lewis acid catalysts on structural and chemical properties of lignins. *Ind. Crops Prod.* **2015**, *65*, 180–189. [CrossRef]
39. Thielemans, W.; Can, E.; Morye, S.S.; Wool, R.P. Novel applications of lignin in composite materials. *J. Appl. Polym. Sci.* **2002**, *83*, 323–331. [CrossRef]
40. Košíková, B.; Gregorová, A.; Osvald, A.; Krajčovičová, J. Role of lignin filler in stabilization of natural rubber-based composites. *J. Appl. Polym. Sci.* **2007**, *103*, 1226–1231. [CrossRef]
41. Guerrero, P.P.; Lisperguer, J.; Navarrete, J.; Rodrigue, D. Effect of modified Eucalyptus Nitens lignin on the morphology and thermo-mechanical properties of recycled polystyrene. *BioResources* **2014**, *9*, 6514–6526.
42. Jagadale, S.C.; Chavan, R.P.; Rajkumar, K.; Shinde, D.N.; Patil, C.L. Lignin as plasticizer in nitrile rubber, it's effect on properties. *Int. J. Res. Eng. Appl. Sci.* **2016**, *6*, 78–84.
43. Jiang, C.; He, H.; Jiang, H.; Ma, L.; Jia, D.M. Nano-lignin filled natural rubber composites: Preparation and characterization. *Express Polym. Lett.* **2013**, *7*, 480–493. [CrossRef]
44. Gu, J.; Chen, W.J.; Lin, L.; Luo, Y.F.; Jia, D.M. Effect of nanocrystalline cellulose of the curing characteristics and aging resistance properties of carbon black reinforced natural rubber. *Chin. J. Polym. Sci.* **2013**, *31*, 1382–1393. [CrossRef]
45. Yu, P.; He, H.; Jiang, C.; Wang, D.; Jia, Y.; Zhou, L.; Jia, D.M. Reinforcing styrene butadiene rubber with lignin-novalac epoxy resin networks. *Express Polym. Lett.* **2015**, *9*, 36–48. [CrossRef]
46. Tangudom, P.; Thongsang, S.; Sombatsompop, N. Cure and mechanical properties and abrasive wear behaviour of natural rubber, styrene–butadiene rubber and their blends reinforced with silica hybrid fillers. *Mater. Des.* **2014**, *53*, 856–864. [CrossRef]
47. Maciejewska, M.; Walkiewicz, F.; Zaborski, M. Novel ionic liquids as accelerators for the sulfur vulcanization of butadiene-styrene elastomer composites. *Ind. Eng. Chem. Res.* **2013**, *52*, 8410–8415. [CrossRef]
48. Rattanasom, N.; Prasertsri, S.; Ruangritnumchai, T. Comparison of the mechanical properties at similar hardness level of natural rubber filled with various reinforcing—Fillers. *Polym. Test.* **2009**, *28*, 8–12. [CrossRef]
49. Mostafa, A.; Abouel-Kasem, A.; Bayoumi, M.R.; El-Sebaie, M.G. Effect of carbon black loading on the swelling and compression set behavior of SBR and NBR rubber compounds. *Mater. Des.* **2009**, *30*, 1561–1568. [CrossRef]

50. Goldberg, A.; Lesuer, D.R.; Patt, J. Fracture morphologies of carbon-black loaded SBR subjected to low-cycle, high-stress fatigue. *Rubber Chem. Technol.* **1989**, *62*, 272–287. [CrossRef]
51. Furtado, C.R.G.; Leblanc, J.L.; Nunes, R.C.R. Fatigue resistance of mica-carbon black-styrene butadiene rubber (SBR) compounds. *Eur. Polym. J.* **1999**, *35*, 1319–1325. [CrossRef]
52. Bhowmick, A.K.; Nando, G.B.; Basu, S.; De, S.K. Scanning electron microscopy studies of fractured natural rubber surfaces. *Rubber Chem. Technol.* **1979**, *53*, 327–334. [CrossRef]
53. Mars, W.V.; Fatemi, A. Factors that affect the fatigue life of rubber: A literature survey. *Rubber Chem. Technol.* **2004**, *77*, 391–412. [CrossRef]
54. Cataldo, F. On the ozone protection of polymers having non-conjugated unsaturation. *Polym. Degrad. Stab.* **2001**, *72*, 287–296. [CrossRef]
55. Rattanasom, N.; Thammasiripong, U.; Suchiva, K. Mechanical properties of deproteinized natural rubber in comparison with synthetic *cis*-1,4 polyisoprene vulcanizates: Gum and black-filled vulcanizates. *J. Appl. Polym. Sci.* **2005**, *97*, 1139–1144. [CrossRef]
56. Hao, P.T.; Ismail, H.; Hashim, A.S. Study of two types of styrene butadiene rubber in tire tread compounds. *Polym. Test.* **2001**, *20*, 539–544. [CrossRef]
57. Gusev, A.A. Micromechanical mechanism of reinforcement and losses in filled rubbers. *Macromolecules* **2006**, *39*, 5960–5962. [CrossRef]
58. Kar, K.K.; Bhowmick, A.K. Hysteresis loss in filled rubber vulcanzates and its relationship with heat generation. *J. Appl. Polym. Sci.* **1997**, *64*, 1541–1555. [CrossRef]
59. Kakroodi, A.R.; Sain, M. Lignin in reinforced rubber composites. In *Lignin in Polymer Composites*; Faruk, O., Sain, M., Eds.; Matthew Deans: Oxford, UK, 2016; pp. 195–206.
60. Liu, S.; Cheng, X. Applications of lignin as antioxidant in styrene butadiene rubber composite. *AIP Conf. Proc.* **2010**, *1251*, 344. [CrossRef]
61. De Paoli, M.A.; Furlan, L.T. Lignin for reinforcing rubber. *Quim. Nova* **1983**, *7*, 121–122.
62. Crabtree, J.; Kemp, A.R. Weathering of soft vulcanized rubber. *Ind. Eng. Chem.* **1946**, *38*, 278–295. [CrossRef]
63. Muniandy, K.; Ismail, H.; Othman, N. Studies on natural weathering of rattan powder-filled natural rubber composites. *BioResources* **2012**, *7*, 3999–4011.
64. Choi, S.S.; Han, D.H.; Ko, S.W.; Lee, H.S. Thermal aging behaviors of elemental sulfur-free polyisoprene vulcanizates. *Korean Chem. Soc.* **2005**, *26*, 1853–1855.
65. Seidelt, S.; Müller-Hagedorn, M.; Bockhorn, H. Description of tire pyrolysis by thermal degradation behaviour of main components. *J. Anal. Appl. Pyrolysis* **2006**, *75*, 11–18. [CrossRef]
66. Calvo, F.; Francisco, G.; Dobado, J.A.; Isac-Garcia, J.; Martin-Martínez, F.J. *Lignin and Lignans as Renewable Raw Materials: Chemistry, Technology and Applications*; John Wiley & Sons: Hoboken, NJ, USA, 2015.
67. Brazier, D.W.; Nickel, G.H. Thermoanalytical methods in vulcanizate analysis II. derivative thermogravimetric analysis. *Rubber Chem. Technol.* **1975**, *48*, 661–667. [CrossRef]

*Article*

# Nitroxide-Mediated Copolymerization of Itaconate Esters with Styrene

Sepehr Kardan [1], Omar Garcia Valdez [1,2], Adrien Métafiot [1] and Milan Maric [1,*]

[1] Department of Chemical Engineering, McGill University, Montreal, QC H3A 0C5, Canada; sepehr.kardan@mail.mcgill.ca (S.K.); ogarciavaldez@greencentrecanada.com (O.G.V.); adrien.metafiot2@mcgill.ca (A.M.)

[2] Green Center Canada, 945 Princess Street, Suite 105, Kingston, ON K7L 0E9, Canada

* Correspondence: milan.maric@mcgill.ca; Tel.: +1-514-398-4272

Received: 19 March 2019; Accepted: 27 April 2019; Published: 1 May 2019

**Abstract:** Replacing petro-based materials with renewably sourced ones has clearly been applied to polymers, such as those derived from itaconic acid (IA) and its derivatives. Di-n-butyl itaconate (DBI) was (co)polymerized via nitroxide mediated polymerization (NMP) to impart elastomeric (rubber) properties. Homopolymerization of DBI by NMP was not possible, due to a stable adduct being formed. However, DBI/styrene (S) copolymerization by NMP at various initial molar feed compositions $f_{DBI,0}$ was polymerizable at different reaction temperatures (70–110 °C) in 1,4 dioxane solution. DBI/S copolymerizations largely obeyed first order kinetics for initial DBI compositions of 10% to 80%. Number-average molecular weight ($M_n$) versus conversion for various DBI/S copolymerizations however showed significant deviations from the theoretical $M_n$ as a result of chain transfer reactions (that are more likely to occur at high temperatures) and/or the poor reactivity of DBI via an NMP mechanism. In order to suppress possible intramolecular chain transfer reactions, the copolymerization was performed at 70 °C and for a longer time (72 h) with $f_{DBI,0} = 50\%$–80%, and some slight improvements regarding the dispersity (Đ = 1.3–1.5), chain activity and conversion (~50%) were observed for the less DBI-rich compositions. The statistical copolymers produced showed a depression in $T_g$ relative to poly(styrene) homopolymer, indicating the effect of DBI incorporation.

**Keywords:** nitroxide mediated polymerization; itaconate esters; copolymerization

---

## 1. Introduction

The limited supply of fossil resources has forced the exploration of renewable feedstocks as an alternative route to materials such as polymers [1]. One such feedstock for polymers is itaconic acid (IA), which was first isolated from the pyrolysis of citric acid [2,3] and is now made by fermentation from fungi [3,4]. IA has been used historically in coatings, adhesives, binders and thickeners [1,5–8]. Itaconic acid's high availability, low cost, structural similarity with acrylates and methacrylates, and its dicarboxylic acid functionality have motivated research on the development of polymeric materials from IA and derivatives like dialkyl itaconates or β monoalkyl itaconates [9,10]. Itaconic acid is indeed listed as one of the most promising bio-based feedstocks according to a report from the Biomass Program of the US Department of Energy [9]. The dual functionality of IA is particularly appealing as it makes it possible to polymerize it via free radical mechanisms (Marvel and Shepherd first described it in 1959 [11]), step-wise polymerization mechanisms (such as using ring-opening step-wise polymerization of itaconic anhydride [12–14]), ring-opening metathesis polymerizations (ROMP) [15,16] and acyclic diene metathesis (ADMET) [17].

To impart a wider array of properties, itaconic acid derivatives such as poly(dialkyl itaconate)s have provided flexibility in blends or in statistical or block copolymers. For example, di-n-butyl itaconate (DBI) can impart lower glass transition temperatures in copolymers and consequently similar related dialkyl itaconates have been polymerized via conventional free radical polymerization (FRP) [2,11,18–31] and more recently via reversible de-activation radical polymerization (RDRP), also known as controlled radical polymerization (CRP), specifically reversible addition fragmentation chain transfer polymerization (RAFT) [32,33] and atom transfer radical polymerization (ATRP) [34,35]. In one case, NMP has been reported using TEMPO-based initiators but no molecular weight distributions were provided, ascribed to the styrene/DBI copolymers adsorbing onto the gel permeation chromatography (GPC) columns [36]. Itaconate esters with stiffer substituents like itaconic anhydride have been similarly polymerized by conventional radical polymerization [37–39] and controlled radical polymerizations like RAFT [40] and would be expected to behave similarly to copolymerizations of styrene with maleic anhydride to provide alternating monomer sequences in the chain.

In this report, we present the nitroxide-mediated polymerization (NMP) of DBI/S using the BlocBuilder family of unimolecular initiators to obtain statistical copolymers with enhanced elastomeric properties. These initiators improved upon TEMPO-based initiators in permitting the homopolymerization of acrylates, acrylamides and methacrylates (with a small concentration of controlling co-monomer ~ 1mol%–10 mol%), which was not possible with first-generation nitroxides [41]. NMP has often been overlooked compared to RAFT and ATRP as an RDRP process, as witnessed by the case with the itaconate esters. Unlike the RAFT and ATRP systems, very little post-polymerization work-up is required without removal of transition metal ligands or odorous chain transfer agents [41], although these issues have been enormously reduced recently [42–44]. The advantages associated with NMP thus make it worthwhile to evaluate its ability to polymerize itaconate-based monomers, which should be challenging due to the secondary vinylic bond in its structure. Thus, our goal here is to apply NMP to obtain (co)polymers with active chain ends and controllable molecular weight (Scheme 1). We thus studied the polymerizations as a function of temperature, initial compositions and effect of solvent.

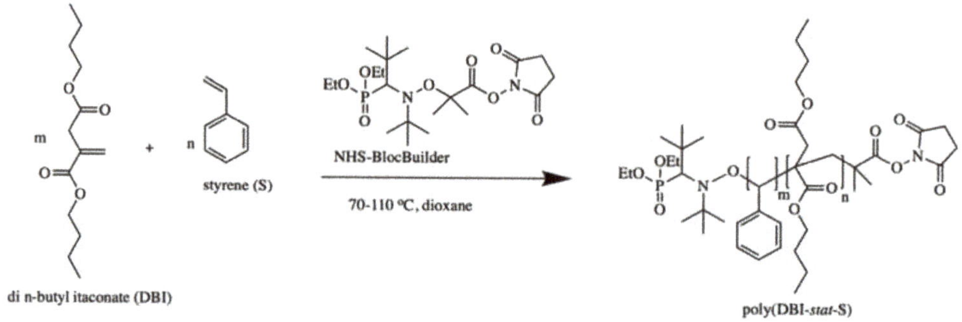

**Scheme 1.** Reaction scheme to copolymerize DBI with S via NMP using NHS-BlocBuilder as initiator.

## 2. Materials and Methods

### 2.1. Materials

$N$-(2-Methylpropyl)-$N$-(1-diethylphosphono-2,2-dimethylpropyl)-$O$-(2-carboxylprop-2-yl) hydroxylamine (99%, BlocBuilder-MA™) was received from Arkema. $N,N'$-Dicyclohexylcarbodiimide (DCC, 99%) was received from Sigma-Aldrich and used in conjunction with BlocBuilder-MA™ to synthesize the succinimidyl ester terminated alkoxyamine NHS-BlocBuilder following a procedure previously reported [45]. Tetrahydrofuran (THF, 99.9% HPLC grade), dioxane (99%), methanol (99%) and 1,4-dioxane (99%), were purchased from Fisher Scientific. Styrene (99%, S) was purified to remove the inhibitor by passing through a column of basic alumina mixed with 5 weight% calcium hydride and then stored in a sealed flask under a head of nitrogen in a refrigerator until needed. Di-n-butyl itaconate (96%, DBI), dimethyl itaconate (96%, DMI), isobutyramide (99%), calcium hydride (90–95%

reagent) and basic alumina (Brockmann, Type 1, 150 mesh) were purchased from Sigma-Aldrich and used as received. Chloroform-D (99.8%) was obtained from Cambridge Isotope Laboratories.

### 2.2. DBI/S Copolymerization

The copolymerizations were performed in a 15 mL three-neck round bottom glass flask equipped with a vertical flux condenser, a thermal well and a magnetic Teflon stir bar. The flask was placed on a heating mantle and the equipment was placed on a magnetic stirrer. The condenser was connected to a chilling unit that used a glycol/water mixture (10/90 *v/v* %) to prevent loss of the monomers and solvent due to evaporation. To the reactor is added the initiator (NHS-BlocBuilder, 0.10 g), 50 wt% solvent (1,4-dioxane), and varying compositions of DBI and S. Once stirring started and the chiller is set to 4 °C, an ultra-pure nitrogen flow was introduced to purge the system for 30 min. An example is used for DBI/S-110-20. In this case, 0.10 g ($2.09 \times 10^{-4}$ mol) of NHS-BlocBuilder was added to 1.59 g DBI (0.066 mol) and 2.73 g (0.0262 mol) of previously purified S along with 5.11 mL of dioxane solvent. A thermocouple was inserted into the temperature well and connected to a controller. The reactor was then heated to the chosen reaction temperature while maintaining the nitrogen purge and stirring with a magnetic stir bar. Table 1 shows the different formulations that were studied. Samples were taken periodically, and once the last sample was taken, 50 mL of methanol was added to the remaining solution to precipitate the polymer. The precipitated polymer was dried overnight in a vacuum oven at 45 °C to remove any remaining solvents or unreacted volatile monomers (styrene). The composition of the copolymer was determined by $^1$H NMR using the methyl end groups of DBI and the aromatic styrene protons (ppm, CDCl$_3$): (t, 0.8–1.05, $CH_3$-CH$_2$-CH$_2$-), (m, 1.2–1.6, backbone and CH$_3$-$CH_2$-CH$_2$-, CH$_3$-CH$_2$-$CH_2$-), (s, 2.7, ≡C-$CH_2$-COO-), (m, 4.2, COO-$CH_2$-CH$_2$-CH$_2$-CH$_3$), (ar, 6.4–7.7, -$C_6H_5$). The molecular weight according to GPC for the specific example was M$_n$ = 12.3 kg·mol$^{-1}$, Đ = 1.37, relative to PS standards in THF at 40 °C.

**Table 1.** Formulations for dibutyl itaconate/styrene (DBI/S) statistical copolymerizations initiated by NHS-BlocBuilder (NHS-BB) at 70–110 °C in 50 wt% 1,4-dioxane solutions.

| Sample ID [a] | $f_{DBI,0}$ [a] | T (°C) | [DBI] (M) | [S] (M) | [NHS-BB] (M) | [Dioxane] (M) |
|---|---|---|---|---|---|---|
| DBI/S-110-10 | 0.10 | 110 | 0.41 | 3.62 | 0.026 | 5.54 |
| DBI/S-110-20 | 0.20 | 110 | 0.73 | 2.91 | 0.023 | 5.58 |
| DBI/S-110-30 | 0.29 | 110 | 0.99 | 2.33 | 0.021 | 5.61 |
| DBI/S-110-40 | 0.40 | 110 | 1.22 | 1.83 | 0.019 | 5.64 |
| DBI/S-110-50 | 0.51 | 110 | 1.41 | 1.41 | 0.018 | 5.66 |
| DBI/S-110-60 | 0.60 | 110 | 1.58 | 1.05 | 0.017 | 5.68 |
| DBI/S-110-70 | 0.71 | 110 | 1.72 | 0.73 | 0.016 | 5.69 |
| DBI/S-110-80 | 0.80 | 110 | 1.84 | 0.46 | 0.015 | 5.71 |
| DB1/S-100-50 | 0.50 | 100 | 1.41 | 1.41 | 0.018 | 5.66 |
| DBI/S-80-50 | 0.50 | 80 | 1.41 | 1.41 | 0.018 | 5.66 |
| DBI/S-70-50 | 0.50 | 70 | 1.41 | 1.41 | 0.018 | 5.66 |
| DBI/S-70-60 | 0.60 | 70 | 1.58 | 1.05 | 0.017 | 5.68 |
| DBI/S-70-70 | 0.70 | 70 | 1.72 | 0.73 | 0.016 | 5.69 |
| DBI/S-70-80 | 0.80 | 70 | 1.84 | 0.46 | 0.015 | 5.71 |
| DBI/S-70-90 | 0.90 | 70 | 1.96 | 0.22 | 0.014 | 5.72 |

[a] Sample ID is defined as: DBI/S-XXX-YY = dibutyl itaconate (DBI)/styrene (S) statistical copolymerization at XXX = temperature (°C) and YY = % molar composition of DBI in initial mixture with molar fraction given as $f_{DBI,0}$.

### 2.3. Chain Extension Experiments

To determine chain end fidelity, chain extension experiments were performed using two macroinitiators, one rich in DBI (DBI/S-110-80; M$_n$ = 2.60 kg mol$^{-1}$, Đ = 1.36) and the other rich in S (DBI/S-110-20, M$_n$ = 12.3 kg mol$^{-1}$, Đ = 1.37). For DBI/S-110-20 as the macroinitiator, typically 1.00 g of macroinitiator was placed inside the reactor with 6.00 g of 1,4 dioxane solvent and 5.00 g of styrene monomer previously purified and mixing started with a magnetic stir bar. The identical reactors conditions were applied as for the copolymerizations described in Section 2.2. After purging with nitrogen at room temperature for 30 min, the temperature was increased to 110 °C to commence the chain extension. The nitrogen purge remained during the rest of the reaction. Samples were periodically taken to assess the molecular weight distribution. At the conclusion of the polymerization after cooling to < 40 °C, the contents were precipitated into 50 mL of methanol. The precipitated polymer was dried overnight in a vacuum oven at 45 °C to remove any remaining solvents or unreacted volatile monomers (styrene). For DBI/S-110-20 product (DBI-S-110-20-*b*-S) the $M_n$ = 22 100 kg mol$^{-1}$, Đ = 1.61. A similar procedure was followed using DBI/S-110-80 as the macroinitiator.

*2.4. Characterization*

Gel permeation chromatography (GPC) was used to obtain molecular weight distributions (MWDs) of the different copolymer samples using HPLC grade THF as the mobile phase. The GPC was calibrated relative to linear PS standards with THF as the eluent at 40 °C. A Waters Breeze GPC system was used at a mobile phase flow rate of 0.3 mL·min$^{-1}$ equipped with three Styragel HR columns (HR1 with a molecular weight measurement range of $10^2$ to $5 \times 10^3$ g mol$^{-1}$, HR2 with a molecular weight measurement range of $5 \times 10^2$ to $2 \times 10^4$ g mol$^{-1}$ and HR4 with a molecular weight measurement range of $5 \times 10^3$ to $6 \times 10^5$ g mol$^{-1}$) and a guard column. The GPC was equipped with an RI 2410 differential refractive index (RI) detector. For these experiments, the RI detector was used. DBI/S conversion and copolymer composition were determined by $^1$H NMR in CDCl$_3$. A 300 MHz Varian Gemini 2000 spectrometer was used for the $^1$H NMR measurements. Samples were placed in 5 mm up NMR tubes using CDCl$_3$ as a solvent. After injecting and shimming, the samples were scanned 32 times. Individual conversions were calculated using the integrated areas at $\delta$ = 6.4–7.7 ppm for the aromatic protons of styrene and $\delta$ = 0.8–1.05 ppm for DBI and taking the reference peaks at 5.3 and 6.1 ppm for the vinyl protons of styrene and DBI, respectively. Once the individual conversion of each monomer was determined, the overall molar conversion was calculated by the following equation: $X_{overall} = X_S\, f_{S,0} + X_{DBI}\, f_{DBI,0}$, where $f_{S,0}$ and $f_{DBI,0}$ are the initial molar fractions of the monomers and $X_S$ and $X_{DBI}$ are the individual monomer conversions determined from $^1$H NMR measurements. Differential scanning calorimetry (DSC) was used to determine the glass transition temperature ($T_g$) of the various copolymers. A cycle of heat/cool/heat with a heating rate of 10 °C min$^{-1}$ was performed on the samples and $T_g$ was determined by observing the change in slope in the heat flow (W g$^{-1}$) versus temperature plot and finding the inflection point using TA Universal Analysis software.

## 3. Results and Discussion

We first attempted to determine the best conditions for the controlled polymerization of dialkyl itaconates, which is defined here as linear progression of degree of polymerization with monomer conversion, dispersity Đ ($M_w/M_n$) < 1.5 and ability to reinitiate a second batch of monomer. Initially, the homopolymerization of DBI via NMP at different temperatures was examined. Interestingly, although DBI was homopolymerized via conventional radical processes and RDRP processes such as RAFT [32,33] and the related dimethyl itaconate (DMI) by ATRP [34], DBI did not homopolymerize via NMP with no conversion after several hours at elevated temperatures of ~110 °C. We also examined the homopolymerization of dimethyl itaconate (DMI) under the same conditions and again observed no conversion after several hours. We suspected this was due to a stable adduct formed by reaction of the alkoxyamine (NHS-BlocBuilder) and one unit of DBI (see Figure 1). The nature of the C-O-N bond between the alkoxyamine and the monomer is very stable since it is centered on a quaternary carbon, and therefore has a high activation energy barrier to produce the propagating radicals required for subsequent polymerization. During the initiation stage of NMP, when the alkoxyamine is activated for the first time, the initiating radical is produced and reacts with the one monomer unit prior to being deactivated by the nitroxide moiety, rendering the adduct.

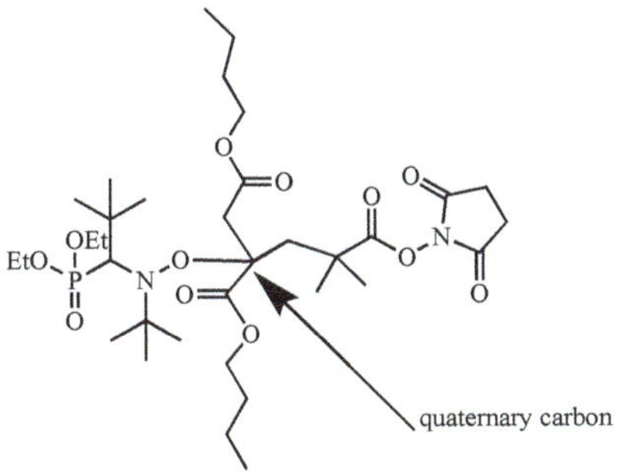

**Figure 1.** Possible adduct formed during the homopolymerization of dibutyl itaconate (DBI) via NMP with NHS-BlocBuilder.

From other radical polymerizations, the stability of this particular radical is reflected in the much lower $k_p$ of dialkyl itaconates compared to styrenic and (meth) acrylic monomers. For example, PLP-SEC measurements revealed the $k_p$ of DBI~40 L mol$^{-1}$ s$^{-1}$ at 110 °C by extrapolation [46], which is much lower, by comparison to styrene, which has a $k_p$~1580 L mol$^{-1}$ s$^{-1}$ at the same temperature [47]. The application of group transfer polymerization (GTP) to dialkyl itaconates revealed only the ability to cap a chain end but no further addition of more monomer units was observed [48,49]. Further, chain transfer to monomer is prevalent in systems with sterically hindered monomers like DBI and were argued to be related to the very low $k_p$ of dialkyl itaconates [50]. In ATRP systems, for DMI with copper halides teamed with various ligands such as methyl 2-bromopropionate (MBrP), p-toluene 2-sulfonyl chloride, pentamethyl diethylenetriamine (PMDETA) and 2,2'bipyridine (bpy) at temperatures of 100 °C and 120 °C, were controlled up to about 50% conversion, with an abrupt decrease in polymerization rates at that juncture [34]. These variations did not enhance polymerization rate or control of the polymerization. Later, Hirano et al. employed ATRP of DBI at 60 °C; higher temperatures resulted in intramolecular chain transfer [25]. This same group continued to try to increase the rate of reaction by using hydrogen bonding and Lewis acids [26,27]. In RAFT systems, a variety of CTAs were applied, and poly(DBI) could be homopolymerized in a controlled fashion at low temperatures of ~20 °C, which likely suppressed many transfer reactions, although rate of polymerization was compromised (e.g., 150 h to obtain 50% conversion) and generally fairly low molecular weights resulted [33]. Thus, our experiments generally matched that observed previously, but NMP of dialkyl itaconates did not result in polymer with appreciable molecular weight, compared to ATRP and RAFT processes. We thus turned our attention to binary copolymerization systems with a monomer that is easily polymerizable by NMP: styrene.

We performed copolymerizations of DBI with S ($f_{DBI,0}$ = 0.1 to 0.8) at 110 °C, as NMP of styrenic-based monomers is well controlled and relatively fast at such temperatures [51]. We attempted to see if the polymerization kinetics we observed would approach those predicted for NMP of model copolymerizations. Semi-logarithmic kinetic plots of the different copolymerizations ($\ln(1/(1 - X))$ versus time) were used to extract the apparent rate constant, $\langle k_p \rangle$[P·] from the slope, where $\langle k_p \rangle$ is the compositionally averaged propagation rate constant and [P·] is the concentration of the active chains. Such a plot would be expected to be linear.

$$ln\left(\frac{1}{1-X}\right) = k_p[P\cdot]t \tag{1}$$

We would expect the propagation rate to vary as a function of the copolymerization feed. The propagation rate constant that we are measuring is actually an average rate constant that is dependent on the individual propagation rate constants and the reactivity ratios as exemplified for a terminal kinetic model (since reactivity ratio data was available for the DBI/S pair) in Equation (2) where $r_1$ and $r_2$ are the reactivity ratios of monomers 1 and 2, $f_1$ and $f_2$ are the respective molar fractions of the monomer mixture and $k_{11}$ and $k_{22}$ refer to the individual homopropagation rate constants [52].

$$\langle k_p \rangle = \frac{r_1 f_1^2 + 2 f_1 f_2 + r_2 f_2^2}{\frac{r_1 f_1}{k_{11}} + \frac{r_2 f_2}{k_{22}}} \tag{2}$$

The steady-state concentration of radical ended chains, is provided by Fischer's expression [53]:

$$[P\cdot] = \left( \frac{\langle K \rangle [I]_0}{3 \langle k_t \rangle} \right)^{1/3} t^{-1/3} \tag{3}$$

where $\langle K \rangle$ is the average equilibrium constant, $[I]_0$ is the initiator concentration, $\langle k_t \rangle$ is the average termination rate constant and time is given by $t$. Further, $\langle K \rangle$ assuming a terminal model, was provided by Charleux and co-workers in Equation (4) [54].

$$\langle K \rangle = \frac{\frac{r_1 f_1}{k_{11}} + \frac{r_2 f_2}{k_{22}}}{\frac{r_1 f_1}{k_{11} K_1} + \frac{r_2 f_2}{k_{22} K_2}} \tag{4}$$

Here, the individual reactivity ratios, monomer molar fractions and individual homopropagation rate constants are defined as above while $K_1$ and $K_2$ are the individual equilibrium constants. The termination rate constant is provided by the following [55]:

$$\langle k_t \rangle = \left( p_1 k_{t,1}^{1/2} + p_2 k_{t,2}^{1/2} \right)^2, \tag{5}$$

where:

$$p_1 = \frac{\frac{r_1 f_1}{k_{11}}}{\frac{r_1 f_1}{k_{11}} + \frac{r_2 f_2}{k_{22}}} \quad \text{and} \quad p_2 = 1 - p_1 \tag{6}$$

For the terminal model, all of the parameters are available, with the exception of the K for DBI, to predict the apparent rate constant $\langle k_p \rangle [P\cdot]$, which can be compared to our experimental values. Thus, reactivity ratios are used from DBI/S conventional radical copolymerizations at 60 °C in benzene ($r_{DBI}$ = 0.38 ± 0.02 and $r_S$ = 0.40 ± 0.05; [19] while the propagation rate constants at 110 °C for DBI and S being $k_{p,DBI}$ = 40 L·mol$^{-1}$·s$^{-1}$ [46] and $k_{p,S}$ = 1580 L·mol$^{-1}$·s$^{-1}$ [47], respectively. The $K_S$ is estimated to be $1.1 \times 10^{-9}$ s$^{-1}$ at 110 °C [51,56] while individual termination rate constants at 110 °C are estimated to be $k_{t,DBI}$ = $1.4 \times 10^6$ L·mol$^{-1}$·s$^{-1}$ [57] and $k_{t,S}$ = $1.5 \times 10^8$ L·mol$^{-1}$·s$^{-1}$ [58]. As $f_{DBI,0}$ is increased, the experimental $\langle k_p \rangle [P\cdot]$ did not vary much. Iterating on $K_{DBI}$ to try to fit the data only resulted in good agreement between the experimental and predicted relationship at high $f_{DBI,0}$ (Figure 2). There are several possible explanations for the disagreement. First, there could indeed be a strong penultimate effect as was suggested by Yee et al. [59]. For DMI/S copolymerizations, the terminal model for the $\langle k_p \rangle$ did not match the experimental data very well. We suspect that a similar effect was occurring in the DBI/S system. Further, Yee et al. found that the terminal group was nearly always DMI up to very high styrene content (~80 mol%) [59]. They found that this particular pair had a strong penultimate unit effect on the copolymerization propagation rate constant and the measured average propagation rate constant was about three times higher than that predicted by the terminal model. In RAFT and ATRP, the same stability of the dialkyl itaconate was noted and further it was stated that the polymerization of DBI or DMI was not a truly reversible deactivation polymerization process and was indeed a mixture between conventional and controlled polymerization [60].

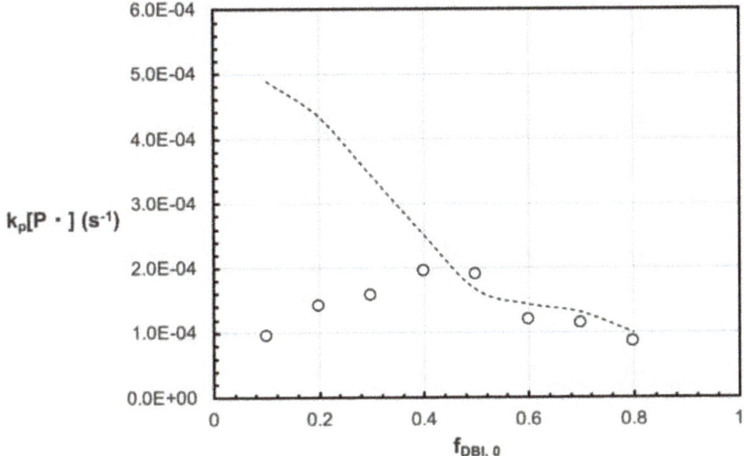

**Figure 2.** Apparent rate constant $k_p[P\cdot]$ as a function of initial molar composition of dibutyl itaconate, $f_{DBI,0}$, for DBI/S copolymerizations at 110 °C in 50 wt% dioxane solutions. The dashed lines indicate the prediction for $k_p[P\cdot]$ based on the terminal model for the copolymerization.

The $M_n$ versus conversion plots are shown in Figure 3 where Figure 3a plots polymerizations with initial feed compositions of $f_{DBI,0} = 0.1$–$0.4$ while Figure 3b plots polymerizations with initial feed compositions of $f_{DBI,0} = 0.5$–$0.8$. For low initial loadings of DBI such as for $f_{DBI,0} = 0.10$ and $0.20$, $M_n$ increased linearly with X. However, experiments with $f_{DBI,0} > 0.20$ (Figure 3b), plateauing of the $M_n$ became increasingly evident. (full data set of $M_n$, $M_w$, individual conversion of the copolymerizations carried out at 110 °C are shown in Table S1 of the Supporting Information).

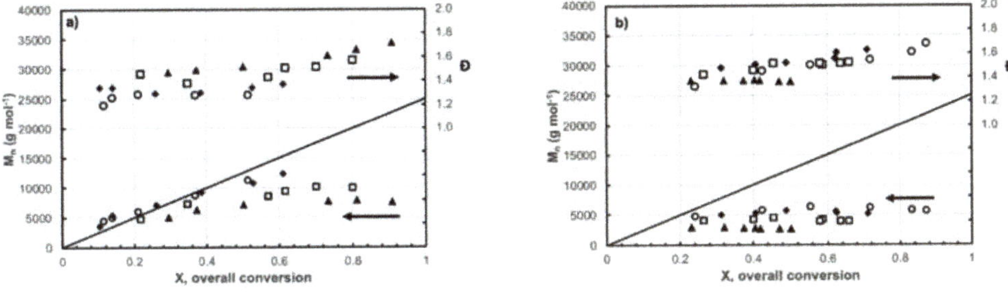

**Figure 3.** Number average molecular weight $M_n$ and dispersity Đ versus overall conversion (X) for DBI/S copolymerization at (**a**) initial molar compositions of DBI $f_{DBI,0} = 0.1$–$0.4$ ($f_{DBI,0} = 0.1$ are open circles ○, $f_{DBI,0} = 0.2$ are solid diamonds ◆, $f_{DBI,0} = 0.3$ are open squares □, and $f_{DBI,0} = 0.4$ are solid triangles ▲) and at (**b**) initial molar compositions of DBI $f_{DBI,0} = 0.5$–$0.8$ ($f_{DBI,0} = 0.5$ are open circles ○, $f_{DBI,0} = 0.6$ are solid diamonds ◆, $f_{DBI,0} = 0.7$ are open squares □, and $f_{DBI,0} = 0.8$ are solid triangles ▲).

One of the reasons for this behavior can be attributed to the high affinity of the itaconate monomer to be involved in chain transfer reactions [18,25]. For a DBI conventional free radical polymerization, the chain-transfer reaction occurs when the propagating radical abstracts a hydrogen atom from a monomer unit, yielding an end-saturated polymer chain and a new monomeric radical [18]. Later, Hirano and Takayoshi suggested that at temperatures higher than 60 °C, intramolecular chain-transfer reactions are favored [25]. They argued that this happens due to the formation of a less stable secondary radical from a more stable tertiary radical, which requires a higher activation energy than the propagating reaction [25]. In an NMP process, intramolecular chain-transfer reactions might occur when the chains are active. Hirano and Takayoshi also suggested that chain-transfer reactions can be suppressed by a reversible deactivation process such as ATRP [25]. However, in the NMP process, the reactivation of the polymer chains depends on the homolysis capability of the C-O-N bond formed between the alkoxyamine and the DBI monomer unit as previously discussed. Together with the irreversible chain termination reactions that occur inevitably, the chain transfer/termination becomes more prominent at higher DBI initial compositions. As illustrated in Figure 3b, when $f_{DBI,0} > 0.5$, the $M_n$ is almost constant at any conversion and as $f_{DBI,0}$ the plateau $M_n$ decreases (as these reactions happen more frequently and earlier). These same trends are shown in the GPC chromatograms in Figure 4. At $f_{DBI,0} < 0.3$, the molecular weight distributions shift steadily to higher molecular weight while propagation is essentially stopped for $f_{DBI,0} > 0.6$ with only a few monomeric units added on as the $M_n$ listed in Table 2 (from GPC relative to PS standards) indicate.

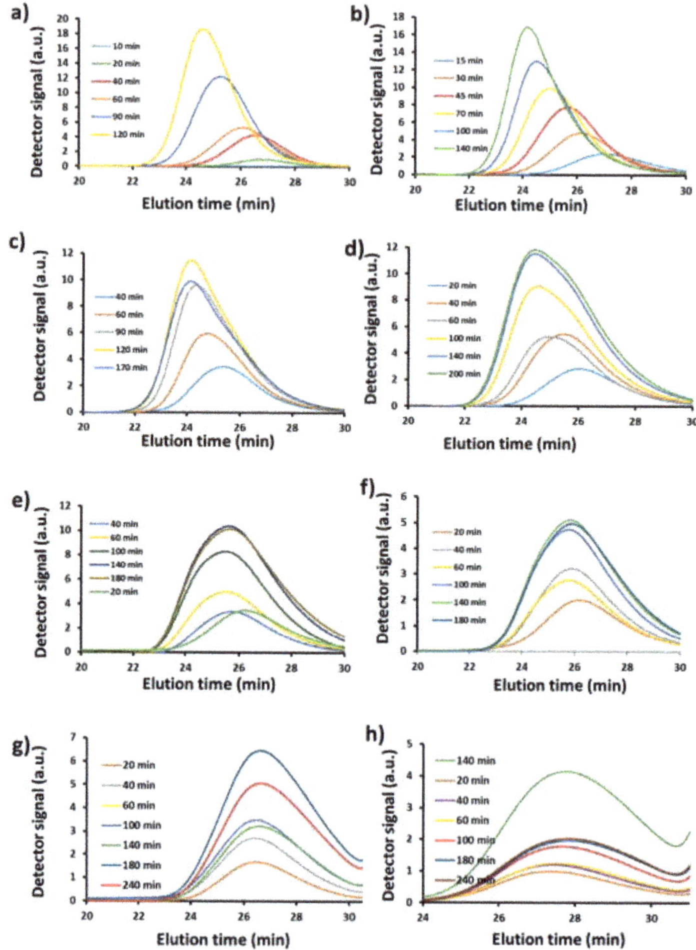

**Figure 4.** GPC traces (elution time) of DBI-S copolymerizations done at done at 110 °C.: (**a**) $f_{DBI,0} = 0.1$, (**b**) $f_{DBI,0} = 0.2$, (**c**) $f_{DBI,0} = 0.3$, (**d**) $f_{DBI,0} = 0.4$, (**e**) $f_{DBI,0} = 0.5$, (**f**) $f_{DBI,0} = 0.6$, (**g**) $f_{DBI,0} = 0.7$, (**h**) $f_{DBI,0} = 0.8$.

**Table 2.** Summary of dibutyl itaconate/styrene copolymerizations in 50 wt% dioxane solution at 110 °C.

| Sample ID [a] | $f_{DBI,0}$ | Time (min) | X [b] | $F_{DBI}$ [c] | $M_n$ (kg mol$^{-1}$) [d] | Đ [d] |
|---|---|---|---|---|---|---|
| DBI/S-110-10 | 0.10 | 120 | 0.51 | 0.13 | 11.2 | 1.28 |
| DBI/S-110-20 | 0.20 | 140 | 0.61 | 0.25 | 12.3 | 1.37 |
| DBI/S-110-30 | 0.29 | 170 | 0.80 | 0.30 | 9.9 | 1.57 |
| DBI/S-110-40 | 0.40 | 200 | 0.91 | 0.40 | 7.6 | 1.72 |
| DBI/S-110-50 | 0.51 | 180 | 0.87 | 0.45 | 5.5 | 1.68 |
| DBI/S-110-60 | 0.60 | 180 | 0.71 | 0.50 | 5.0 | 1.62 |
| DBI/S-110-70 | 0.71 | 230 | 0.75 | 0.60 | 3.7 | 1.52 |
| DBI/S-110-80 | 0.80 | 240 | 0.50 | 0.65 | 2.6 | 1.36 |

[a] Sample ID is defined as: DBI/S-XXX-YY = dibutyl itaconate (DBI)/styrene (S) statistical copolymerization at XXX = temperature (°C) and YY = % molar composition of DBI in initial mixture with molar fraction given as $f_{DBI,0}$. [b] X = overall molar conversion. [c] The final copolymer composition with respect to DBI is given as $F_{DBI}$ and determined by $^1$H NMR. [d] Number average molecular weight $M_n$ and dispersity Đ determined by GPC relative to poly(styrene) standards in THF at 40 °C.

With the apparent loss of control with copolymerizations richer in DBI, we carried out a study of copolymerizations of DBI/S (50/50 mol%) at lower temperatures of 100, 80 and 70 °C and compared the results observed at the same composition at 110 °C in order to determine if lower temperature might reduce the possible intramolecular chain transfer reactions. Figure 5 shows the $M_n$ versus overall conversion plots for the DBI/S copolymerizations ($f_{DBI,0}$ = 0.5) done at 110, 100, 80 and 70 °C. Table 3 also summarizes the various properties for these equimolar copolymerizations. Indeed, temperature is a factor that affects the overall conversion and it was therefore necessary to run reactions for longer at lower temperatures. For example, conversion was close to 90% in only 180 min at 110 °C but at 70 °C, overall conversion was only 47% after three days. There seemed to be only slight improvement, but the molecular weight did not plateau as early as with polymerizations done at higher temperatures. This observation is supported by the GPC traces in Figure 6, which shows the polymerization at 70 °C continually growing whereas at 110 °C, the polymerization was effectively stopped.

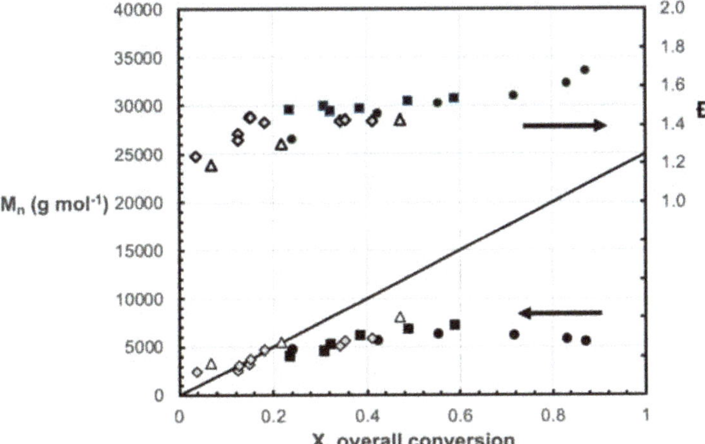


**Figure 5.** Number average molecular weight $M_n$ and dispersity Đ versus overall conversion (X) for DBI/S copolymerizations at composition $f_{DBI,0}$ = 0.5 at various temperatures: at (**a**) 110 °C (filled circles ●), (**b**) 100 °C (filled squares ■), (**c**) 80 °C (open diamonds ◇) and (**d**) 70 °C (open triangles △).

**Table 3.** Summary of equimolar dibutyl itaconate/styrene copolymerizations in 50 wt% dioxane solution at various temperatures.

| Sample ID [a] | $f_{DBI,0}$ | Time (min) | X [b] | $M_n$ (kg mol$^{-1}$) [c] | Đ [c] |
|---|---|---|---|---|---|
| DBI/S-110-50 | 0.51 | 180 | 0.87 | 5.5 | 1.68 |
| DBI/S-100-50 | 0.50 | 180 | 0.59 | 7.1 | 1.53 |
| DBI/S-80-50 | 0.50 | 340 | 0.41 | 5.8 | 1.42 |
| DBI/S-70-50 | 0.50 | 4320 | 0.47 | 8.0 | 1.42 |

[a] Sample ID is defined as: DBI/S-XXX-YY = dibutyl itaconate (DBI)/styrene (S) statistical copolymerization at XXX = temperature (°C) and YY = % molar composition of DBI in initial mixture with molar fraction given as $f_{DBI,0}$. [b] X = overall molar conversion. [c] Number average molecular weight $M_n$ and dispersity Đ determined by GPC relative to poly(styrene) standards in THF at 40 °C.

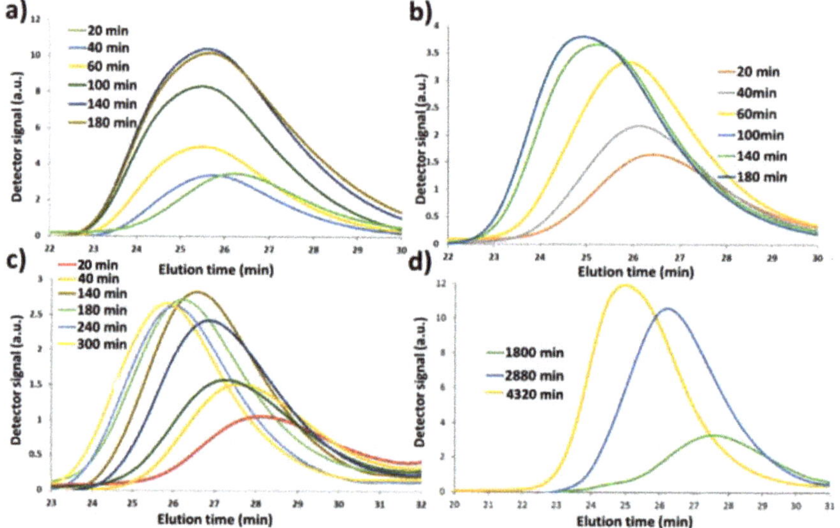


**Figure 6.** GPC traces of DBI-S copolymerizations ($f_{DBI,0} = 0.50$ in all cases) done at (**a**) 110 °C, (**b**) 100 °C, (**c**) 80 °C and (**d**) 70 °C.

Table S2 in the Supporting Information shows the individual conversion of DBI and S, $M_n$, $M_w$ and Đ for DBI-S copolymerizations (50–50 mol%) done at 110, 100, 80 and 70 °C. In every case, the conversion of S was higher than DBI, meaning that the final composition of the copolymers is richer in S than DBI. The GPC traces of the copolymerizations for DBI-S (50–50 mol%) done at 110, 100, 80 and 70 °C are shown in Figure 6. Although at 110 °C almost full conversion was achieved at relatively shorter times, the GPC traces at this temperature (Figure 6a) showed little growth towards higher chain lengths [41] and the same molecular weight distribution from the early stages of the reaction just became broader during the course of the reaction. Although the conversion did increase linearly with time, this might be due to the chain transfer and termination of chains with the formation of low molecular weight chains. The GPC traces of the reactions at 100 °C and 80 °C moved towards higher molecular weight only at the early stages of the reactions (low conversion) and low molecular weight tailing was still observed, followed by no further increase in molecular weight after about 30% conversion. At 70 °C, although the reaction took longer to reach 40% conversion ~ 2 days, it showed shifting towards higher molecular weight and very little tailing. As noted earlier for the radical polymerization of DBI, intramolecular chain transfer side reactions take place at high temperatures [25] which was corroborated by our DBI/S copolymerizations at $f_{DBI,0} = 0.50$. We thus focused on copolymerizations at the lower temperature of 70 °C for three days to see if the mixtures with higher DBI loadings led to higher molecular weight polymers.

As observed from Figure 5, with decreasing temperature, the $M_n$ versus conversion slightly shifts up toward the theoretical line which shows intramolecular chain transfer reactions can be mitigated to some degree by decreasing temperature. The control of the polymerizations seems to be better at 70 °C. At this temperature, as the compositions become richer in DBI, the $M_n$ versus conversion plots do rise but still become noticeably flatter with conversion (Table 4 summarizes the copolymerizations while Figure 7 shows $M_n$ versus conversion and individual conversions are listed in Table S3 of Supporting Information). The GPC chromatograms for experiments with $f_{DBI,0}$ = 0.60, 0.70 and 0.80 at 70 °C show clear shifts towards higher molecular weights (Figure 8). Only in the case of $f_{DBI,0}$ = 0.90 do the GPC traces show no growth. Also, it is worth mentioning that the obtained molecular weight is still much lower than the expected value. The main reason could be that the maximum temperature of the reaction which is suggested by Hirano and Tayoshi is 60 °C, which was not tested in our experiment due to the minimum temperature required for NHS-BlocBuilder dissociation (T ≥ 65 °C) [61]. Thus, DBI/S copolymerizations can be pushed to higher initial DBI compositions > 50 mol%, provided the polymerization temperature does not decrease < 65 °C. Another issue regarding the differences between the actual and theoretical molecular weights is the difference in hydrodynamic volume from GPC measurements. All of the GPC measurements here were measured relative to PS standard in THF. For poly(DBI), no Mark-Houwink-Sakurada (MHS) parameters were available for THF but MHS parameters for poly(DBI) in THF were reasonably approximated by those for poly(DBI) in toluene at similar temperatures (K = 5.7 × 10⁻³ mL g⁻¹, a = 0.70 at 25 °C) [62]. These MHS parameters are similar to PS in THF at 25 °C (K = 14.1 × 10⁻³ mL g⁻¹, a = 0.70 at 25 °C) [63]. The differences in hydrodynamic volume would not be very significant between PS and poly (DBI) and would not be solely responsible for the flattening of the $M_n$ versus conversion plots for the copolymerizations performed.

**Table 4.** Summary of dibutyl itaconate/styrene copolymerizations in 50 wt% dioxane solution at 70 °C at different initial compositions.

| Sample ID [a] | $f_{DBI,0}$ | Time (min) | X [b] | $M_n$ (kg mol⁻¹) [c] | Đ [c] |
|---|---|---|---|---|---|
| DBI/S-70-60 | 0.60 | 4320 | 0.56 | 6.9 | 1.63 |
| DBI/S-70-70 | 0.70 | 4320 | 0.58 | 7.2 | 1.41 |
| DBI/S-70-80 | 0.80 | 4320 | 0.66 | 5.9 | 1.33 |
| DBI/S-70-90 | 0.90 | 4320 | 0.38 | 3.2 | 1.38 |

[a] Sample ID is defined as: DBI/S-XXX-YY = dibutyl itaconate (DBI)/styrene (S) statistical copolymerization at XXX = temperature (°C) and YY = % molar composition of DBI in initial mixture with molar fraction given as $f_{DBI,0}$. [b] X = overall molar conversion. [c] Number average molecular weight $M_n$ and dispersity Đ determined by GPC relative to poly(styrene) standards in THF at 40 °C.

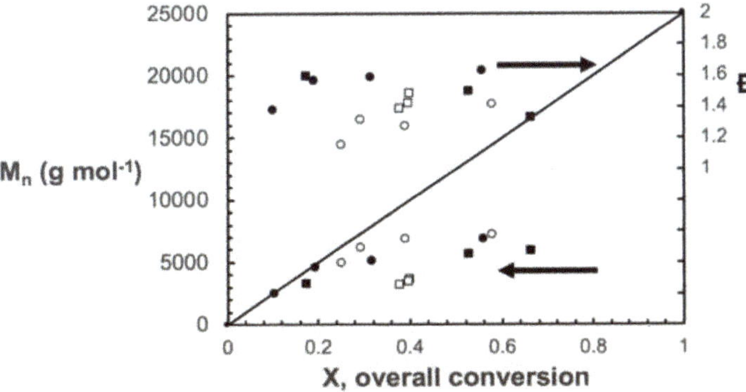


**Figure 7.** Number average molecular weight $M_n$ versus overall conversion (X) for DBI/S copolymerization with initial DBI feed compositions $f_{DBI,0}$ = 0.60–0.90 done at 70 °C: $f_{DBI,0}$ = 0.6 (filled circles, ●); $f_{DBI,0}$ = 0.7 (open circles, ○); $f_{DBI,0}$ = 0.8 (filled squares, ■), $f_{DBI,0}$ = 0.9 (open squares, □).

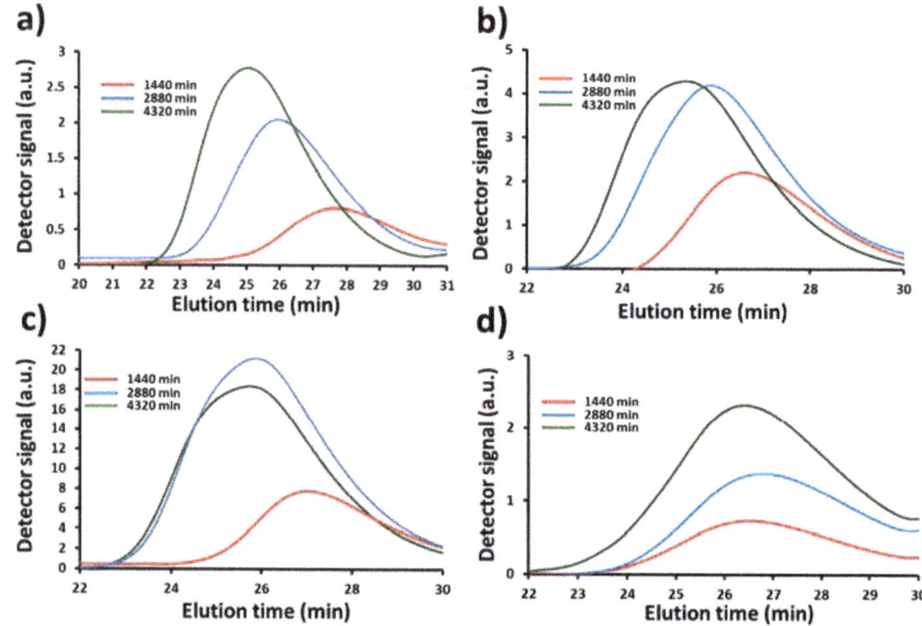

**Figure 8.** GPC traces (elution time) of DBI-S copolymerizations at 70 °C at (**a**) $f_{DBI,0} = 0.60$, (**b**) $f_{DBI,0} = 0.70$, (**c**) $f_{DBI,0} = 0.80$ and (**d**) $f_{DBI,0} = 0.90$.

It should be stated that we attempted further DBI copolymerizations with methyl methacrylate (MMA) using NHS-BlocBuilder with little avail at temperatures between 90 and 110 °C. It is likely that the nitroxide adduct with DBI is too stable even when using a monomer like MMA that has a higher $k_p$ compared to S at similar temperatures.

Finally, chain extension experiments with S using a P(DBI/S) macroinitiator were done: S-rich ($f_{DBI,0} = 0.2$) or DBI-rich ($f_{DBI,0} = 0.8$) macroinitiators were used from the copolymerizations done at 110 °C (Table 5 summarizes the chain extensions). In a typical NMP system, it is well-established that <10% of the polymer chains are irreversibly terminated, which means than >90% of the chains are end-functionalized with nitroxide moieties capable of reactivation to allow extension of the polymer chains [41]. The main objective of these experiments was to confirm our previous discussions, specifically if the DBI-rich macroinitiator could be re-activated and add the second batch of monomer. Figure 9a shows the GPC traces of the chain extension experiment using a macroinitiator with $f_{DBI,0} = 0.2$.

**Table 5.** Summary of chain extension reactions of dibutyl itaconate/styrene (DBI/S) macroinitiators with styrene at 110 °C in 50 wt% dioxane solution.

| Experiment | Macroinitiator | | Product | |
|---|---|---|---|---|
| | $M_n$ (kg mol$^{-1}$) | Đ | $M_n$ (kg mol$^{-1}$) | Đ |
| DBI/S-110-20-b-S | 12.3 | 1.37 | 22.1 | 2.99 |
| DBI/S-110-80-b-S | 2.6 | 1.36 | 19.2 | 1.61 |

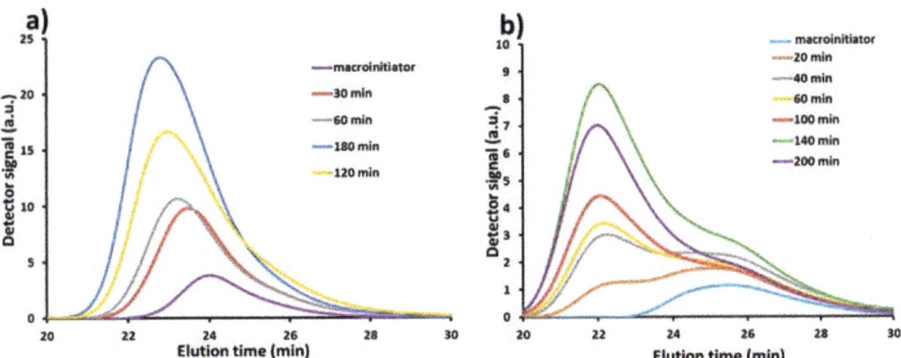

**Figure 9.** GPC traces (elution time) of DBI-S chain extension experiments with S at 110 °C using a macroinitiator P(DBI-S) a) $f_{DBI,0} = 0.20$ and b) $f_{DBI,0} = 0.80$.

As expected, shifts towards higher molecular weights were observed with time, but also some tailing, meaning that most of the polymer chains end-capped with SG1 did extend. When the macroinitiator from a feed of $f_{DBI,0} = 0.8$ was used, the GPC traces (Figure 9b) exhibited some shifting to higher molecular weight with time, but each trace had a very pronounced shoulder, confirming that most of the polymer chains are either terminated or are not capable to be reactivated again.

One of the potential applications of poly (dialkyl itaconates) is as a component in thermoplastic elastomers. They can be copolymerized in much the same way as dienes are with monomers like styrene to afford tough but flexible materials. The reactivity ratios for the DBI/S system by conventional radical polymerization were reported previously in benzene at 60 °C ($r_{DBI} = 0.38 \pm 0.02$ and $r_S = 0.40 \pm 0.05$ [19]). In the same study, reactivity ratios were similar for other di-n-alkyl itaconates (DXI where X = methyl, n-ethyl, n-propyl, n-amyl and n-octyl) ($r_{DXI} = 0.25–0.60$, $r_S = 0.25–0.40$). More recently, others have extended the analysis to higher di-n-alkyl itaconates (n = 12, 14, 16, 18 and 22) in bulk conventional free radical polymerization at 60 °C using AIBN initiator [28]. For n = 12, 14, 16, reactivity ratios for di-n-alkyl itaconate ranged from 0.22–0.28 and $r_S = 0.19–0.39$ while for n = 18 and 22, $r_{DXI} = 0.42–0.50$ and $r_S = 0.37–0.47$. Similar ranges in reactivity ratios were observed for DMI/S copolymerizations although there was considerable spread in the data (Davis and co-workers found that DMI was contaminated with poly (DMI) [59]. Our data, after fitting with Fineman-Ross and Kelen-Tudos methods, revealed $r_{DBI} = 0.29$ and $r_S = 0.59$ and $r_{DBI} = 0.32$ and $r_S = 0.77$, respectively. A non-linear least square fitting to the Mayo-Lewis equation provided $r_{DBI} = 0.31$ and $r_S = 0.61$.

Given that conditions for obtaining higher loadings of DBI are accessible, it would be possible now to tune the desired glass transition temperature ($T_g$) of the copolymers, which was measured using differential scanning calorimetry (DSC). The variation in $T_g$ with composition is bracketed by the $T_g$s of the homopolymers ($T_{g,PS}$ are 100 °C [64] while the $T_{g,PDBI}$ has been reported to vary between 5–17 °C [22,30,65,66]). Figure 10 shows the $T_g$ of the various statistical copolymers as function of final copolymer composition as well as the Fox equation prediction, which is given in Equation (7) where $w_{DBI}$ and $w_S$ are the weight fractions of DBI and S and $T_{g,PDBI}$ and $T_{g,PS}$ are the experimental $T_g$s of PDBI and PS homopolymers.

$$1/T_{g,th} = w_{DBI}/T_{g,PDBI} + w_S/T_{g,PS} \tag{7}$$

Although the fit to the Flory-Fox equation is not particularly good, it is still clear that even having a low fraction of DBI in the copolymer was able to dramatically reduce the $T_g$. For example, a copolymer with composition $F_{DBI} = 18$ mol% had a $T_g$ of 45 °C, much lower than the $T_g$ of the PS homopolymer (~100 °C). This is due to the flexibility that DBI provided to the polymer chains, giving additional free volume and thus a dramatic decrease in the $T_g$. The copolymers were noticeably more flexible with increasing DBI composition.

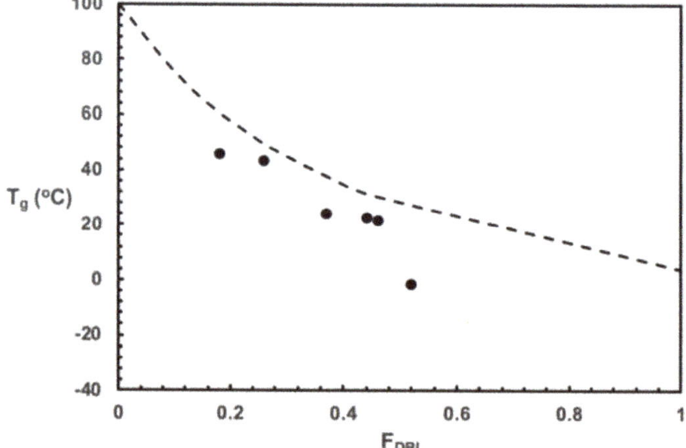

**Figure 10.** $F_{DBI}$ effects on $T_g$ in P(DBI-S) statistical copolymers. The Fox equation prediction of the P(DBI-S) $T_g$ are represented by the dotted line while the experimental data is presented by the solid circles.

## 4. Conclusions

In this study, nitroxide-mediated copolymerization of DBI with S was studied using the succcinimdyl functionalized NHS-BlocBuilder at different temperatures. It was not possible to obtain homopolymers of PDBI by NMP, likely due to the formation of a stable adduct, effectively blocking further propagation. When DBI was copolymerized with styrene-rich initial compositions ($f_{DBI,0} < 0.2$) at 110 °C, $M_n$ versus conversion plots were relatively linear up to fairly high conversion ~ 0.6 with relatively narrow molecular weight distributions ($Đ = 1.3$–$1.4$). At $f_{DBI,0} > 0.2$, $M_n$ versus conversion plots flattened with increasing conversion and $M_n$ was much lower than theoretical predictions, indicating the high tendency of itaconate monomers to generate intramolecular chain transfer side reactions and the increased probability for formation of stable adducts. A change in reaction temperature to 70 °C indicated a slight improvement in terms of control (narrowed molecular weight distributions and $M_n$ versus conversion remained linear up to higher conversions). This however was accompanied by excessively long polymerization times of around three days. DSC was used to measure the $T_g$ of the copolymers. As expected, $T_g$ decreased as composition of DBI in the copolymer increased, leading the material to become less rigid than the PS homopolymer.

**Supplementary Materials:** The following are available online at http://www.mdpi.com/2227-9717/7/5/254/s1, Tables S1, S2 and S3 indicating individual monomer conversions and molecular weight data are listed.

**Author Contributions:** Conceptualization, M.M., O.G.V. and A.M.; methodology, O.G.V., M.M.; formal analysis, S.K.; O.G.V., A.M., M.M.; investigation, S.K., O.G.V., A.M.; data curation, S.K.; writing—Original draft preparation, S.K.; writing—Review and editing, M.M., A.M.; supervision, M.M.; project administration, M.M.; funding acquisition, M.M.

**Funding:** This research was funded by an NSERC Discovery Grant, grant number 288125 (M.M).

**Acknowledgments:** We would like to thank Arkema (Mickael Havel and Noah Macy) particularly for their help in providing the BlocBuilder and SG1 initiators described in this work.

**Conflicts of Interest:** The authors declare no conflict of interest.

## References

1. Robert, T.; Friebel, S. Itaconic acid—A versatile building block for renewable polyesters with enhanced functionality. *Green Chem.* **2016**, *18*, 2922–2934. [CrossRef]
2. Tate, B.E. Polymerization of itaconic acid and derivatives. *Adv. Poly. Sci.* **1967**, *5*, 214–232.

3. Velada, J.; Hernáez, E.; Cesteros, L.C.; Katime, I. Study of the thermal degradation of several poly (monoalkylaryl itaconates). *Polym. Degrad. Stab.* **1996**, *52*, 273–282. [CrossRef]
4. Calam, C.T.; Oxford, A.E.; Raistrick, H. Studies in the biochemistry of micro-organisms: Itaconic acid, a metabolic product of a strain of Aspergillus terreus Thom. *Biochem. J.* **1939**, *33*, 1488. [CrossRef] [PubMed]
5. Willke, T.; Vorlop, K.-D. Biotechnological production of itaconic acid. *Appl. Microbiol. Biotechnol.* **2001**, *56*, 289–295. [CrossRef]
6. Willke, T.; Vorlop, K.-D. Industrial bioconversion of renewable resources as an alternative to conventional chemistry. *Appl. Microbiol. Biotechnol.* **2004**, *66*, 131–142. [CrossRef]
7. Dai, J.; Liu, X.; Ma, S.; Wang, J.; Shen, X.; You, S.; Zhu, J. Soybean oil-based UV-curable coatings strengthened by crosslink agent derived from itaconic acid together with 2-hydroxyethyl methacrylate phosphate. *Prog. Org. Coat.* **2016**, *97*, 210–215. [CrossRef]
8. Katime, I.; Rodríguez, E. Absorption of metal ions and swelling properties of poly (acrylic acid-co-itaconic acid) hydrogels. *J. Macromol. Sci. Part A Pure Appl. Chem.* **2001**, *38*, 543–558. [CrossRef]
9. Werpy, T.; Petersen, G.; Aden, A.; Bozell, J.; Holladay, J.; White, J.; Manheim, A. *Top Value Added Chemicals from Biomass Vol. I—Results of Screening for Potential Candidates from Sugars and Synthesis Gas*; Report DOE/GO-102004-1992; U.S. Department of Energy, Office of Scientific and Technical Information: Washington, DC, USA, 2004. Available online: https://www.nrel.gov/docs/fy04osti/35523.pdf (accessed on 30 April 2019).
10. Klement, T.; Büchs, J. Itaconic Acid—A Biotechnological Process in Change. *Bioresour. Technol.* **2013**, *135*, 422–431. [CrossRef]
11. Marvel, C.S.; Shepherd, T.H. Polymerization Reactions of Itaconic Acid and Some of Its Derivatives. *J. Org. Chem.* **1959**, *24*, 599–605. [CrossRef]
12. Takasu, A.; Ito, M.; Inai, Y.; Hirabayashi, T.; Nishimura, Y. Synthesis of biodegradable polyesters by ring-opening copolymerization of cyclic anhydrides containing a double bond with 1,2-epoxybutane and one-pot preparation of the itaconic acid-based polymeric network. *Polym. J.* **1999**, *31*, 961–969. [CrossRef]
13. Tang, T.; Moyori, T.; Takasu, A. Isomerization-free polycondensations of cyclic anhydrides with diols and preparation of polyester gels containing cis or trans carbon double bonds via photo-cross-linking and isomerization in the gels. *Macromolecules* **2013**, *46*, 5464–5472. [CrossRef]
14. Farmer, T.; Castle, R.; Clark, J.; Macquarrie, D. Synthesis of unsaturated polyester resins from various bio-derived platform molecules. *Int. J. Mol. Sci.* **2015**, *16*, 14912–14932. [CrossRef] [PubMed]
15. Winkler, M.; Lacerda, T.M.; Mack, F.; Meier, M.A.R. Renewable polymers from itaconic acid by polycondensation and ring-opening-metathesis polymerization. *Macromolecules* **2015**, *48*, 1398–1403. [CrossRef]
16. Bai, Y.; De bruyn, M.; Clark, J.H.; Dodson, J.R.; Farmer, T.J.; Honore, M.; Ingram, I.D.V.; Naguib, M.; Whitwood, A.C.; North, M. Ring opening metathesis polymerization of a new bio-derived monomer from itaconic anhydride and furfuryl alcohol. *Green Chem.* **2016**, *18*, 3945–3948. [CrossRef]
17. Lv, A.; Li, Z.-L.; Du, F.-S.; Li, Z.-C. Synthesis, functionalization, and controlled degradation of high molecular weight polyester from itaconic acid via ADMET polymerization. *Macromolecules* **2014**, *47*, 7707–7716. [CrossRef]
18. Nagai, S.; Yoshida, K. Polymerization of itaconic acid derivatives. Part III: Rate of polymerization of dialkyl itaconates. *Kobunshi Kagaku* **1960**, *17*, 79–82. [CrossRef]
19. Braun, V.D.; Ahn, T.-O. Copolymerisation von Itaconsauredialkylestern mit Styrol. *Kolloid Z.* **1963**, *188*, 1–4. [CrossRef]
20. Sato, T.; Morita, N.; Tanaka, H.; Ota, T. Solvent effect on the radical polymerization of di-n-butyl itaconate. *J. Polym. Sci. A Polym. Chem.* **1989**, *27*, 2497–2508. [CrossRef]
21. Otsu, T.; Watanabe, H. Radical polymerization reactivity of dialkyl itaconates and characterization of their polymers. *Eur. Polym. J.* **1993**, *29*, 167–174. [CrossRef]
22. Fernandez-Garcia, M.; Madruga, E.L. Glass transition in dimethyl and di-n-butyl poly(itaconate ester)s and their copolymers with methyl methacrylate. *Polymer* **1997**, *38*, 1367–1371. [CrossRef]
23. Madruga, E.L.; Fernandez-Garcia, M. Free-radical homopolymerization and copolymerization of di-n-butyl itaconate. *Polymer* **1994**, *35*, 4437–4442. [CrossRef]
24. Veličković, J.; Filipović, J.; Djakov, D.P. The synthesis and characterization of poly(itaconic) acid. *Polym. Bull.* **1994**, *32*, 169–172. [CrossRef]

25. Hirano, T.; Takeyoshi, R.; Seno, M.; Sato, T. Chain-transfer reaction in the radical polymerization of di-n-butyl itaconate at high temperatures. *J. Polym. Sci. A Polym. Chem.* **2002**, *40*, 2415–2426. [CrossRef]
26. Hirano, T.; Higashi, K.; Seno, M.; Sato, T. Radical polymerization of di-n-butyl itaconate in the presence of Lewis acids. *Eur. Polym. J.* **2003**, *39*, 1801–1808. [CrossRef]
27. Hirano, T.; Higashi, K.; Seno, M.; Sato, T. Reaction control in radical polymerization of di-n-butyl itaconate utilizing a hydrogen-bonding interaction. *J. Polym. Sci. A Polym. Chem.* **2004**, *42*, 4895–4905. [CrossRef]
28. Rangel-Rangel, E.; Torres, C.; Rincon, L.; Loteich-Khatib, S.; Lopez-Carrasquero, F. Copolymerizations of long side chain di n-alkyl itaconates and methyl n-alkyl itaconates with styrene: Determination of monomers reactivity ratios by NMR. *Rev. Latinoam. Metal. Mater.* **2012**, *32*, 79–88.
29. Lopez-Carrasqero, F.; Rangel-Rangel, E.; Cardenas, M.; Torres, C.; Dugarte, N.; Laredo, E. Copolymers of long-side-chain di-n-alkyl itaconates or methyl m-alkyl itaconates with styrene: Synthesis, characterization, and thermal properties. *Polym. Bull.* **2013**, *70*, 131–146. [CrossRef]
30. Sarkar, P.; Bhowmick, A.K. Green approach toward sustainable polymer: Synthesis and characterization of poly(myrcene-co-dibutyl itaconate). *ACS Sustain. Chem. Eng.* **2016**, *4*, 2129–2141. [CrossRef]
31. Lei, W.; Russell, T.P.; Hu, L.; Zhou, X.; Qiao, H.; Wang, W.; Wang, R.; Zhang, L. Pendant Chain effect on the synthesis, characterization, and structure–property relations of poly (di-n-alkyl itaconate-co-isoprene) biobased elastomers. *ACS Sustain. Chem. Eng.* **2017**, *5*, 5214–5223. [CrossRef]
32. Szablan, Z.; Toy, A.A.; Terrenoire, A.; Davis, T.P.; Stenzel, M.H.; Müller, A.H.E.; Barner-Kowollik, C. Living free-radical polymerization of sterically hindered monomers: Improving the understanding of 1,1 di-substituted monomer systems. *J. Polym. Sci. A Polym. Chem.* **2006**, *44*, 3692–3710. [CrossRef]
33. Satoh, K.; Lee, D.-H.; Nagai, K.; Kamigaito, M. Precision Synthesis of Bio-Based Acrylic Thermoplastic Elastomer by RAFT Polymerization of Itaconic Acid Derivatives. *Macromol. Rapid Commun.* **2014**, *35*, 161–167. [CrossRef] [PubMed]
34. Fernandez-Garcia, M.; Fernandez-Sanz, M.; De la Fuente, J.L.; Madruga, E.L. Atom-transfer radical polymerization of dimethyl itaconate. *Macromol. Chem. Phys.* **2001**, *202*, 1213–1218. [CrossRef]
35. Okada, S.; Matyjaszewski, K. Synthesis of bio-based poly(*N*-phenylitaconimide) by atom transfer radical polymerization. *J. Polym. Sci. A Polym. Chem.* **2015**, *53*, 822–827. [CrossRef]
36. Inciarte, H.; Orozco, M.; Fuenmayor, M.; López-Carrasquero, F.; Oliva, H. Comb-like copolymers of n-alkyl monoitaconates and styrene. *e-Polymers* **2006**, *6*. [CrossRef]
37. Nagai, S. The polymerization and polymers of itaconic acid derivatives. VI. The polymerization and copolymerization of itaconic anhydride. *Bull. Chem. Soc. Jpn.* **1964**, *37*, 369–373. [CrossRef]
38. Wallach, J.A.; Huang, S.J. Copolymers of itaconic anhydride and methacrylate-terminated poly(lactic acid) macromonomers. *Biomacromolecules* **2000**, *1*, 174–179. [CrossRef]
39. Shang, S.; Huang, S.J.; Weiss, R.A. Synthesis and characterization of itaconic anhydride and stearyl methacrylate copolymers. *Polymer* **2009**, *50*, 3119–3127. [CrossRef]
40. Javakhishvili, I.; Kasama, T.; Jankova, K.; Hvilsted, S. RAFT copolymerization of itaconic anhydride and 2-methoxyethyl acrylate: A multi-functional scaffold for preparation of "clickable" gold nanoparticles. *Chem. Commun.* **2013**, *49*, 4803–4805. [CrossRef]
41. Nicolas, J.; Guillaneuf, Y.; Lefay, C.; Bertin, D.; Gigmes, D.; Charleux, B. Nitroxide mediated polymerization. *Prog. Polym. Sci.* **2013**, *38*, 63–235. [CrossRef]
42. Willcock, H.; O'Reilly, R.K. End group removal and modification of RAFT polymers. *Polym. Chem.* **2010**, *1*, 149–157. [CrossRef]
43. Su, X.; Jessop, P.G.; Cunningham, M.F. ATRP catalyst removal and ligand recycling using $CO_2$-switchable materials. *Macromolecules* **2018**, *51*, 8156–8164. [CrossRef]
44. Wang, Y.; Lorandi, F.; Fantin, M.; Chmielarz, P.; Isse, A.A.; Gennaro, A.; Matyjaszewski, K. Miniemulsion ARGET ATRP via interfacial and ion-pair catalysis: From ppm to ppb of residual copper. *Macromolecules* **2017**, *50*, 8417–8425. [CrossRef]
45. Vinas, J.; Chagneux, N.; Gigmes, D.; Trimaille, T.; Favier, A.; Bertin, D. SG1-based alkoxyamine bearing a N-succinimidyl ester: A versatile tool for advanced polymer synthesis. *Polymer* **2008**, *49*, 3639–3647. [CrossRef]
46. Szablan, Z.; Stenzel, M.H.; Davis, T.P.; Barner, L.; Barner-Kowollik, C. Depropagation kinetics of sterically demanding monomers: A pulsed laser size exclusion chromatography study. *Macromolecules* **2005**, *38*, 5944–5954. [CrossRef]

47. Buback, M.; Gilbert, R.G.; Hutchinson, R.A.; Klumperman, B.; Kuchta, F.D.; Manders, B.G.; O'Driscoll, K.F.; Russell, G.T.; Schweer, J. Critically evaluated rate coefficients for free-radical polymerization, 1. Propagation rate coefficient for styrene. *Macromol. Chem. Phys.* **1995**, *196*, 3267–3280. [CrossRef]

48. Kassi, E.; Loizou, E.; Porcar, L.; Patrickios, C.S. Di(n-butyl) itaconate end-functionalized polymers: Synthesis by group transfer polymerization and solution characterization. *Eur. Polym. J.* **2011**, *47*, 816–822. [CrossRef]

49. Kassi, E.; Constantinou, M.S.; Patrickios, C.S. Group transfer polymerization of bio-based monomers. *Eur. Polym. J.* **2013**, *49*, 761–767. [CrossRef]

50. Tomić, S.L.; Filipović, J.M.; Velicković, J.S.; Katsikas, L.; Popović, I.G. The polymerisation kinetics of lower dialkyl itaconates. *Macromol. Chem. Phys.* **1999**, *200*, 2421–2427. [CrossRef]

51. Benoit, D.; Grimaldi, S.; Robin, S.; Finet, J.P.; Tordo, P.; Gnanou, Y. Kinetics and mechanism of controlled free-radical polymerization of styrene and n-butyl acrylate in the presence of an acyclic-phosphonylated nitroxide. *J. Am. Chem. Soc.* **2000**, *122*, 5929–5939. [CrossRef]

52. Fukuda, T.; Ma, Y.-D.; Inagaki, H. Free radical copolymerization. 3. Determination of rate constants of propagation and termination for styrene/methyl methacrylate system. A critical test of terminal-model kinetics. *Macromolecules* **1985**, *18*, 17–26. [CrossRef]

53. Fischer, H. The persistent radical effect in controlled radical polymerizations. *J. Polym. Sci. A Polym. Chem.* **1999**, *37*, 1885–1901. [CrossRef]

54. Charleux, B.; Nicolas, J.; Guerret, O. Theoretical expression of the average activation-deactivation equilibrium constant in controlled/living free-radical copolymerization operating via reversible termination. Application to a strongly improved control in nitroxide-mediated polymerization of methyl methacrylate. *Macromolecules* **2005**, *38*, 5484–5492.

55. Ma, Y.D.; Sung, K.S.; Tsujii, Y.; Fukuda, T. Free-radical copolymerization of styrene and diethyl fumarate. Penultimate-unit effects on both propagation and termination processes. *Macromolecules* **2001**, *34*, 4749–4756. [CrossRef]

56. Lessard, B.; Marić, M. Effect of acrylic acid neutralization on 'livingness' of poly[styrene-ran-(acrylic acid)] macro-initiators for nitroxide-mediated polymerization of styrene. *Polym. Int.* **2008**, *57*, 1141–1151. [CrossRef]

57. Buback, M.; Egorov, M.; Junkers, T.; Panchenko, E. Termination kinetics of dibutyl itaconate free-radical polymerization studied via the SP–PLP–ESR technique. *Macromol. Chem. Phys.* **2005**, *206*, 333–341. [CrossRef]

58. Buback, M.; Kuchta, F.D. Termination kinetics of free-radical polymerization of styrene over an extended temperature and pressure range. *Macromol. Chem. Phys.* **1997**, *198*, 1455–1480. [CrossRef]

59. Yee, L.H.; Heuts, J.A.P.; Davis, T.P. Copolymerization propagation kinetics of dimethyl itaconate and styrene: Strong entropic contributions to the penultimate unit effect. *Macromolecules* **2001**, *34*, 3581–3586. [CrossRef]

60. Szablan, Z.; Toy, A.A.; Davis, T.P.; Hao, X.; Stenzel, M.H.; Barner-Kowollik, C. Reversible addition fragmentation chain transfer polymerization of sterically hindered monomers: Toward well-defined rod/coil architectures. *J. Polym. Sci. A Polym. Chem.* **2004**, *42*, 2432–2443. [CrossRef]

61. Marque, S.; Le Mercier, C.; Tordo, P.; Fischer, H. Factors influencing the C–O–bond homolysis of trialkylhydroxylamines. *Macromolecules* **2000**, *33*, 4403–4410. [CrossRef]

62. Veličković, J.; Ćoseva, S.; Fort, R.J. Solution properties of poly(dicyclohexyl itaconate). *Eur. Polym. J.* **1975**, *11*, 377–380. [CrossRef]

63. Strazielle, C.; Benoit, H.; Vogl, O. Preparation et characterisation des polymers tete-a-tete—VI: Proprieties physicochimiques du polystyrene tete-a-tete en solution diluee. Comparaison avec des polystyrenes de structure differente. *Eur. Polym. J.* **1978**, *14*, 331–334. [CrossRef]

64. Andrews, R.J.; Grulke, E.A. Glass transition temperatures of polymers. In *Polymer Handbook*, 4th ed.; Brandrup, J., Immergut, E.H., Grulke, E.A., Abe, A., Bloch, D.R., Eds.; Wiley: New York, NY, USA, 1999; Volume 1, pp. 181–308.

65. Arrighi, V.; Holmes, P.F.; McEwen, I.J.; Qian, H.; Terrill, N.J. Order in amorphous di-n-alkyl itaconate polymers, copolymers, and blends. *J. Polym. Sci. B Polym. Phys.* **2004**, *42*, 4000–4016. [CrossRef]

66. Cowie, J.M.G. Physical properties of polymers based on itaconic acid. *Pure Appl. Chem.* **1979**, *51*, 2331–2343. [CrossRef]

*Article*

# Chemical Recycling of Used Printed Circuit Board Scraps: Recovery and Utilization of Organic Products

Se-Ra Shin [†] [iD], Van Dung Mai [†] [iD] and Dai-Soo Lee *[iD]

Department of Semiconductor and Chemical Engineering, Chonbuk National University, 567 Baekje-daero, Deokjin-gu, Jeonju 54896, Korea; srshin89@jbnu.ac.kr (S.-R.S.); dungmv1983@gmail.com (V.D.M.)
* Correspondence: daisoolee@jbnu.ac.kr; Tel.: +82-63-270-2310
† These authors contributed equally to this work.

Received: 6 December 2018; Accepted: 28 December 2018; Published: 4 January 2019

**Abstract:** The disposal of end-of-life printed circuit boards (PCBs) comprising cross-linked brominated epoxy resins, glass fiber, and metals has attracted considerable attention from the environmental aspect. In this study, valuable resources, especially organic material, were recovered by the effective chemical recycling of PCBs. Pulverized PCB was depolymerized by glycolysis using polyethylene glycol (PEG 200) with a molecular weight of 200 g/mol under basic conditions. The cross-linked epoxy resins were effectively decomposed into a low-molecular species by glycolysis with PEG 200, followed by the effective separation of the metals and glass fibers from organic materials. The organic material was modified into recycled polyol with an appropriate viscosity and a hydroxyl value for rigid polyurethane foams (RPUFs) by the Mannich reaction and the addition polymerization of propylene oxide. RPUFs prepared using the recycled polyol exhibited superior thermal and mechanical properties as well as thermal insulation properties compared to conventional RPUFs, indicating that the recycled polyol obtained from the used PCBs can be valuable as RPUF raw materials for heat insulation.

**Keywords:** chemical recycling; glycolysis; used printed circuit board; recycled polyol; rigid polyurethane foam

---

## 1. Introduction

The development of various electronic and electrical equipment continuously generates a large amount of electronic and electrical equipment wastes [1–6]. Printed circuit boards (PCBs) constitute one of the essential components in electronic and electrical products. PCBs contain cross-linked brominated epoxy resins, metals, and glass fibers, which are difficult to decompose and reuse due to the inherent insolubility of the epoxy resins comprising a cross-linked network structure [1,7–11]. Thus, the appropriate disposal of used PCBs (UPCBs) has become an ongoing important issue from the environmental aspect [2,3,12].

UPCBs can be disposed by various methods, including landfills, incineration, pyrolysis, and chemical recycling [5]. Landfills and incineration can cause several environmental problems due to the presence of heavy metals and hazardous components in UPCBs [7]. On the other hand, thermal and chemical recycling is efficient for disposing UPCBs because of the separation and recovery of valuable materials from UPCBs. Several studies have reported the recovery of valuable resources from UPCBs by various methods [3,4,7,13–16]. Veti et al. reported the recycling of UPCBs to recover valuable metals [16] and found that lead (Pb), tin (Sn), and copper (Cu) are efficiently recovered by magnetic and electrostatic separation, followed by additional recovery via the electrowinning process. Zhu et al. reported a new technology for the recovery of valuable materials from UPCBs [13] by using 1-ethyl-3-methylimidazolium tetrafluoroborate [EMIM+] [BF$_4$−] at 260 °C and found that solders, Cu, glass fibers, and brominated epoxy resins are completely separated from UPCBs. Quan et al. examined the pyrolysis of UPCBs and successfully obtained the separated pyrolysis products comprising metals, glass fibers, and pyrolysis oil from UPCBs [3]. The pyrolysis oil derived from brominated epoxy resins contains a high concentration of phenol and phenol derivatives. Thus, pyrolysis oil is reused as a phenolic derivative to prepare a new phenol–formaldehyde resin. In addition, the authors expected that the glass fibers obtained from the pyrolysis residue can be applied as reinforcement filler for sheet molding compound and bulk molding compound.

The pyrolysis of UPCBs is one of the efficient recycling methods for the rapid recovery of valuable metals and glass fibers [1,3,9,14,15,17]. However, pyrolysis generally requires high temperature (>300 °C) and generates toxic organic materials [2–4,7]. In particular, hazardous gases including bromine compounds are released [10,18,19]. Nevertheless, owing to its high efficiency and short residence time, pyrolysis has been frequently employed for the recycling of UPCB to obtain valuable materials [18,20]. Chemical recycling, including glycolysis, aminolysis, and alcoholysis, demonstrates promise for the depolymerization of thermoplastic and thermosetting resins [11,21–25]. Typically, the chemical recycling of polymers involves the chain scission into small molecules by using hydroxyl or amine groups of solvolytic agents via transesterification [24]. Therefore, UPCBs mostly comprising brominated epoxy resins as the organic material are depolymerized, and the metallic and inorganic materials (i.e., metals and glass fiber) can be separated from the organic material (mostly brominated epoxy resin) during the process. Besides, organic products comprising hydroxyl-terminated groups and abundant aromatic rings can be converted to cost-effective raw chemicals to prepare a new polymer, especially polyurethane, and inorganics can be reused as reinforcements or fillers [12,13,21,26,27]. However, there are few reports on the utilization of the organic product from UPCBs for polyurethanes [21,28,29].

In this study, valuable resources, especially organic material were recovered from UPCBs via glycolysis. The pulverized UPCBs were depolymerized using polyethylene glycol (PEG) with a molecular weight of 200 g/mol under basic conditions. To optimize the depolymerization of UPCBs, the UPCB to PEG 200 ratio, reaction temperature, and reaction time were varied. Nonmetallic components such as organic material (depolymerization product) and glass fibers were collected and characterized. The organic product recovered after glycolysis contains phenolic and aliphatic hydroxyl end groups; thus, it can be used as a polyol to prepare rigid polyurethane foams (RPUFs). However, a solid organic product was obtained, and the hydroxyl value was not appropriate to prepare RPUFs. Thus, the organic product is liquefied and modified into recycled polyol for RPUFs via the Mannich reaction and the addition polymerization of propylene oxide. The recycled polyol obtained after the modification of the organic product was incorporated for manufacturing RPUFs, replacing 60 wt% of conventional polypropylene glycol (PPG) for RPUF. Besides, the RPUFs prepared using the recycled polyol exhibited improved thermal and mechanical properties compared to conventional RPUF. A flow chart for the chemical recycling of PCBs is given in Figure 1. To the best of our knowledge, this is the first report on the utilization of the organic product from UPCBs to prepare a recycled polyol for rigid polyurethane foams.

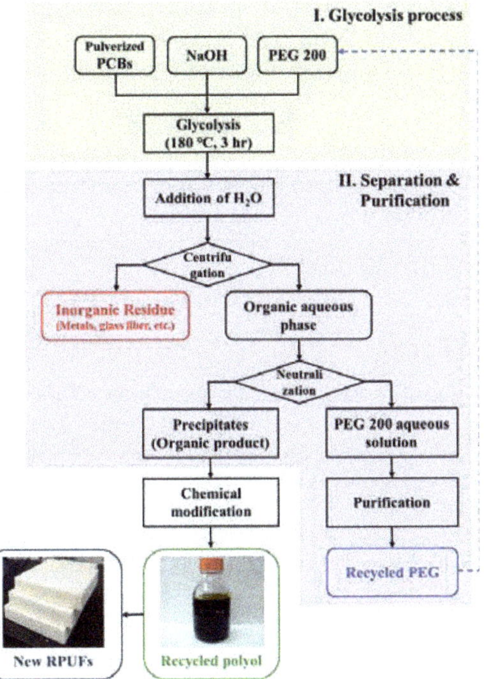



**Figure 1.** Flow chart for chemical recycling of pulverized printed circuit boards (PCBs).

## 2. Materials and Methods

### 2.1. Materials

UPCBs were collected from end-of-life computers and pulverized to remove metallic components by the density difference. PEG with a molecular weight of 200 g/mol (PEG 200) was purchased from Aldrich (Yong-In, Korea). Sodium hydroxide (NaOH) and potassium hydroxide (KOH), as well as a 0.5 M hydrochloric acid solution, were purchased from Daejung Chemicals & Metals (Si-Heung, Korea). Diethanolamine (Aldrich), a 37% formaldehyde solution (Aldrich), and propylene oxide (PO) (SKC chemicals, Ul-San, Korea) were used to modify the pre-polyol. A conventional PPG based on sugar/glycerin with a hydroxyl number of 450 mg KOH/g (JOP-0585) was supplied by Jungwoo Fine Chem Co., Ltd. (Ik-San, Korea). Polymeric 4,4′-diphenylmethane diisocyanate (pMDI, Cosmate M200) was purchased from Kumho Mitsui Chemicals (Yeo-Su, Korea). Dimethylcyclohexylamine (Polycat® 8, PC-8) from Air Products and Chemicals (Allentown, PA, USA) was used as amine catalysts. A silicone surfactant (B-8462) and tris(1-chloro-2-propyl) phosphate (TCPP) were purchased from Evonik (Essen, Germany) and Aldrich, respectively. As blowing agents, SOLKANE® 365/227 from Solvay (Brussels, Belgium) and distilled water were used. JOP-0585 was dried at 80 °C for 24 h under vacuum prior to use. All chemicals were used as received.

### 2.2. Glycolysis of Pulverized UPCB

The glycolysis of pulverized PCBs after the removal of metallic components was carried out under basic conditions. NaOH was dissolved in PEG 200 (NaOH:PEG 200 = 1:11 by mole) at 80 °C in a four-neck round-bottom flask, equipped with a reflux condenser and mechanical stirrer, under nitrogen, followed by the addition of a determined amount of PCB powder. The reaction temperature was allowed to increase to 180 °C. After 3 h, the system was cooled to room temperature, and distilled water was added to the mixture to decrease its viscosity. The inorganic residue, mostly glass fibers, was separated from the aqueous solution by centrifugation. The washing of the residue by water and separation were repeated several times until the pH of the aqueous solution reached 7. The aqueous solution containing the glycolysis product and residual PEG 200 was neutralized by the addition of 0.5 M HCl, followed by the precipitation of the glycolysis product from the aqueous phase. The precipitates were collected, washed with water, and dried in a convection oven at 100 °C to remove water. After complete drying, a solid glycolysis product was obtained (referred to as pre-polyol). To optimize the efficiency of depolymerization: (i) the PCB to PEG 200 mass ratio was changed by 1/6, 1/7, and 1/9; (ii) glycolysis temperature was varied to 160, 180, and 200 °C; and (iii) glycolysis time was varied by 1, 2, 3, 4, and 5 h. The yield of the glycolysis product obtained from each experiment was estimated by thermogravimetric analysis.

### 2.3. Preparation of the Recycled Polyol

The solid pre-polyol was liquefied by the Mannich reaction and addition polymerization of PO, affording recycled polyol for RPUFs. The pre-polyol was obtained by glycolysis at 180 °C for 3 h, with a hydroxyl value of 226.9 mg KOH/g (Table 1). Briefly, the Mannich reaction and alkoxylation of pre-polyol were carried out as follows. A predetermined amount of pre-polyol (1.0 mol), diethanolamine (2.2 mol), and water were added to a round-bottom reactor equipped with a mechanical stirrer, thermometer, and nitrogen inlet. The reaction temperature was increased to 90 °C, and a formaldehyde solution (2.2 mol) was slowly added into the flask for the Mannich reaction. After 3 h at 120 °C, dehydration was carried out under reduced pressure until the water content decreased to less than 0.1%. Subsequently, the obtained Mannich adduct was modified by addition polymerization with PO, affording recycled polyol with a suitable hydroxyl value for RPUFs in autoclave. A dark brown liquid as the final product, recycled polyol, was collected and dehydrated at 80 °C under reduced pressure. The hydroxyl value of the product was measured by titration following ASTM D 4704. Table 1 summarizes the characteristics of conventional PPG (JOP-0585), pre-polyol, and recycled polyol. Fourier-transform infrared spectrum (FTIR) presented the following (Figure S1, cm$^{-1}$): 3376 ($-$OH), 2966 ($sp^3$ C$-$H), 2932 ($sp^3$ C$-$H), 2874 ($sp^3$ C$-$H), 1658 (Aromatic ring), 1509 (Aromatic ring), 1458, 1374, 1302, 1119, 1080, 934, 876, 840, and 584.

**Table 1.** Characteristics of JOP-0585, pre-polyol, and recycled polyol.

| | JOP-0585 | Pre-Polyol | Recycled Polyol |
|---|---|---|---|
| Viscosity (Pa·s) | 5.0 | - | 2.7 |
| Hydroxyl value (mg KOH/g) | 450.0 | 226.9 | 460.0 |
| Acid value (mg KOH/g) | >0.1 | >1.0 | 1.2 |
| Color | Light yellow | Dark brown | Dark brown |
| Br% | 0 | 15.0 ± 1.0 [a]<br>14.12 ± 1.04 [b] | 1.80 ± 0.13 [b] |

[a] Br% of pre-polyol was determined by EDS analysis. [b] Br% was determined by relative ratio of peak area% from pyrolysis gas chromatography–mass spectroscopy–electron capture detection (Py–GC/MS/ECD).

### 2.4. Preparation of RPUFs

RPUFs with different contents of recycled polyol were prepared in two steps. First, predetermined amounts of the surfactant, catalyst, phosphate, and blowing agents were added to the polyol and homogeneously mixed under mechanical stirring. Second, a precalculated amount of pMDI was added to the polyol mixture and vigorously mixed under mechanical stirring at 6000 rpm for 7 s. The mixture was rapidly poured into a steel mold (300 mm × 300 mm × 50 mm) with a lid. After curing at 60 °C for 20 min, the RPUFs were demolded and stored for at least 24 h at room temperature before the characterization. The NCO index was maintained at a constant value, 120. The recycled polyol contents with respect to the total polyol weight were increased from 0% to 60%. For comparison, the RPUF prepared using only conventional PPG was designated as CON. Table 2 summarizes the formulations of the RPUFs with different contents of the recycled polyol.

**Table 2.** Sample code and formulation for rigid polyurethane foams (RPUFs).

| Sample Code | CON | P20 | P40 | P60 |
|---|---|---|---|---|
| | (Composition by wt%) | | | |
| *Polyol part* | | | | |
| JOP-0585 | 100.0 | 80.0 | 60.0 | 40.0 |
| Recycled polyol | - | 20.0 | 40.0 | 60.0 |
| B-8462 | 2.0 | 2.0 | 2.0 | 2.0 |
| PC-8 | 3.0 | 3.0 | 2.0 | 1.5 |
| TCPP | 15.0 | 15.0 | 15.0 | 15.0 |
| Water | 1.5 | 1.5 | 1.5 | 1.5 |
| 365/227 | 35.0 | 35.0 | 35.0 | 35.0 |
| *Isocyanate part* | | | | |
| NCO Index | 120 | 120 | 120 | 120 |

*2.5. Characterization*

The chemical structures of the pre-polyol and the recycled polyol were examined by $^1$H and $^{13}$C NMR spectroscopy (600 MHz, JNM-ECA600, JEOL Ltd., Tokyo, Japan) in DMSO-$d_6$ at room temperature. Fourier transform infrared (FTIR) spectra (FTIR 2000, JASCO, Easton, MD, USA) were recorded in the wavenumber range from 4000 to 500 cm$^{-1}$ at a resolution of 4 cm$^{-1}$. Scanning electron microscopy (SEM) images (JSM 6400, JEOL Ltd., Akishima, Tokyo, Japan) were recorded to examine the morphology of the glass fiber after depolymerization at an accelerating voltage of 20 kV. The content of bromine in the pre-polyol and recycled polyol was determined by energy-dispersive spectroscopy (EDS) (JSM-6400, JEOL Ltd., Akishima, Tokyo, Japan), pyrolysis gas chromatography–mass spectroscopy (Py–GC/MS) (QP2010 plus, Shimadzu, Kyoto, Japan), and gas chromatography–electron capture detection (GC/ECD) (6890N, Agilent Technologies, Santa Clara, CA, USA). EDS analysis was carried out with gold-coated pre-polyol powder at an accelerating voltage of 20 kV. Py–GC/MS with medium polar column (DB–624, 30 m × 0.251 mm × 1.40 mm, Agilent Technologies) was employed for the identification of the overall compounds in the pre-polyol and the recycled polyol. The pyrolysis temperature was 670 °C and the oven was held at 40 °C for 3 min, and then ramped to 260 °C at 10 °C/min. The brominated phenolic compounds thereof were identified by GC/ECD with medium polar column (DB–624, 30 m × 0.251 mm × 1.40 mm, Agilent Technologies). The analysis conditions of GC/ECD were identical to GC. The content of metals in recycled polyol such as Al, Ni, Cu, Cd, and Pb was determined by inductively coupled plasma–mass spectrometry (ICP/MS).

To characterize the reactivity of RPUFs, the characteristic times, including the cream time, gel time, and tack-free time, were estimated. The compressive strength of the RPUFs was estimated by using a universal testing machine according to ASTM D 1621. Cubic samples with a size of 40 mm × 40 mm × 40 mm were prepared, and the blowing direction and direction perpendicular to blowing were evaluated. The compressive strength of five specimens per sample was estimated, and the average values were calculated. To eliminate the effect of density, the measured compressive strengths were normalized for the blowing direction and direction perpendicular to blowing as follows (Equation (1)):

$$\sigma_n = \sigma(40/\rho)^2 \left[\{1 + \sqrt{(40/\rho_s)}\}/\{1 + \sqrt{(\rho/\rho_s)}\}\right]^2 \tag{1}$$

where $\sigma_n$ represents the normalized compressive strength; $\sigma$ represents the measured compressive strength; $\rho$ represents the density of RPUF (kg/m$^3$); and $\rho_s$ represents the density of the solid polyurethane matrix, which is given as 1200 kg/m$^3$ [30].

Thermal conductivities of the RPUFs were evaluated using a heat flow meter (HFM 436 Lambda, Netzsch, Selb, Germany) with two plane plates maintained at different temperatures (ASTM C 518).

Thermal conductivities of three specimens per sample were estimated, and the average values were obtained. The closed-cell content of the RPUFs with dimensions of 25 mm × 25 mm × 25 mm was estimated on an ULTRAPYC 1200e (Quantachrome, Boynton Beach, FL, USA) pycnometer according to ASTM D 6226. The closed-cell contents of five specimens per sample were estimated and averaged. The cell morphology of the RPUFs was examined by SEM (JSM-6400, JEOL Ltd., Akishima, Tokyo, Japan) at an accelerating voltage of 20 kV. The samples were coated with gold to avoid the charging of the electrons. Thermogravimetric analysis (TGA) was carried out on a Q50 system from TA Instruments (New Castle, DE, USA). Approximately 10 mg of sample placed on a platinum pan was heated from ambient temperature to 800 °C at a heating rate of 20 °C/min under nitrogen atmosphere. TGA measurements were carried out three times per sample and representative data were used for analysis. Dynamic mechanical property measurements of the RPUFs were carried out on a dynamic mechanical analyzer (DMA, Q800, New Castle, DE, USA) from TA Instruments in the tension mode from 30 °C to 250 °C at a heating rate of 3 °C/min (a frequency of 1 Hz and an amplitude of 15%).

## 3. Results and Discussion

### 3.1. Glycolysis of UPCBs

PEG is not only the solvent but also the co-catalyst for the glycolysis of pulverized PCBs. A sufficient amount of PEG is required to disperse PCB powders and combine with sodium hydroxide to promote glycolysis. To identify the effect of the PCB content on the efficiency of glycolysis, the PCB to PEG mass ratio was varied from 1:6 to 1:9 (Table 3). Figure 2 shows the effect of the PEG content on the weight loss of the glass fiber under the same glycolysis conditions (at 180 °C after 3 h). With the decrease of PEG content in the decomposition of PCB from 86.1 wt% to 84.4 wt%, the weight loss of glass fiber at 800 °C increased from 4% to 6%, respectively, and the yield increased from 89.74% to 92.47% (Table 3). However, with the increase in the PCB content from 86.1 wt% to 88.5 wt%, the weight loss of the glass fiber after glycolysis was almost the same.

**Table 3.** Composition and yield of the decomposition of used PCBs (UPCB) with polyethylene glycol (PEG).

| Codes | Before Decomposition | | | | | | | | After Decomposition | | | | Yield |
|---|---|---|---|---|---|---|---|---|---|---|---|---|---|
| | PCB | | PEG | | NaOH | | Glass Fiber | | Decomposed Product | | Recycled PEG | | |
| | g | wt% | g | wt% | g | wt% | g | wt% | g | wt% | g | wt% | % |
| 1:6 | 20.01 | 14.08 | 120.01 | 84.42 | 2.14 | 1.51 | 13.79 | 9.91 | 6.21 | 4.46 | 119.19 | 85.63 | 89.74 |
| 1:7 | 20.04 | 12.33 | 140.03 | 86.14 | 2.49 | 1.53 | 13.61 | 8.54 | 6.43 | 4.04 | 139.26 | 87.42 | 92.08 |
| 1:9 | 25.12 | 9.91 | 224.31 | 88.50 | 4.02 | 1.58 | 16.98 | 6.84 | 8.13 | 3.28 | 223.05 | 89.88 | 92.47 |

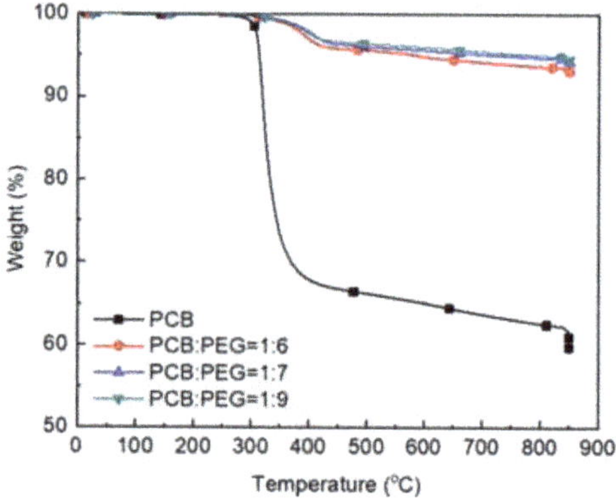


**Figure 2.** Thermogravimetric analysis (TGA) curves of PCB and the glycolysis products at different ratios of PCB to polyethylene glycol (PEG 200).

Figure 3 shows the effect of temperature on the glycolysis of pulverized PCBs. The weight loss values for the glass fiber at 800 °C after glycolysis at 180 °C and 200 °C for a reaction time of 3 h were almost the same (~4.50%, Figure 3a). However, with the decrease in the glycolysis temperature to 160 °C, the weight loss at 800 °C considerably increased (29.28%). Based on the weight loss at 800 °C, the glycolysis yield was estimated. Figure 3b shows the effect of the glycolysis temperature on the glycolysis yield. With the increase in the glycolysis temperature from 160 °C to 180 °C, the glycolysis yield sharply increased. Moreover, similar glycolysis yields were observed at 180 °C and 200 °C. Hence, the optimal reaction temperature was determined to be 180 °C. The effect of temperature can be explained in terms of thermodynamics. At low temperature, the thermal energy is not sufficient for breaking chemical bonds, thereby affording a low yield.

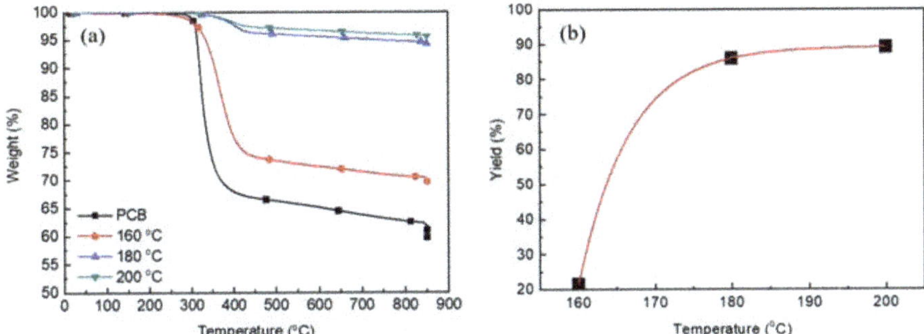

**Figure 3.** TGA analyses of PCB and glass fibers after the reaction at difference reaction temperatures (**a**) and the glycolysis yield for difference reaction temperature (**b**).

Figure 4 shows the effect of the glycolysis time on the efficiency of glycolysis at 180 °C. The weight loss was clearly ~5% after 2 h at 180 °C (Figure 4a). After 3 h, the weight loss did not decrease further because the decomposition reaction reached equilibrium. Figure 4b shows the effect of the reaction time on the glycolysis yield. The glycolysis yield sharply increased with the reaction time until 2 h at 180 °C, following which the yield did not significantly vary with the further increase in the reaction time.

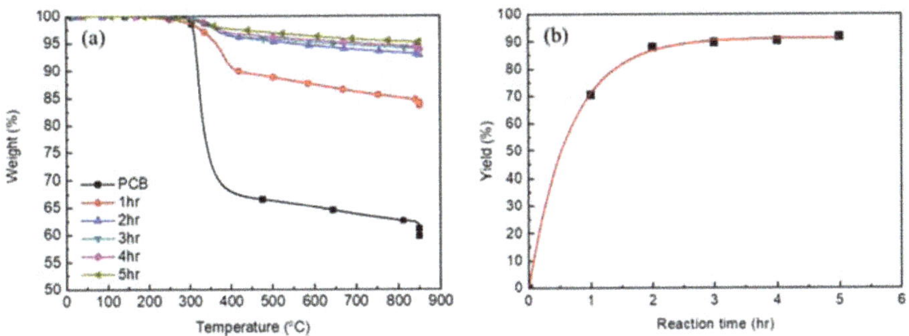

**Figure 4.** TGA analyses of PCB and glass fiber after the reaction for difference reaction time (**a**) and the glycolysis yield for difference reaction time (**b**).

SEM images also revealed the effects of the reaction time on the glycolysis efficiency (Figure 5). After the decomposition for 1 h, the glass fibers retained a considerable amount of polymer on their surfaces, and the glass fibers were still stuck. After 2 h, the glass fibers were separated, but few polymer particles were still retained on the glass fiber surface. Hence, the brominated epoxy resin was not completely decomposed, and a 100% yield was not achieved. After 3 h of glycolysis, the surface of the glass fibers appeared smooth and clean. Therefore, it was concluded that the optimum reaction time for glycolysis is 3 h. The glass fibers obtained after the glycolysis can be used as reinforcing fillers for polymer composites as many previous researchers reported on the reutilization of glass fibers recycled from PCBs [21,27,31,32]. Zheng et al. studied the reutilization of nonmetals recycled from UPCBs as reinforcing fillers in polypropylene composites [27]. They demonstrated that the incorporation of nonmetals recycled from PCBs significantly improved the tensile and flexural properties of the polypropylene composites. Sun et al. investigated sound absorption performance of glass fibers recycled from UPCBs [32]. They found that the recycled glass fibers showed excellent absorption ability in broad-band frequency range, implying the promising candidates for sound absorption materials. In this study, we concentrated on the utilization of organic products from UPCB.

**Figure 5.** Scanning electron microscopy (SEM) images of the PCB powder and glass fiber after the reaction for 1 h, 2 h, and 3 h.

PCBs are comprised of reinforcing glass fibers, metals, and epoxy resins containing brominated flame retardants. Bisphenol A and diglycidyl ether of bisphenol A epoxy resin cured using an anhydride hardener has been widely used for manufacture of PCBs. To limit the possible hazards associated with scarce fire resistance and the inherent flammability of these materials, flame retardants are typically added in their formulation. Typical flame retardants include tetrabromobisphenol A, which is substituted for bisphenol A in the epoxy resin. After the reaction, the epoxy resin was decomposed and converted to monomers in the final product. Figure 6 shows the FTIR spectrum of the final glycolysis product: A broad peak corresponding to the O–H stretching vibrations was observed at 3296 cm$^{-1}$. Sharp absorption peaks were observed between 2871 and 2965 cm$^{-1}$, which were assigned to C–H stretching vibrations. A peak observed at 1705 cm$^{-1}$ corresponds to the stretching vibrations of C=O in the ester bonds of the curing agent. Absorption peaks between 1475 and 1609 cm$^{-1}$ were attributed to the stretching vibration of the aromatic rings. Peaks were observed at 1246 cm$^{-1}$ and 1101 cm$^{-1}$ corresponding to C–O and O–C stretching vibrations, respectively. The peak observed at 832 cm$^{-1}$ corresponds to H–C$_{arom}$ bending vibrations.

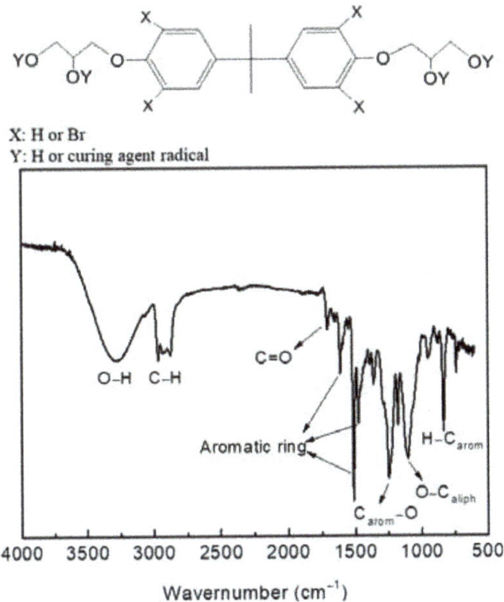



**Figure 6.** Fourier-transform infrared spectrum (FTIR) of the glycolysis product (pre-polyol).

In addition, the $^1$H NMR and $^{13}$C NMR spectra were recorded to examine the chemical structures of the final product as shown in Figure 7. In the $^1$H NMR (Figure 7a), peaks observed between 6.6 and 7.3 ppm correspond to the aromatic ring protons. The signals ranging from 3.3 to 4 ppm correspond to the aliphatic chain protons. The peak at 1.6 ppm is derived from the methyl group proton of the bis-phenol A. Figure 7b shows the $^{13}$C NMR of the final product. Peaks observed between 112 and 155 ppm correspond to the aromatic ring carbons. Peaks observed at 60–72 ppm are associated with the aliphatic carbons of bisphenol A diglycidyl ether. Peaks observed at 30 ppm and 43 ppm are identical to the methyl group and methane carbons, respectively, verifying the presence of bisphenol A diglycidyl ether and tetrabromobisphenol A, while the curing agent structures in the final product are hardly identified.

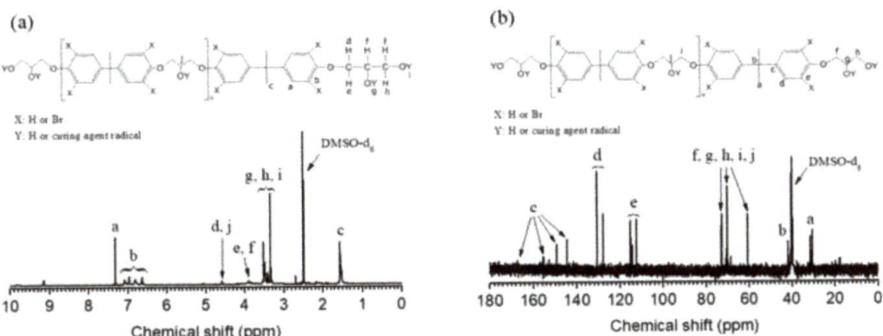



**Figure 7.** $^1$H NMR (**a**); and $^{13}$C NMR (**b**) spectra of the glycolyses product (pre-polyol) in DMSO-$d_6$.

A model of the epoxy resin based on bisphenol A diglycidyl ether and tetrahydrophthalic anhydride was successfully decomposed using a PEG 200/NaOH system. Scheme 1 shows the plausible reaction mechanism for glycolysis. The ester bonds of the cured epoxy resin are broken by hydrolysis, affording an organic sodium salt. The organic salt structure includes bisphenol A diglycidyl ether, tetrabromobisphenol A, and the curing agent, which is soluble in a mixture of PEG and water. After the addition of HCl, the sodium cations in organic salts are replaced by protons, and the organic salts are converted to polyol ethers, which are not soluble in water because of their high molecular weight and branched structure; hence, the polyol ethers precipitate to form a solid phase.

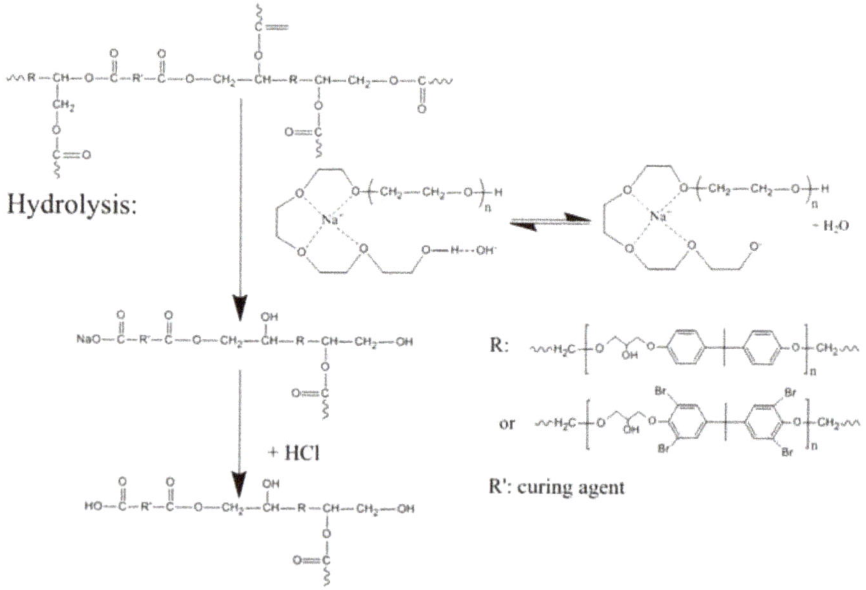

**Scheme 1.** Plausible reaction mechanism for the glycolysis.

PEG collected after the first recycling step by using virgin PEG 200 was continuously used for the decomposition of PCBs under the optimal reaction conditions of a PCB to PEG mass ratio of 1:7, a reaction temperature of 180 °C, and a reaction time of 3 h. Recycled PEG was purified by the removal of the sodium chloride formed during the reaction between the sodium organic salt and HCl. The addition of tetrahydrofuran (THF) led to the removal of sodium chloride from PEG. During the addition of THF, sodium chloride precipitated to a mixture of recycled PEG due to the different solubilities of PEG and sodium chloride in THF. The precipitates were separated using a filter, and the filtrate contained PEG and THF, followed by the extraction of the recycled PEG using a rotary evaporator based on the different boiling points of PEG and THF. This process was repeated until sodium chloride was completely removed. Finally, PEG was further dried overnight in a vacuum oven at 60 °C to completely remove THF. In addition, PEG after the second and third recycling steps was collected by the above process after reusing the PEG obtained from the first and second recycling steps, respectively. Figure 8a shows the TGA curves of the glass fiber collected after the reaction with virgin PEG 200, as well as the first, second, and third recycled PEG. The weight loss of the glass fiber after glycolysis of PCBs with the PEG recycled from the first step slightly decreased in comparison to virgin PEG 200, while those of the glass fiber after the glycolysis of PCBs with the second and third recycle PEG were considerably reduced. The increased weight loss implied the reduction of the glycolysis yield (Figure 8b). With increasing number of cycles for recycling PEG, the glycolysis yield decreased because of the degradation of PEG 200 and formation of a lower-molecular-weight PEG during glycolysis. Lower- or higher-molecular-weight PEG compared to PEG 200 did not coordinate with sodium ions to form a catalyst for glycolysis.

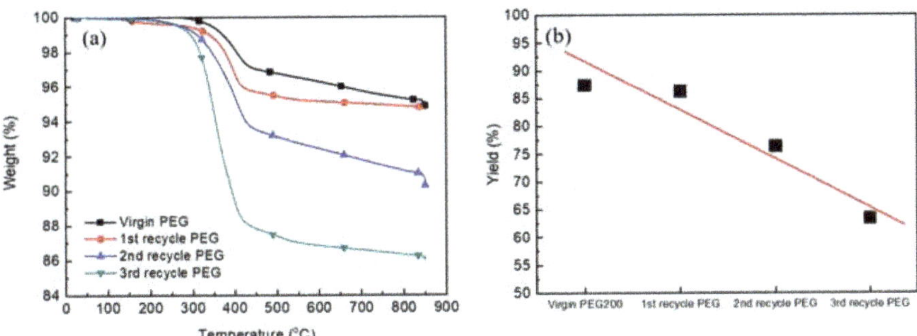

**Figure 8.** TGA curves of the glass fiber after the decomposition with virgin PEG 200 and recycled PEG (**a**) and the glycolysis yield of the recycled PEG (**b**).

### 3.2. Characteristics of RPUFs Prepared from the Recycled Polyol Based on the Glycolysis Product of UPCBs

The glycolysis product and recycled polyol were composed of phenol and phenolic derivatives, and brominated compounds as confirmed by Py−GC/MS (Figures S2 and S3). The representative chemical structures detected in Py−GC/MS are summarized in Tables S1–S3. As shown in Tables S1 and S2, the glycolysis product showed similar chemical structure to conventional brominated epoxy resin. It demonstrates that the organic parts of UPCBs are comprised of diglycidyl ether of bisphenol A containing brominated bisphenol A as flame retardant. The representative brominated compounds in glycolysis products and recycled polyol identified from Py−GC/MS/ECD are summarized in Tables S5 and S6. The bromine content determined by EDS and GC/ECD was estimated as the relative ratio by examining the conventional brominated epoxy resin with already known bromine content. Table 1 summarizes the bromine content of the pre-polyol and recycled polyol. The bromine content of pre-polyol determined by EDS and GC/ECD was similar, i.e., 15.0% and 14.2%, respectively. The recycled polyol was a viscous liquid; thus, the bromine content was only measured by GC/ECD. The bromine content of the recycled polyol was 1.8%, indicating a considerably lower value compared to pre-polyol, probably related to the detachment of the bromine atoms from the benzene ring during the chemical modification. Furthermore, the increased molecular weight after the modification of pre-polyol led to the dilution of the bromine atom concentration of the overall molecules. Thus, the brominated phenolic derivatives were hardly detected in Py−GC/MS/ECD spectrum of recycled polyol (Table S2). Table 4 summarizes the content of metals in recycled polyol as determined by ICP/MS. The content of the representative elements, such as Al, Ni, Cu, Cd, and Pb, were evaluated. As Al and Cu were the major components of PCB; their concentrations were greater than those of various metal elements. On the other hand, concentrations of heavy metals, such as Ni, Cd, and Pb, were almost not detected (~0). Notably, the recycled polyol obtained after chemical recycling and modification can replace the petrochemical polyols for manufacturing new RPUFs without environmental hazards. Furthermore, the recycled polyol containing abundant phenol derivatives is thought to contribute to the enhanced physical properties of the resulting foams.

**Table 4.** Content of metals in the recycled polyol.

| Element | Concentration (%) |
| --- | --- |
| Al | 0.0266 |
| Ni | 0.0000 |
| Cu | 0.0066 |
| Cd | 0.0000 |
| Pb | 0.0010 |

Table 5 summarizes the reactivities of recycled polyol during foaming. All of the characteristic times (i.e., cream time, gel time, and tack-free time) were more rapid with the increase in the recycled polyol content despite the low content of the amine catalyst (PC-8). The accelerated reactivities were related to the catalytic effect of the tertiary amine in the recycled polyol formed during the Mannich reaction.

**Table 5.** Characteristic times and density of RPUFs.

| Sample | CON | P20 | P40 | P60 |
|---|---|---|---|---|
| Cream time (s) | 19 | 15 | 14 | 13 |
| Gel time (s) | 108 | 70 | 55 | 48 |
| Tack-free time (s) | 170 | 120 | 95 | 85 |
| Density (kg/m$^3$) | 44.1 | 44.0 | 44.7 | 45.0 |

Figure 9 shows the cross-sectional SEM images of RPUFs with different contents of recycled polyol. All RPUFs exhibited a polyhedral and uniform closed-cell morphology without shrinkage or collapse even at a high recycled polyol content (60 wt%). With the increase in the recycled polyol content, the average cell size clearly decreased considerably. The rapid reactivity of the recycled polyol should restrict the expansion and coalescence of bubbles during foaming, leading to a smaller cell size of the resulting foams. Recycled polyol is the PO adduct with an abundant benzene ring; hence, it may exhibit a high affinity toward the raw materials for RPUFs, especially pMDI. High compatibility among the components contributed to the formation of a finer, homogenous cell structure.

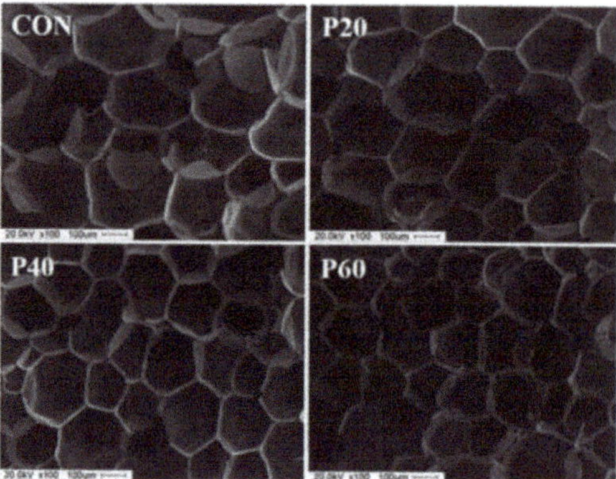


**Figure 9.** SEM micrographs of the RPUFs prepared from recycled polyol.

In addition, the closed-cell content of RPUF is one of the important parameters for thermal insulating materials [33,34]. The thermal conductivities of carbon dioxide (0.0150 W/m°C) and HFA-365/227 (0.0107 W/m°C) are considerably less than that of air (0.0240 W/m°C). Thus, blowing gas entrapped in a closed cell cannot escape from the cell, maintaining a low thermal conductivity. Figure 10 shows the closed-cell content of RPUFs prepared with different contents of recycled polyol. The closed-cell content increased to the maximum of 93.2% for P40 from 89.1% for CON. Although the closed-cell content for RPUF containing 60 wt% of the recycled polyol was slightly decreased, those of all of the RPUFs prepared using recycled polyol were greater than that of CON, revealing that the incorporation of recycled polyol for preparing RPUFs does not cause issues in the cell structure, including opening and breaking.

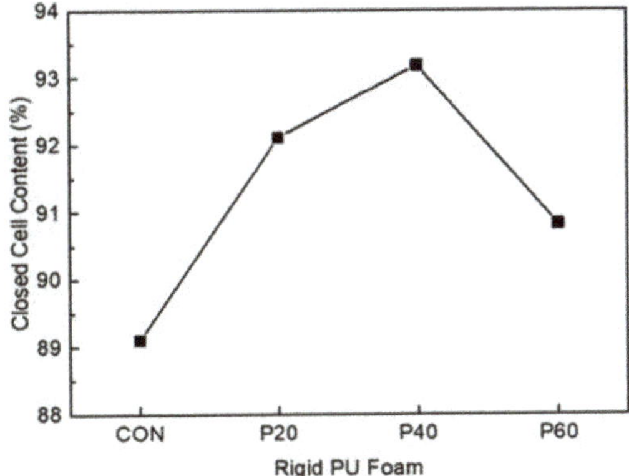

**Figure 10.** Closed-cell content of the RPUFs prepared from recycled polyol.

Figure 11 shows the thermal conductivities of RPUFs with different contents of recycled polyol. Typically, the thermal conductivity of RPUFs is strongly affected by the cell size, closed-cell content, thermal conductivity of the blowing gas entrapped in the cells, and density [30,35–37]. In this study, RPUFs prepared using recycled polyol exhibited a smaller cell size and higher closed-cell content compared to CON. From these contributions, the thermal conductivity of recycled-polyol-based RPUF is less than that of CON. In particular, P40 exhibited the minimum thermal conductivity of 0.0184 kcal/mh°C, which was in good agreement with the highest closed-cell content.

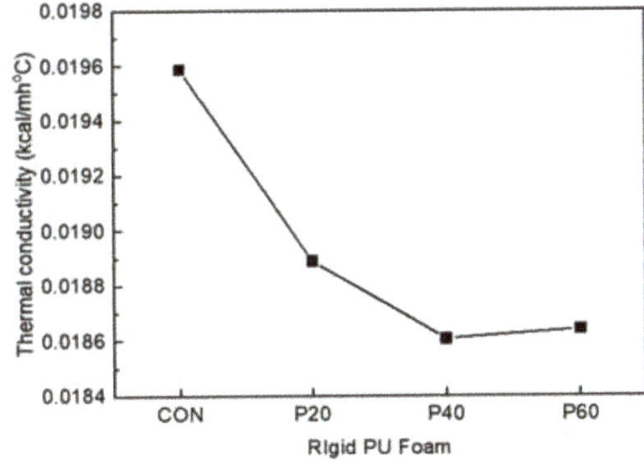

**Figure 11.** Thermal conductivity (K-factor) of RPUFs with different contents of recycled polyol.

Figure 12 shows the normalized compressive strength in the direction of blowing (B) and in the direction perpendicular to blowing (P) at a 10% strain of RPUFs with different contents of recycled polyol. Recycled-polyol-based RPUFs exhibited superior compressive strength in both the measured directions compared to CON due to the smaller cell size [38]. Besides, a significant amount of the aromatic ring in the recycled polyol would render additional strength to the cell, leading to the improved compressive strength of RPUFs. The maximum compressive strengths were observed in both directions for the RPUF containing 40 wt% of the recycled polyol. The compressive strengths of P40 in the direction of blowing and in the direction perpendicular to blowing were improved by 62% and 18%, respectively, compared to CON. However, with the further increase in the recycled polyol content of up to 60 wt%, the compressive strength decreases, possibly related to the presence of slight defects in the cell structure; this decreased strength was already confirmed by the decrease in the closed-cell content of P60. However, its compressive strengths (B: 0.128 MPa and P: 0.169 MPa) were comparable to CON (B: 0.112 MPa and P: 0.178 MPa).

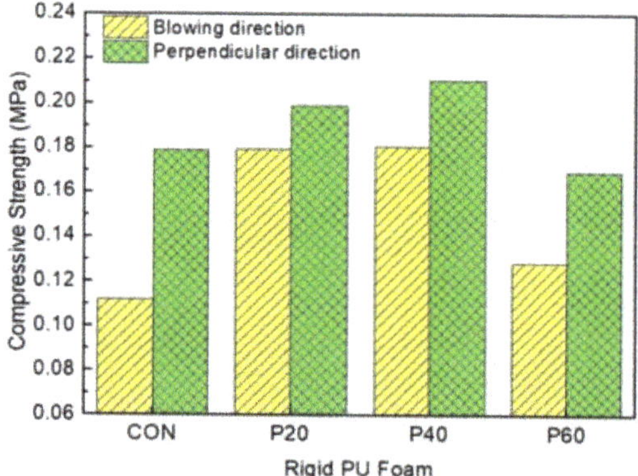


**Figure 12.** Compressive strengths measured in the blowing direction (B) and direction perpendicular to the blowing (P) of RPUFs with different contents of recycled polyol.

The thermal degradation behavior of RPUFs with different contents of recycled polyol was evaluated by TGA under nitrogen. Thermogravimetry (TG) and derivative TG (DTG) thermograms of RPUFs are shown in Figure 13. The TGA curve of CON exhibited a typical thermal degradation behavior for RPUF. Three major thermal decomposition steps were observed. The first step corresponded to the thermal degradation of the urethane unit at 190–220 °C; the second step corresponded to the thermal degradation of polyol at 300–360 °C; and the third step observed at 480–520 °C corresponded to the thermal degradation of aromatic isocyanate and hydrocarbons [39–41]. On the other hand, recycled-polyol-based RPUFs exhibited additional thermal degradation at 429 °C, corresponding to the degradation of recycled polyol. Thus, the weight loss at this temperature increases with the recycled polyol content. $T_{max3}$ corresponding to the thermal decomposition of aromatic groups shifted to lower temperature, eventually overlapping possibly with the thermal degradation of recycled polyol. Table 6 summarizes the maximum degradation temperature ($T_{max}$) at each degradation stage and residue% after the degradation of RPUFs. The thermal degradation of the hard segment occurred at lower temperature with increasing content of recycled polyol. However, in the major thermal degradation stage at which the most weight loss ($T_{max2}$) occurred, the RPUF with more recycled polyol was degraded at the higher temperature. In particular, the $T_{max2}$ of the polyol increased by 25 °C from 314.0 °C for CON to 339.2 °C for P60. In addition, RPUFs based on the recycled polyol retained a high amount of residual char after heating in TGA. Clearly, the incorporation of recycled polyol can enhance the thermal stability due to the presence of thermally stable bromine atoms and the aromatic ring [42–46].

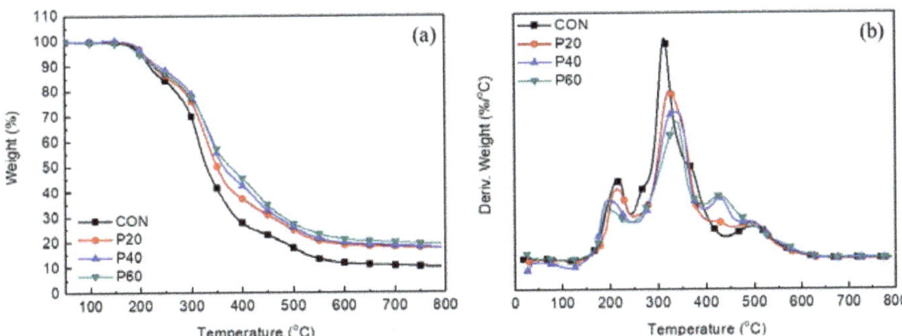

**Figure 13.** Thermogravimetry (TG) (**a**); and derivative TG (DTG) (**b**) thermograms of RPUFs with different content of recycled polyol under nitrogen atmosphere.

**Table 6.** Maximum degradation temperature ($T_{max}$), percentage of residue after degradation in TGA, and $T_g$ of RPUFs with different contents of recycled polyol.

| Sample | $T_{max1}$ (°C) | $T_{max2}$ (°C) | $T_{max3}$ (°C) | Residue (%) | $T_g$ (°C) |
|--------|-----------------|-----------------|-----------------|-------------|------------|
| CON | 216.8 | 314.0 | 496.6 | 10.1 | 134.6 |
| P20 | 211.7 | 324.5 | 494.6 | 17.6 | 149.4 |
| P40 | 200.2 | 338.1 | 488.2 | 18.0 | 165.9 |
| P60 | 195.0 | 339.2 | - | 19.1 | 167.6 |

Figure 14 shows the results obtained by DMA (i.e., storage modulus (G′), loss modulus (G″), and tan delta) of RPUFs with different contents of the recycled polyol. Table 6 summarizes the glass transition temperature ($T_g$), which is determined by the maximum point of the loss modulus and $T_g$ values. The G′ (rubbery region) and $T_g$ values of RPUFs increased with the recycled polyol content due to the high amount of the aromatic ring in the recycled polyol. At a temperature greater than $T_g$ at which the molecular chain was softened to move, the presence of a rigid aromatic ring in the polymer network restricted the movement of the molecular chain, leading to increased G′ and $T_g$. With the increase in the recycled polyol content, the tan delta curves broadened, and the maximum point was hardly observed for the RPUFs containing greater than 40 wt% of the recycled polyol content (Figure 14c). Typically, the width of a tan delta curve is closely related to the phase separation between the soft and hard segments in a polyurethane network. The broadening of the tan delta curve revealed glass transition of the crosslinked polyurethanes [47–49]. As mentioned above, the recycled polyol used in this study is an aromatic ring-rich PO adducts; thus, it exhibits good compatibility with pMDI. It would induce the phase mixing of soft and hard segments in the polyurethane network, leading to an increase of loss modulus peak or tan delta peak temperatures for RPUF loaded with a high content of recycled polyol.

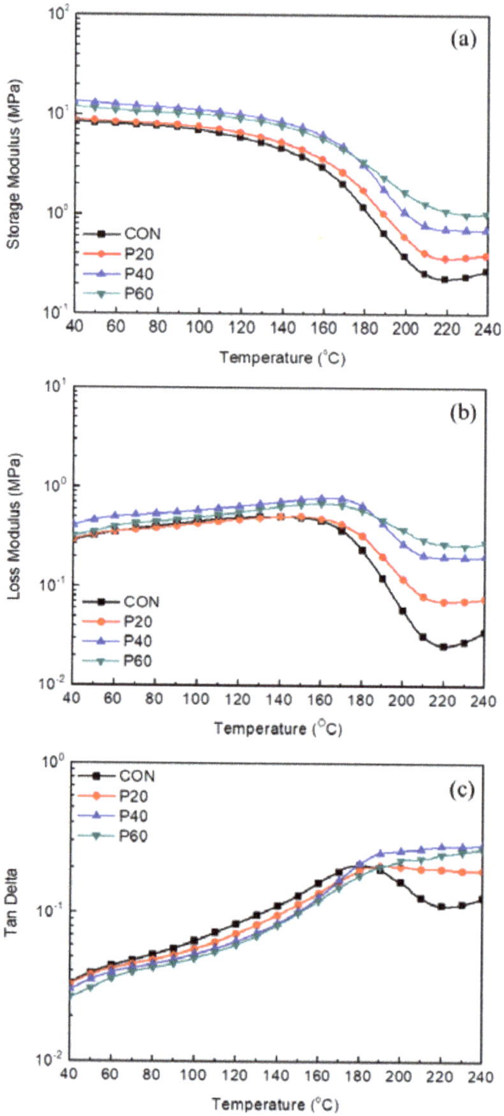

**Figure 14.** Dynamic mechanical analyzer (DMA) results of RPUFs with different content of recycled polyol as a function of temperature: storage modulus (G′) (**a**); loss modulus (G″) (**b**); and Tan delta (**c**).

## 4. Conclusions

In this study, pulverized UPCBs were successfully depolymerized by glycolysis with PEG 200 under basic conditions, and organic materials and glass fibers were recovered. In addition, the glycolysis conditions, including the PCB to PEG 200 mass ratio, reaction temperature, and reaction time, were optimized. The glass fibers obtained after glycolysis exhibited a smooth surface and short length without destruction, indicating that the glass fibers can be reused as reinforcements. Furthermore, the bromine content of the glycolysis products was ~15% with a significant amount of the aromatic groups, which was derived from the brominated epoxy resins of PCBs. Almost no heavy metals were detected in the organic components. PEG 200 recovered after glycolysis was collected and reused again for the glycolysis of PCBs, and the glycolysis efficiency of recovered PEG 200 was obtained by the weight loss of the organic materials remaining in the glass fiber after glycolysis. The glycolysis yield slightly decreased with increasing the number of recycling PEG 200. The organic product (pre-polyol) obtained after glycolysis was converted to a recycled polyol via Mannich reaction and addition polymerization of PO to prepare RPUFs for thermal insulation. Besides, the recycled polyol could replace 60 wt% of conventional PPG for RPUFs without difficulties and deterioration of foam properties. The RPUFs based on recycled polyol exhibited superior compressive strength, thermal insulation property, and thermal stability to the conventional RPUF. The superior performance of RPUFs based on recycled polyol was attributed to the fine cell morphology as well as the presence of aromatic rings in the recycled polyol. It is inferred that the recycled polyol obtained from glycolysis and chemical modification of UPCBs can be promising feedstock for thermal insulating RPUFs.

**Supplementary Materials:** The following are available online at http://www.mdpi.com/2227-9717/7/1/22/s1, Figure S1: FTIR spectrum of recycled polyol obtained by modification of glycolysis product of UPCBs, Figure S2: Py–GC/MS of conventional brominated epoxy resin of tetrabromobisphenol A, Figure S3: Py–GC/MS of glycolysis product obtained from chemical recycling of UPCBs, Figure S4: Py–GC/MS of recycled polyol obtained from modification of glycolysis product, Table S1: Representative chemical structure of standard brominated epoxy resin from Py–GC/MS, Table S2: Representative chemical structure of glycolysis product by from Py–GC/MS, Table S3: Representative chemical structure of recycled polyol from Py–GC/MS, Table S4: Representative brominated compounds in conventional brominated epoxy resin by using Py–GC/MS/ECD, Table S5: Representative brominated compounds in glycolysis product by using Py–GC/MS/ECD, Table S6: Representative brominated compounds in recycled polyol by using Py–GC/MS/ECD.

**Author Contributions:** S.R.S., V.D.M., and D.S.L. contributed to conceptualization, methodology, design, and writing draft. S.R.S. and V.D.M. performed the investigation and data analysis.

**Funding:** This study was supported by the R & D Center for Valuable Recycling (Global-Top R & BD Program) of the Ministry of Environment (Project No: 2016002240004).

**Conflicts of Interest:** The authors declare no conflict of interest.

## References

1. Yin, J.; Li, G.; He, W.; Huang, J.; Xu, M. Hydrothermal decomposition of brominated epoxy resin in waste printed circuit boards. *J. Anal. Appl. Pyrolysis* **2011**, *92*, 131–136. [CrossRef]

2. Zhan, Z.H.; Qiu, K.Q. Pyrolysis kinetics and TG-FTIR analysis of waste epoxy printed circuit boards. *J. Cent. South Univ. Technol.* **2011**, *18*, 331–336. [CrossRef]

3. Quan, C.; Li, A.; Gao, N.; dan, Z. Characterization of products recycling from PCB waste pyrolysis. *J. Anal. Appl. Pyrolysis* **2010**, *89*, 102–106. [CrossRef]

4. De Marco, I.; Caballero, B.M.; Chomón, M.J.; Laresgoiti, M.F.; Torres, A.; Fernández, G.; Arnaiz, S. Pyrolysis of electrical and electronic wastes. *J. Anal. Appl. Pyrolysis* **2008**, *82*, 179–183. [CrossRef]

5. Hadi, P.; Ning, C.; Ouyang, W.; Xu, M.; Lin, C.S.; McKay, G. Toward environmentally-benign utilization of nonmetallic fraction of waste printed circuit boards as modifier and precursor. *Waste Manag.* **2015**, *35*, 236–246. [CrossRef] [PubMed]

6. Guo, J.; Guo, J.; Xu, Z. Recycling of non-metallic fractions from waste printed circuit boards: A review. *J. Hazard. Mater.* **2009**, *168*, 567–590. [CrossRef] [PubMed]

7. Chien, Y.-C.; Wang, H.P.; Lin, K.-S.; Huang, Y.-J.; Yang, Y.-W. Fate of bromine in pyrolysis of printed circuit board wastes. *Chemosphere* **2000**, *40*, 383–387. [CrossRef]

8. Grause, G.; Furusawa, M.; Okuwaki, A.; Yoshioka, T. Pyrolysis of tetrabromobisphenol-A containing paper laminated printed circuit boards. *Chemosphere* **2008**, *71*, 872–878. [CrossRef]

9. Kim, Y.-M.; Han, T.U.; Watanabe, C.; Teramae, N.; Park, Y.-K.; Kim, S.; Hwang, B. Analytical pyrolysis of waste paper laminated phenolic-printed circuit board (PLP-PCB). *J. Anal. Appl. Pyrolysis* **2015**, *115*, 87–95. [CrossRef]

10. Ortuno, N.; Conesa, J.A.; Molto, J.; Font, R. Pollutant emissions during pyrolysis and combustion of waste printed circuit boards, before and after metal removal. *Sci. Total Environ.* **2014**, *499*, 27–35. [CrossRef]

11. El Gersifi, K.; Destais-Orvoën, N.; Durand, G.; Tersac, G. Glycolysis of epoxide-amine hardened networks. I. Diglycidylether/aliphatic amines model networks. *Polymer* **2003**, *44*, 3795–3801. [CrossRef]

12. Jin, Y.Q.; Tao, L.; Chi, Y.; Yan, J.H. Conversion of bromine during thermal decomposition of printed circuit boards at high temperature. *J. Hazard. Mater.* **2011**, *186*, 707–712. [CrossRef]

13. Zhu, P.; Chen, Y.; Wang, L.Y.; Qian, G.Y.; Zhou, M.; Zhou, J. A new technology for separation and recovery of materials from waste printed circuit boards by dissolving bromine epoxy resins using ionic liquid. *J. Hazard. Mater.* **2012**, *239–240*, 270–278. [CrossRef] [PubMed]

14. Zhou, Y.; Wu, W.; Qiu, K. Recovery of materials from waste printed circuit boards by vacuum pyrolysis and vacuum centrifugal separation. *Waste Manag.* **2010**, *30*, 2299–2304. [CrossRef]

15. Guan, J.; Li, Y.-S.; Lu, M.-X. Product characterization of waste printed circuit board by pyrolysis. *J. Anal. Appl. Pyrolysis* **2008**, *83*, 185–189. [CrossRef]

16. Veit, H.M.; Diehl, T.R.; Salami, A.P.; Rodrigues, J.S.; Bernardes, A.M.; Tenorio, J.A. Utilization of magnetic and electrostatic separation in the recycling of printed circuit boards scrap. *Waste Manag.* **2005**, *25*, 67–74. [CrossRef] [PubMed]

17. Akonda, M.H.; Lawrence, C.A.; Weager, B.M. Recycled carbon fibre-reinforced polypropylene thermoplastic composites. *Compos. Part A* **2012**, *43*, 79–86. [CrossRef]

18. Soler, A.; Conesa, J.A.; Ortuno, N. Emissions of brominated compounds and polycyclic aromatic hydrocarbons during pyrolysis of E-waste debrominated in subcritical water. *Chemosphere* **2017**, *186*, 167–176. [CrossRef]

19. Luda, M.P.; Balabanovich, A.I.; Hornung, A.; Camino, G. Thermal degradation of a brominated bisphenol a derivative. *Polym. Adv. Technol.* **2003**, *14*, 741–748. [CrossRef]

20. Guo, X.; Qin, F.G.F.; Yang, X.; Jiang, R. Study on low-temperature pyrolysis of large-size printed circuit boards. *J. Anal. Appl. Pyrolysis* **2014**, *105*, 151–156. [CrossRef]

21. Dang, W.; Kubouchi, M.; Sembokuya, H.; Tsuda, K. Chemical recycling of glass fiber reinforced epoxy resin cured with amine using nitric acid. *Polymer* **2005**, *46*, 1905–1912. [CrossRef]

22. Molero, C.; de Lucas, A.; Rodríguez, J.F. Recovery of polyols from flexible polyurethane foam by "split-phase" glycolysis with new catalysts. *Polym. Degrad. Stab.* **2006**, *91*, 894–901. [CrossRef]

23. Lee, S.-H.; Ohkita, T.; Teramoto, Y. Polyol recovery from biomass-based polyurethane foam by glycolysis. *J. Appl. Polym. Sci.* **2005**, *95*, 975–980. [CrossRef]

24. Destais-Orvoën, N.; Durand, G.; Tersac, G. Glycolysis of epoxide–amine hardened networks II—Aminoether model compound. *Polymer* **2004**, *45*, 5473–5482. [CrossRef]

25. Yang, P.; Zhou, Q.; Yuan, X.-X.; van Kasteren, J.M.N.; Wang, Y.-Z. Highly efficient solvolysis of epoxy resin using poly(ethylene glycol)/NaOH systems. *Polym. Degrad. Stab.* **2012**, *97*, 1101–1106. [CrossRef]

26. Zhou, Y.; Qiu, K. A new technology for recycling materials from waste printed circuit boards. *J. Hazard. Mater.* **2010**, *175*, 823–828. [CrossRef]

27. Zheng, Y.; Shen, Z.; Cai, C.; Ma, S.; Xing, Y. The reuse of nonmetals recycled from waste printed circuit boards as reinforcing fillers in the polypropylene composites. *J. Hazard. Mater.* **2009**, *163*, 600–606. [CrossRef]

28. Hadi, P.; Xu, M.; Lin, C.S.; Hui, C.-W.; McKay, G. Waste printed circuit board recycling techniques and product utilization. *J. Hazard. Mater.* **2015**, *283*, 234–243. [CrossRef]

29. Guo, J.; Li, J.; Rao, Q.; Xu, Z. Phenolic molding compound filled with nonmetals of waste PCBs. *Environ. Sci. Technol.* **2007**, *42*, 624–628. [CrossRef]

30. Yan, D.; Xu, L.; Chen, C.; Tang, J.; Ji, X.; Li, Z. Enhanced mechanical and thermal properties of rigid polyurethane foam composites containing graphene nanosheets and carbon nanotubes. *Polym. Int.* **2012**, *61*, 1107–1114. [CrossRef]

31. Hu, D.; Jia, Z.; Li, J.; Zhong, B.; Fu, W.; Luo, Y.; Jia, D. Characterization of Waste Printed Circuit Boards Nonmetals and its Reutilization as Reinforcing Filler in Unsaturated Polyester Resin. *J. Polym. Environ.* **2017**, *26*, 1311–1319. [CrossRef]

32. Sun, Z.; Shen, Z.; Ma, S.; Zhang, X. Sound absorption application of fiberglass recycled from waste printed circuit boards. *Mater. Struct.* **2013**, *48*, 387–392. [CrossRef]

33. Lee, C.S.; Ooi, T.L.; Chuah, C.H.; Ahmad, S. Rigid Polyurethane Foam Production from Palm Oil-Based Epoxidized Diethanolamides. *J. Am. Oil Chem. Soc.* **2007**, *84*, 1161–1167. [CrossRef]

34. Javni, I.; Zhang, W.; Petrović, Z. Soybean-oil-based polyisocyanurate rigid foams. *J. Polym. Environ.* **2004**, *12*, 123–129. [CrossRef]

35. Kim, Y.H.; Choi, S.J.; Kim, J.M.; Han, M.S.; Kim, W.N.; Bang, K.T. Effects of organoclay on the thermal insulating properties of rigid polyurethane poams blown by environmentally friendly blowing agents. *Macromol. Res.* **2007**, *15*, 676–681. [CrossRef]
36. Kim, S.H.; Kim, B.K.; Lim, H. Effect of isocyanate index on the properties of rigid polyurethane foams blown by HFC 365mfc. *Macromol. Res.* **2008**, *16*, 467–472. [CrossRef]
37. Han, M.S.; Choi, S.J.; Kim, J.M.; Kim, Y.H.; Kim, W.N.; Lee, H.S.; Sung, J.Y. Effects of silicone surfactant on the cell size and thermal conductivity of rigid polyurethane foams by environmentally friendly blowing agents. *Macromol. Res.* **2009**, *17*, 44–50. [CrossRef]
38. Kim, Y.H.; Kang, M.J.; Park, G.P.; Park, S.D.; Kim, S.B.; Kim, W.N. Effects of liquid-type silane additives and organoclay on the morphology and thermal conductivity of rigid polyisocyanurate-polyurethane foams. *J. Appl. Polym. Sci.* **2012**, *124*, 3117–3123. [CrossRef]
39. Bakhshi, H.; Yeganeh, H.; Yari, A.; Nezhad, S.K. Castor oil-based polyurethane coatings containing benzyl triethanol ammonium chloride: Synthesis, characterization, and biological properties. *J. Mater. Sci.* **2014**, *49*, 5365–5377. [CrossRef]
40. Gu, R.; Sain, M.M. Effects of Wood Fiber and Microclay on the Performance of Soy Based Polyurethane Foams. *J. Polym. Environ.* **2012**, *21*, 30–38. [CrossRef]
41. Pawar, M.S.; Kadam, A.S.; Dawane, B.S.; Yemul, O.S. Synthesis and characterization of rigid polyurethane foams from algae oil using biobased chain extenders. *Polym. Bull.* **2015**, *73*, 727–741. [CrossRef]
42. Lakshmi, M.S.; Narmadha, B.; Reddy, B.S.R. Enhanced thermal stability and structural characteristics of different MMT-Clay/epoxy-nanocomposite materials. *Polym. Degrad. Stab.* **2008**, *93*, 201–213. [CrossRef]
43. Dominguez-Rosado, E.; Liggat, J.; Snape, C.; Eling, B.; Pichtel, J. Thermal degradation of urethane modified polyisocyanurate foams based on aliphatic and aromatic polyester polyol. *Polym. Degrad. Stab.* **2002**, *78*, 1–5. [CrossRef]
44. Lee, S.C.; Sze, Y.W.; Lin, C.C. Polyurethanes synthesized from polyester polyols derived from PET waste. II. Thermal properties. *J. Appl. Polym. Sci.* **1994**, *52*, 869–873. [CrossRef]
45. Du, H.; Zhao, Y.; Li, Q.; Wang, J.; Kang, M.; Wang, X.; Xiang, H. Synthesis and characterization of waterborne polyurethane adhesive from MDI and HDI. *J. Appl. Polym. Sci.* **2008**, *110*, 1396–1402. [CrossRef]
46. Xiong, X.; Zhou, L.; Ren, R.; Ma, X.; Chen, P. Thermal, mechanical properties and shape memory performance of a novel phthalide-containing epoxy resins. *Polymer* **2018**, *140*, 326–333. [CrossRef]
47. Mora-Murillo, L.D.; Orozco-Gutierrez, F.; Vega-Baudrit, J.; González-Paz, R.J. Thermal-Mechanical Characterization of Polyurethane Rigid Foams: Effect of Modifying Bio-Polyol Content in Isocyanate Prepolymers. *J. Renew. Mater.* **2017**, *5*, 220–230. [CrossRef]
48. Paruzel, A.; Michałowski, S.; Hodan, J.; Horák, P.; Prociak, A.; Beneš, H. Rigid Polyurethane Foam Fabrication Using Medium Chain Glycerides of Coconut Oil and Plastics from End-of-Life Vehicles. *ACS Sustain. Chem. Eng.* **2017**, *5*, 6237–6246. [CrossRef]
49. Sharma, C.; Kumar, S.; Unni, A.R.; Aswal, V.K.; Rath, S.K.; Harikrishnan, G. Foam stability and polymer phase morphology of flexible polyurethane foams synthesized from castor oil. *J. Appl. Polym. Sci.* **2014**, *131*, 40668. [CrossRef]

*Article*

# Preparation and Performance of Different Modified Ramie Fabrics Reinforced Anionic Polyamide-6 Composites

Ze Kan [ID], Hao Shi, Erying Zhao and Hui Wang *

Key Laboratory of Biobased Polymer Materials, School of Polymer Science and Engineering, Qingdao University of Science and Technology, Shandong Provincial Education Department, Qingdao 266042, Shandong, China; zkan@qust.edu.cn (Z.K.); shihao0719@163.com (H.S.); eryingzhao123@163.com (E.Z.)
* Correspondence: h54wang@uwaterloo.ca; Tel.: +86-0532-8402-2927

Received: 21 February 2019; Accepted: 16 April 2019; Published: 22 April 2019

**Abstract:** Anionic polyamide-6 (APA-6) composites are prepared by the VARIM process using different modified ramie fabrics to study the structure and properties of different composites. This study can not only evaluate the optimal modification method for the ramie fabrics, but also further explore the interface interaction mechanism between ramie fabrics and APA-6. In this article, the ramie fabrics are modified by a pretreatment, coupling agent and alkali modification. Different modification methods have different effects on the structure, surface properties and mechanical properties of ramie fabrics, which will further affect the impregnation process, interfacial and mechanical properties of the composites. Through the performance analysis of different modified ramie fabrics reinforced APA-6 composites, the conversion, crystallinity and molecular weight of these composites are at a high level, which indicate that the polymerization of these composites is well controlled. The coupling agent modified ramie fabrics composites and the pretreated ramie fabrics composites have higher flexural modulus, tensile strength and dynamic mechanical properties. Alkali-modified ramie fabrics composites have slightly lower mechanical properties, which however have the highest interlaminar shear strength and outperformed interface properties of the composites.

**Keywords:** ramie fabric; anionic polyamide-6; modification; polymerization; properties

---

## 1. Introduction

Natural fibers exhibit many excellent properties and are low-density materials, which probably lead to lightweight composites with a high specific strength and modulus. Natural fibers also offer significant cost advantages and easy processing along with being a highly biodegradable resource, as well as reducing the pollution to the environment [1–3]. Ramie fibers as one of the high-performance natural fibers [4] were selected and used in this work for the purpose of reinforcement of the resin.

Fabric reinforced composites are increasingly used to manufacture the high-performance and lightweight structures in the aerospace and automotive industry [5–7]. However, the wetting of fiber fabrics is still challenging. Thermoplastic resins usually exhibit a high melt viscosity in the range of 100~10,000 Pa·s, which made them much more difficult to impregnate into fiber. According to the previous report, the appropriate viscosity for fabric reinforced composites should normally be lower than 1 Pa·s [8]. Meanwhile, the high melting temperature of the thermoplastic resins would also have an adverse impact on natural fiber. In order to overcome these problems, a vacuum assisted resin infusion molding (VARIM) [9] was adopted by using a reactive thermoplastic polymer, i.e., anionic polyamide-6 (APA-6) [10–13]. The anionic ring opening polymerization of caprolactam is polymerized into a high molecular weight polyamide 6 (PA6) at 130~170 °C. The melted caprolactam monomer has a water-like viscosity, which facilitated it to impregnate into fiber. In addition, this method will provide more opportunities to form a chemical bonding between the fiber and resin [5,14]. Fabric reinforced APA-6 composites processed through VARIM have been extensively investigated by many researchers. Van Rijswijk et al. [5,6,14,15] studied the choice of activator and initiator, the influence of polymerization temperature, the influence of the vacuum infusion process, and the interfacial bond formation in detail. Pillay et al. [16,17] prepared the carbon fiber reinforced APA-6 composites. Yan et al. [18] reported the mechanical properties and microstructure of continuous glass fiber reinforced APA-6 composites.

There is a layer between the fiber reinforcement and the resin matrix, i.e., the interface layer. This layer plays a role of connecting the polymer matrix and the reinforcing fibers and determines the performance of the composites [4,5,15,19]. The strong hydrogen bonding between the natural fiber macromolecular chains makes them exhibit strong polarity and water absorption [20–22]. If it is compounded with a non-polar resin, the wettability and adhesion of the interface will be extremely poor. The fiber will peel off over time, and then the performance of the composites will be deteriorated. In order to obtain the best performance composites, it is very important to modify the surface of the reinforcing fiber. A lot of research has been carried out on the modification of natural fibers [23–29]. The physical modification methods include the steam explosion treatment, heat treatment, alkali treatment, low temperature plasma treatment, laser and high energy radiation treatment and the chemical modification methods include the surface coating, graft copolymerization and interface coupling.

At present, there appears to be no research focusing on the modification of natural fibers for using in the thermoplastic reaction processing. Therefore, in this work, four kinds of modification methods were selected to modify the natural fiber. The structure and properties of the modified natural fiber were studied. The optimum modification method of the ramie fiber suitable for VARIM process for the preparation of APA-6 composites was evaluated by studying the properties of the composites.

## 2. Materials and Methods

### 2.1. Materials

Anionic polymerization grade caprolactam with a low moisture content (<400 ppm) supplied by the BASF Chemical company was used. The sodium hydroxide (≥96% by mass, 'NaOH', KeLong Chemical, Chengdu, China), caprolactam magnesium bromide (1.4 mol/kg concentration in caprolactam, 'C1', Brüggemann Chemical, Heilbronn, Germany) were used as initiators. The di-functional hexamethylene-1,6-dicarbamoyl-caprolactam (2 mol/kg concentration in caprolactam, 'C20', Brüggemann Chemical, Heilbronn, Germany) was used as an activator. In this work, 1.2 mol % initiators were used together with 1.2 mol % activators. The ramie fabrics (warp/weft yarn count: 21 s × 21 s, warp/weft density: 52 × 36) was supplied by Chuan-dong Ramie Fabrics Corporation, Chongqing, China and the silane coupling agent KH550 was supplied by KeLong Chemical, Chengdu, China. The sodium hydroxide, sodium dodecylbenzenesulfonate, sodium carbonate and deionized water were supplied by KeLong Chemical, Chengdu, China.

### 2.2. Preparation of the Composites

#### 2.2.1. Modification of the Ramie Fabric

**Pretreatment**: The surface of natural fiber contains many impurities, including the dust, residual pectin, hemicellulose, lignin and sizing agent. The anionic polymerization will be affected by these impurities. Therefore, the impurities need to be removed from the commercial natural fibers prior to use in the preparation of the composites. First, the commercial ramie fabric was washed by water to remove dust, which was recorded as NF-0. Then, the NF-0 was immersed in an aqueous solution containing the sodium hydroxide, sodium dodecylbenzenesulfonate, sodium carbonate and deionized water according to a specific ratio, and the mixture was heated to 95 °C and stirred for 4 h. The pretreated fiber was recorded as NF-1.

**Coupling agent modification:** First, the ethanol and water were formulated into a solvent with a volume ratio of 1:3, and then the ammonia water was added dropwise to adjust the pH of the solvent. Then, 5% KH550 was added into the solution, and the resulting mixture was ultrasonically dispersed for 20 min and preheated in an oven at 80 °C. The pretreated fiber was then immersed in the preheated solvent for 12 h. After the reaction, the fiber was taken out and washed thoroughly with ethanol, and finally placed in an oven at 80 °C to dry to a constant weight. The treated fiber was recorded as NF-KH550.

**Alkali modification:** First, a 10% aqueous solution of sodium hydroxide was prepared and then the pretreated fiber was placed into it. The resulting mixture was given appropriate agitation, treated at room temperature for 15 min, then regenerated with ethanol or isopropanol for 5 min. Finally, the surface lye was thoroughly cleaned by water, placed on a drying plate, and dried in an oven at 60 °C for more than 24 h. The ethanol-regenerated fiber was recorded as NF-alk-alc, and the isopropanol-regenerated fiber was recorded as NF-alk-iso.

### 2.2.2. Preparation of Reactive Mixtures

The caprolactam was heated to 110 °C to melt. Then, after degassing the flask (15 min at 30 mbar), the initiator was added and mixed using a static mixer. Subsequently, the activator was fed in and mixed sufficiently after degassing the flask again (15 min at 30 mbar). Since APA-6 was sensitive to moisture and easy to oxidize at this temperature, the storage and processing would be carried out under a dry nitrogen environment.

### 2.2.3. VARIM

A balanced and symmetrical preform consisting of ten 200 × 200 mm layers of the textile fibers was laid onto a 300 × 300 × 3 mm$^3$ aluminum plate coated with a release agent (GS213-3, Airtech, Huntington Beach, USA). Then a layer of peel ply and a piece of resin distribution medium (Resinflow 90HT, Airtech, Huntington Beach, USA) were placed above the fibers. Two thermocouples were positioned at the surface of the top and the bottom layers of the preform to record and monitor the temperature. The inlet and outlet assembled by the springs and aluminum tubes were placed at corresponding positions. Then, the preform was bagged and sealed using a polyimide film and a heat resistant sealant tape. After degassing, the mold was rotated by 90 degrees and inserted in two infrared heating panels that were vertically placed. Then, the temperature was increased to 165 °C, and a continuous degassing (20 min at 30 mbar) were conducted to remove the moisture sufficiently. Next, the prepared reactive mixtures were infused into the preform via a heated (110 °C) silicon rubber tube. Once the resin reached the end of the preform, the inlet was clamped off. A pressure control system was used to set the infusion and curing pressure at 500 mbar precisely. After 60 min, the mold cooled down until the temperature reached below 30 °C. After that, it was de-molded directly and a composite with an average thickness of 2 mm was obtained. The diagram and specimen of ramie fabric reinforced APA-6 composites by VARIM are shown in Figure 1.

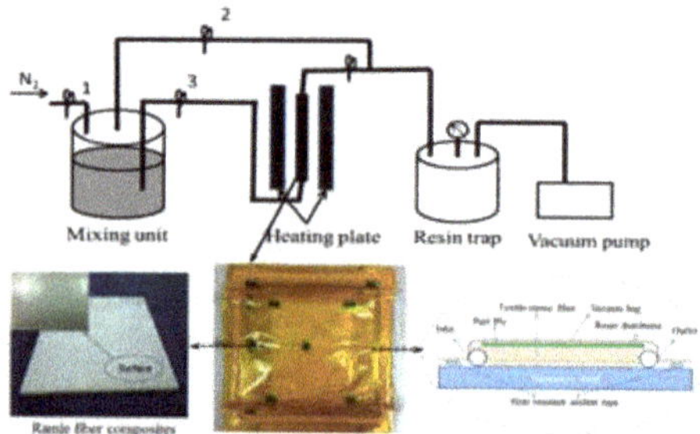



**Figure 1.** Diagram and specimen of ramie fabric reinforced anionic polyamide-6 (APA-6) composites by VARIM.

### 2.3. Characterization Methods

#### 2.3.1. Degree of Conversion

The samples taken near the center were ground into powder, weighed ($m_{tot}$) and put into cellulose thimbles. A Soxhlet extraction was conducted using a heated demineralized water for 12 h at 105 °C. Then, the samples dried were in an oven at 60 °C for an additional 12 h, and weighed again ($m_{pol}$). The monomer caprolactam dissolves easily in water, and APA-6 is water immiscible. Additionally, the fiber weight content ($m_f$) of the composites was determined in a subsequent step by dissolving it into formic acid. The degree of conversion (*DOC*) was then calculated using Equations (1) and (2):

$$DOC = \frac{m_{pol}}{m_{tot}} \times 100\% \tag{1}$$

$$DOC' = \frac{m_{pol} - m_f}{m_{tot} - m_f} \times 100\% \tag{2}$$

where *DOC* is the degree of conversion for pure APA-6, and the *DOC'* is for the composites.

#### 2.3.2. Degree of Crystallinity

The degree of crystallinity ($X_c$) of APA-6 and its composites were measured by the differential scanning calorimeter (DSC, TA Q20, New Castle, USA). The samples, approximately 5 mg, were obtained near the center of the composites and dried overnight at 50 °C in a vacuum oven. During testing, each sample was first held at 25 °C for 2 min before being heated up to 250 °C at 10 °C/min under $N_2$. The $X_c$ was then calculated using Equations (3) and (4):

$$X_c = \left(\frac{\Delta H_m}{\Delta H_{100}}\right)\left(\frac{1}{DOC}\right) \times 100\% \tag{3}$$

$$X_c' = \left(\frac{\Delta H_m}{\Delta H_{100}}\right)\left(\frac{m_{tot}}{m_{tot} - m_f}\right)\left(\frac{1}{DOC'}\right) \times 100\% \tag{4}$$

where $X_c$ is the degree of crystallinity for pure APA-6, and $X_c'$ is the degree of crystallinity for the composites; $\Delta H_m$ is the melting enthalpy of samples, and $\Delta H_{100}$ is the melting enthalpy of fully crystalline PA6: $\Delta H_{100}$ = 190 J/g [30].

#### 2.3.3. Viscosity-Average Molar Mass

The viscosity-average molar mass was determined by the dilute solution viscometry using the Ubbelohde viscometer with a 0.6~0.7 mm capillary diameter. Dried samples (150 mg) were dissolved in an aqueous formic acid (85%, 30 mL) to obtain a transparent solution. In particular, the composites were dissolved in formic acid to remove the fibers through filtration, and then the APA-6 resin was precipitated in deionized water. The viscosity-average molar mass (*Mv*) was calculated according to Equation (5).

$$[\eta] = K'Mv^{\alpha} \tag{5}$$

Here, $K'$ and $\alpha$ is the Mark-Houwink constants which have a specific value for each polymer-solvent combination. In this particular case, $K' = 0.023$ mL/g, $\alpha = 0.82$ [31].

### 2.3.4. Density and Void Content

The densities of pure APA-6 and its composites were measured using the water immersion method as specified in ASTM D-792. The rectangular specimens ($40 \times 10 \times 2$ mm) were measured by using a MH-120 densitometer (MatsuHaku, Taiwan, China). Prior to testing, all specimens were dried at 50 °C in a vacuum oven. The void content of a composite sample was measured according to the ASTM standard D2734-94, which can be calculated by Equation (6):

$$V_v = 100 - D_c\left(\frac{w_r}{d_r} + \frac{w_f}{d_f}\right) \tag{6}$$

where $D_c$, $d_r$, $w_r$ and $w_f$ are the measured composite density, resin density, resin weight fraction (%) and fiber weight fraction (%), respectively. The density of the ramie fiber ($d_f$) is 1.5 g·cm$^{-3}$ [1].

### 2.3.5. Mechanical Properties

The mechanical properties of composites and pure APA-6 were tested using the AGS-J testing machine (Shimadzu, Kyoto, Japan). Flexural properties were obtained according to ASTM D-790. The tensile properties were obtained according to ASTM D-3039. A Dynamic Mechanical Analyze (TA Q800, New Castle, USA) was used to measure the storage modulus via three-point bending at a frequency of 1 Hz with an amplitude control of 25 μm. The glass transition temperature ($T_g$) was obtained at the same time.

### 2.3.6. Morphologies

The morphologies of the fracture surfaces were observed by the polarizing microscope (BX-51, 'PLM', Olympus, Tokyo, Japan) and scanning electron microscope (JSM-5900LV, 'SEM', JEOL, Tokyo, Japan). For the SEM observation, the samples were fractured in liquid nitrogen, polished and covered with gold prior to analysis.

### 2.3.7. Structural Characterizations

The surface chemical groups of the KH550 modified ramie fibers was characterized by the Nicolet 6700 Fourier transform infrared spectrometer (FTIR, Thermo Fisher, Waltham, USA). The content of the elements for the unmodified and KH550 modified ramie fibers were measured by the X-ray photoelectron spectroscopy (XPS, XSAM800, Kratos Company, Manchester, UK).

## 3. Results and Discussion

### 3.1. The Effects on Structure and Properties of Ramie Fabrics by Different Modification Methods

Different modification methods have a great influence on the structure, surface properties and mechanical properties of natural fiber fabrics, and these properties of the fabrics will affect the impregnation process, interface properties and final properties of the composites, respectively.

### 3.1.1. The Effects on the Structure of Different Modified Ramie Fabrics

Different modification methods can impose a different impact on the macroscopic structure of the fiber fabric. Different fabric structures have a great effect on the resin flow in the fiber, which in turn affect the resin impregnation process. In order to characterize the changes of the morphological structure, different modified ramie fabrics were examined by the polarizing microscope, as shown in Figure 2. The fiber bundle width and void area of the above five fibers were counted according to the polarized photograph, shown in Figure 3.

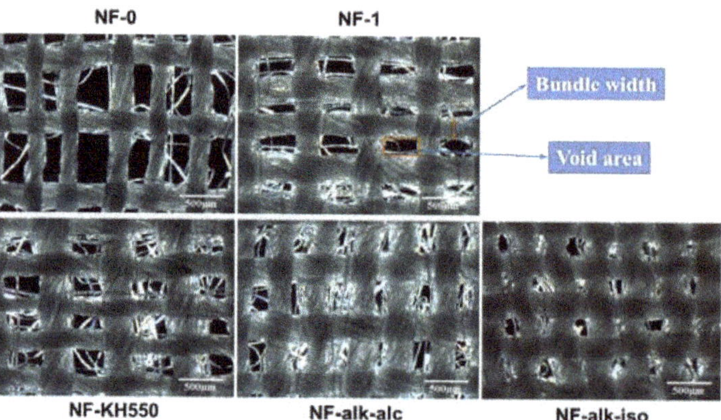


**Figure 2.** PLM observations of different modified ramie fabrics.

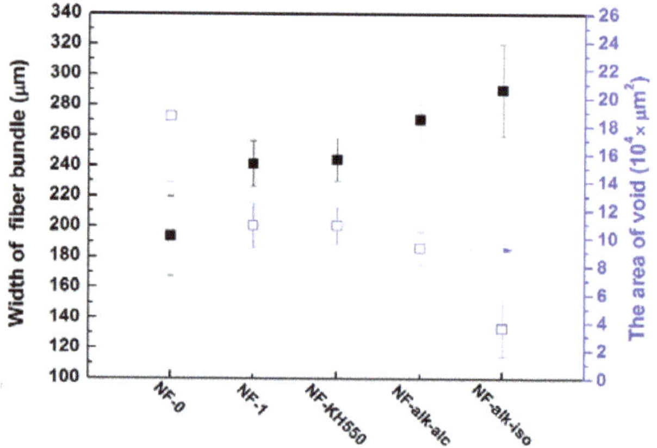


**Figure 3.** The variation of the fiber bundle width (black square) and the void area of different modified ramie fabrics (blue square).

NF-0 is a kind of ramie fabric without any treatment. The unmodified ramie fabric bundle is relatively dense. The uniformity of the ramie fabric is poor and the fiber is plucked. At the same time, the void area is also extremely uneven, and the deviation is large, which will adversely affect the fluidity of the resin and the uniformity of the final product. Moreover, the unmodified fiber contains pectin, hemicellulose, dust adsorbed during storage as well as sizing agents in the weaving process, which will have a negative effect on the anionic polymerization. Therefore, the ramie fibers must be modified before being used in the reaction processing.

NF-1 is a pretreated ramie fabric that has a considerable improvement on uniformity of the fiber fabric structure compared to the unmodified ramie fabric. The fiber bundle width is significantly increased, which indicates that the compactness of the single bundle fiber reduces after modification. This will facilitate the resin impregnation to the fiber. NF-KH550 is a ramie fabric modified with a silane coupling agent, and its structural changes are similar to that of the pretreated fiber.

NF-alk-alc and NF-alk-iso represent fibers that were regenerated from ethanol and isopropanol after the alkali treatment, respectively. It is well known that the surface of natural fiber contains a large amount of hydroxyl groups, which is advantageous for resin impregnation and good interfacial adhesion to the resin. However, most of the hydroxyl groups are in the form of hydrogen bonds, and the free hydroxyl groups are limited. In addition, the strength is high due to the natural fiber that has a high crystallinity and orientation. However, the toughness is poor, and a high crystallinity is also disadvantageous for the resin impregnation. Therefore, the alkali modification (also known as the activation modification) aims to destroy the hydrogen bond structure of cellulose thereby forming more free hydroxyl groups and appropriately reducing the crystallinity and orientation of the cellulose [32]. The mechanism of the alkali modification is that the high concentration of the alkali liquid penetrates into the network structure of the cellulose to induce swelling, which increases the distance between the cellulose molecular chains gradually. This leads to a weakening of the hydrogen bonding between the molecular chains, and even breaking down the hydrogen bonds. The high concentration of the alkali solution will partially penetrate into the crystalline region of the cellulose, thereby destroying the ordered structure of the cellulose crystal region and decreasing the degree of crystallinity and orientation of the fiber. However, the swollen network structure of the fiber is easily aggregated in the subsequent process of removing alkali and drying, and intramolecular hydrogen bonds will be reformed again [24,33,34]. Therefore, the way of the regeneration of the fiber and removal of the solvent is critical to maintain the decrystallizing effect of the cellulose. In this research, ethanol and isopropanol were used to regenerate the fiber, and different methods had different effects on the alkali-treated fiber. The essence of alkali modification is to change the existing aggregate structure of the ramie fiber, by which the effect of the alkali modification can be monitored by characterizing the crystal structure of modified fiber. Figure 4 shows the X-ray diffraction curves of different modified ramie fabrics. It can be concluded that the crystal structures of the pretreatment and coupling agent treated ramie fiber are type I of cellulose, indicating that the coupling agent treatment did not change the crystal type of cellulose and only a part of the amino groups were grafted onto the surface. The X-ray diffraction curves of the ramie fiber after the alkali treatment have different degrees of variations. Among them, the diffraction peak position of the 002 crystal plane of the ramie fiber regenerated by ethanol shifts to a low angle. It means that a small amount of crystal type I of cellulose is converted to crystal type II, and the fiber has two crystal structures. Moreover, the positions of the diffraction peaks of the 101 and 002 crystal plane of the ramie fiber regenerated by isopropanol significantly shift, and the crystal structures of the ramie fiber are all expressed as type II of cellulose. The changes of the crystal structure of the fiber indicate that the crystallinity and orientation of the fiber significantly decrease.

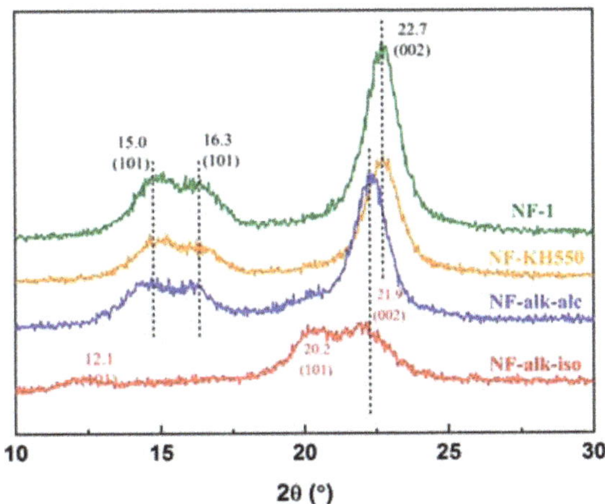


**Figure 4.** X-ray diffraction (XRD) curve of different modified ramie fabrics.

In summary, the alkali treatment had a great influence on the structure of the fiber, in which the fiber bundle width was further increased; the fiber void area was greatly degraded; and the surface fuzzing phenomenon was more serious. This is because the alkali has a swelling effect on the fiber. Among them, the isopropyl alcohol regenerated fiber had serious shrinkage, distortion and uniformity, which might have a negative impact on the preparation of the composites. With the ethanol regeneration, there was no serious deformation and the structure was relatively uniform.

In addition, the surface of the modified fiber has a different fuzzing phenomenon. On the one hand, it is beneficial to the interaction between the resin and the fiber surface, and the wettability and adhesion of the interface will also increase. On the other hand, the surface of the fiber becomes rough and irregular, which hinders the resin flow in the fiber, thereby affecting the wetting process. Therefore, the optimum fibers should have a good interfacial action with the resin and be beneficial to impregnate for the resin, so that the untreated fibers and isopropyl alcohol regenerated fibers after alkali treatment are not suitable.

### 3.1.2. The Effects on Surface Properties of Different Modified Ramie Fiber

The surface properties of the fiber have been known to affect the impregnation process of the resin, they also affect the interfacial adhesion of the resin and fiber, and further influence the properties of the composites. The scanning electronic microscopy (SEM) was used to observe the surface of a different modified ramie fiber as shown in Figure 5. The untreated fiber (NF-0) had a smooth surface, but there were obvious flow marks of the sizing agent. Most of the active groups on the surface of the fiber were covered by the sizing agent, which was not conducive to form a good interface between the fiber and the resin. The flow marks on the surface of the pretreated ramie fiber (NF-1) were found to be reduced, which revealed the characteristics of the fiber, and the surface were slightly rough. Compared with NF-1, NF-KH550 had a small amount of flow marks on the surface due to the surface being coated with a coupling agent. FTIR and XPS tests were used to confirm that KH550 was grafted onto the fiber. As the FTIR spectra shown in Figure 6, although the absorption peaks of Si-O and N-H of KH550 are in the same position with the peaks of C-O and O-H of the ramie fiber at around 1000~1100 cm$^{-1}$ and 3400~3500 cm$^{-1}$, the strength of the peaks of NF-KH550 is significantly increased during these two ranges. The C-H stretching vibration and C-H bending vibration for NF-KH550 are also more obvious at around 2850~2910 cm$^{-1}$ and 1365~1438 cm$^{-1}$. The peak at 1630 cm$^{-1}$ is cellulose-specific and is not affected by the modifier. It can be used as a benchmark. The changes of the other peaks intensity explain the influence of modifiers to some extent. Besides, XPS was adopted to measure the changes of the elements on the surface of the fiber. It can be concluded that the intensity of N and Si peaks for NF-KH550 both increase compared to NF-1. In conclusion, it can be confirmed that the silane was grafted onto the fiber. Both the alkali modified fibers for NF-alk-alc and NF-alk-iso had rougher surfaces, which would be more conducive to resin-fiber bonding.

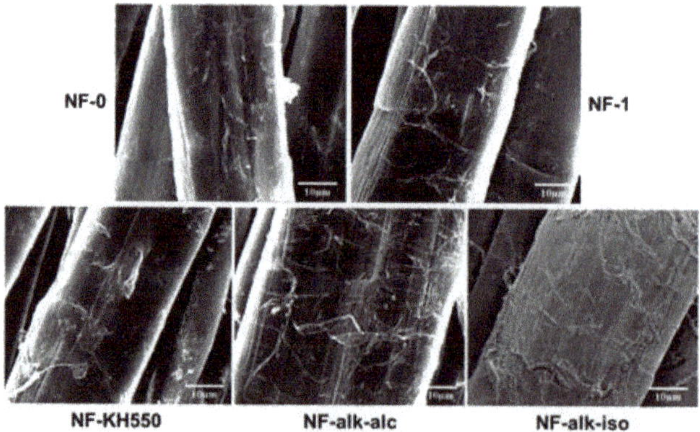


**Figure 5.** SEM imaging of different modified ramie fabrics.

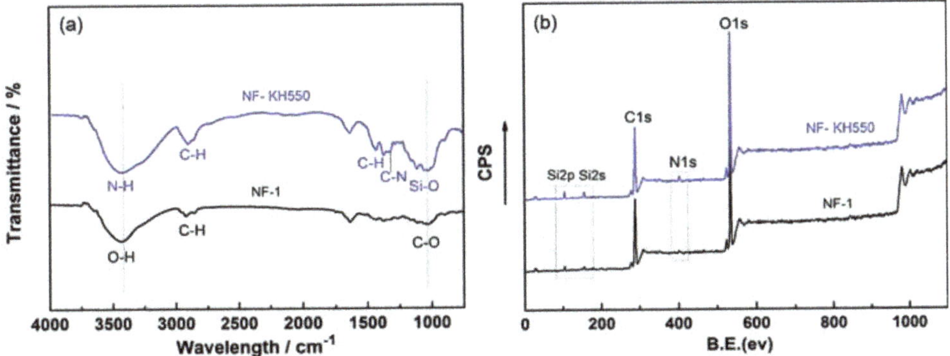



**Figure 6.** FTIR (**a**) and XPS (**b**) spectra of different modified ramie fabrics.

### 3.1.3. The Effects on the Mechanical Properties of Different Modified Ramie Fabrics

The above studies have showed that the different modification methods have different effects on the morphological structure, surface properties and aggregation structure of ramie fabrics, which will inevitably affect the mechanical properties of fabrics and eventually affect the properties of composites. Figure 7 showed the strength values of different modified ramie fabrics, as well as the fabrics strength and elongation curves during stretching. The strength of the pretreated fabrics and silane coupling agent modified fabrics were both higher than that of the untreated sample, because the relative content of cellulose in the modified ramie fabrics was slightly increased after pretreatment. The strength of the two fabrics after the alkali treatments decreased significantly, but also accompanied by a significant increase in elongation at break. This is because the crystallinity and orientation of the fiber after the alkali modification decrease. Therefore, the alkali treated fabrics have an adverse effect on the strength of the composites and are advantageous for the toughness and interfacial adhesion of the composites. In conclusion, the different modified ramie fabrics can be used to prepare composites with different properties to meet the different environmental requirements.

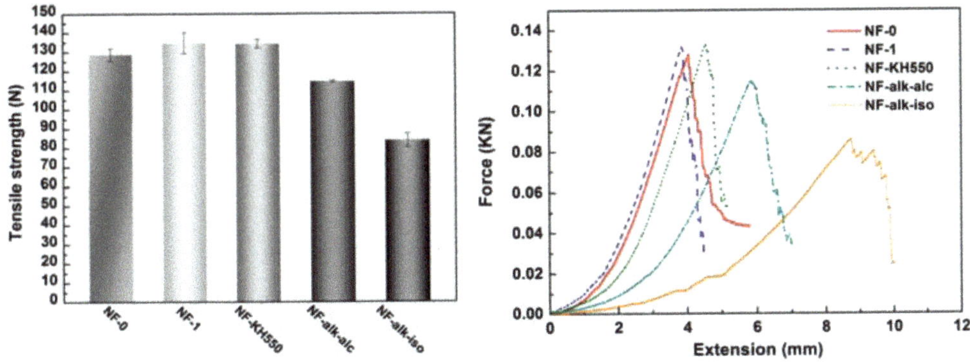



**Figure 7.** The mechanical properties of different modified ramie fabrics.

### 3.2. The Properties of APA-6 Composites Reinforced by Different Modified Ramie Fabrics

#### 3.2.1. Physical Properties

The five modified ramie fabrics mentioned above were used as the reinforcing materials and these composites were prepared by the VARIM process accordingly. It was found that not all modified ramie fabrics were suitable as reinforcing materials in this study. First, the untreated ramie fabrics impeded the anionic polymerization, because the surface of the untreated fiber contained some sizing agents, hemicellulose, lignin, pectin and other impurities, which hindered the occurrence of anionic polymerization. Second, the fabrics regenerated by isopropanol after the alkali modification could not be prepared to obtain composites with uniform wetting and perfect polymerization. The reason was that the alkali treatment seriously changed the morphological structure of the fabric, and the uniformity was deteriorated. The fiber bundle was even severely distorted thereby hindering the impregnation process of the resin.

The composites prepared by NF-1, NF-KH550 and NF-alk-alc all showed a well uniformity. In order to further characterize the polymerization properties of these composites, the conversion, crystallinity and viscosity-average molecular weight were analyzed as shown in Figures 8 and 9. First, the degree of conversion of these three composites showed a different degree of decline compared to the pure APA-6. It meant that the addition of natural fibers had a certain effect on anionic polymerization. Moreover, the effect of the alkali-modified fiber on the anionic polymerization conversion was more obvious. Since the alkali-modified fiber carried more reactive groups, the consumption of the initiator in the polymerization system was increased. On the other hand, the reaction probability of the surface of the fiber with the activator was increased. Therefore, the active center of the polymerization system was also consumed to some extent. The degrees of crystallinity of these three composites were significantly higher than that of the pure APA-6, which is due to the fact that the natural fiber easily induced polymer crystallization. The viscosity-average molecular weights of composites were not much different from that of the pure APA-6, and were much higher than that of nylon 6. However, there were slight differences among these different composites, which related to the reaction between the hydroxyl groups on the surface of the fiber and the activator in the polymerization system. Besides, the degree of conversion and viscosity-average molecular weight of the untreated ramie fabrics reinforced APA-6 composites were also measured, only 18% and 3600 respectively, which demonstrated the inhibition to the anionic polymerization by the unmodified fibers (NF-0). In this system, only a small amount of oligomer was produced, and the polymerization was not reached.

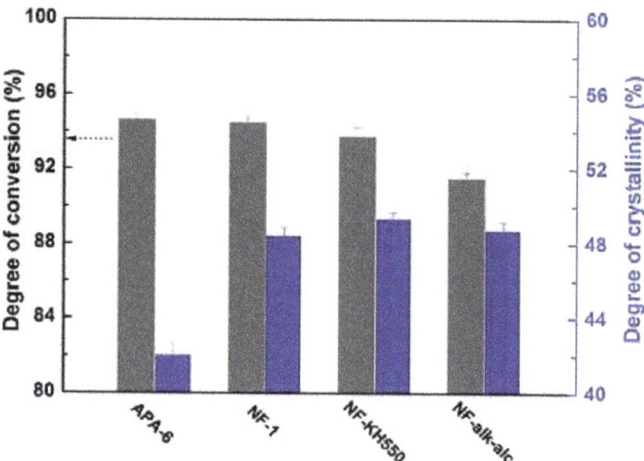

**Figure 8.** The degree of conversion (gray columnar) and crystallinity of different modified ramie fabrics reinforced APA-6 composites (blue columnar).

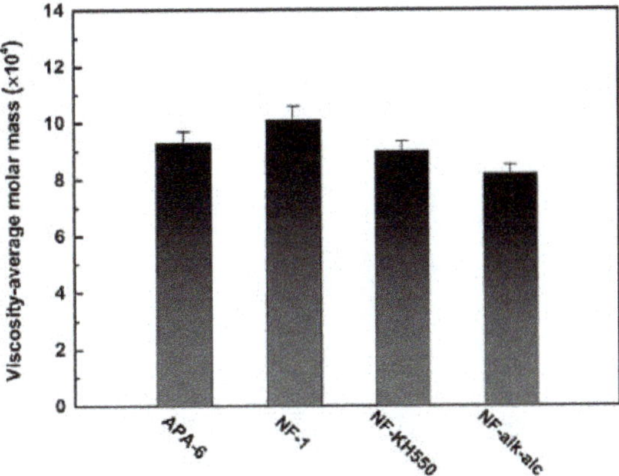

**Figure 9.** Viscosity-average molecular weight of different modified ramie fabrics reinforced APA-6 composites.

Figure 10 showed the fiber content and void content of these three composites. These composites were prepared by the same process, so that the fiber content was the same. The void content of these composites was lower, and the NF-alk-alc composites were the best. The difference in void content indicated that the resin had different wetting effects on different modified fibers, and the alkali-modified ramie fabric had better wettability.

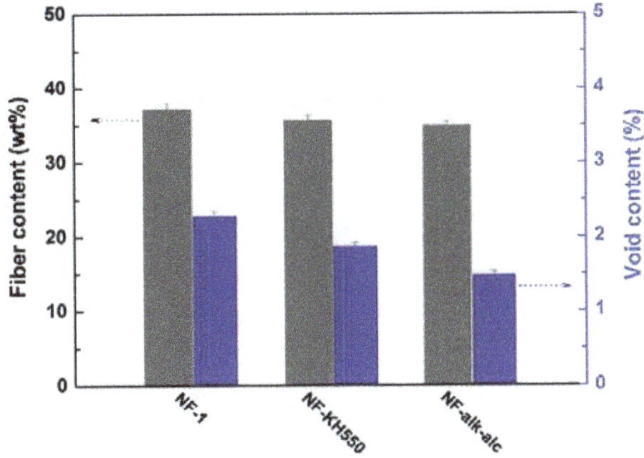

**Figure 10.** The fiber content and the void content of different modified ramie fabrics reinforced APA-6 composites.

### 3.2.2. Mechanical Properties

Figure 11 showed the flexural modulus, tensile strength and elongation at the break of APA-6 and different composites. The flexural modulus of NF-1 and NF-KH550 composites were better than that of the other samples. The tensile strength and elongation at the break of the NF-KH550 composites were larger than that of NF-1 composites, which indicated that the coupling agent modified ramie fiber fabric increased the strength and plastically deform under tension of the composites. The reason may be that the interfacial interaction between the fiber and the matrix becomes better after coating a layer of coupling agent. However, the NF-alk-alc had a great decrease in the mechanical properties, so that the flexural modulus and tensile strength of the composites were much lower than that of other two composites. In addition, the elongation at the break of NF-alk-alc composites was found significantly increased, which indicated that the plastically deform under tension of the NF-alk-alc composites was improved. Their trends are closely related to the changes in the mechanical properties of different modified ramie fabrics. The reasons have been stated in 3.1.3 in this paper.

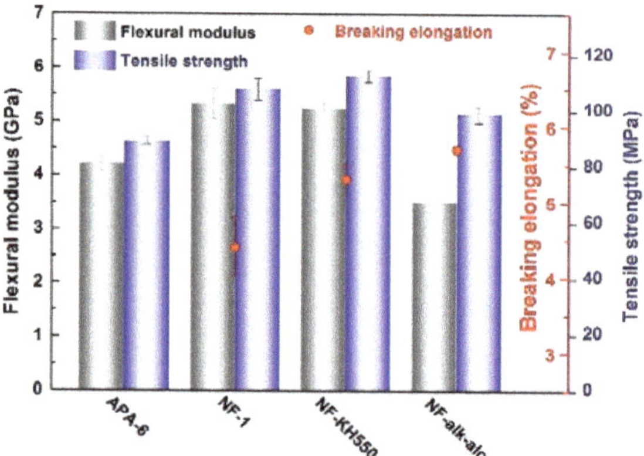

**Figure 11.** The mechanical properties of different modified ramie fabrics reinforced APA-6 composites.

The storage modulus is an important parameter to characterize the resistance to thermal mechanical deformation of the materials. The larger the modulus, the less deformed. Figure 12 showed the results of dynamic mechanical properties of different materials. The NF-1 and NF-KH550 composites had the same variations under the low temperature, but the storage modulus of the NF-1 composites was larger than that of the NF-KH550 composites under the high temperature. Under the high temperature, the chain segments began to move, and the NF-1 composites had a slightly larger molecular weight. NF-alk-alc composites had a lower modulus over the entire temperature range than the other two composites. The reason was that the degree of crystallization and orientation of the fiber decreased during the alkali treatment. In the loss factor curve, the peak at 75 °C represents the α transition of nylon 6, which meant that the values of glass transition the temperature of the composites. It relates to the molecular weight and crystallinity of the materials. The peak at around −60 °C indicates the β relaxation of nylon. This transition peak is related to the moisture and motion units of small-sized chain unit.

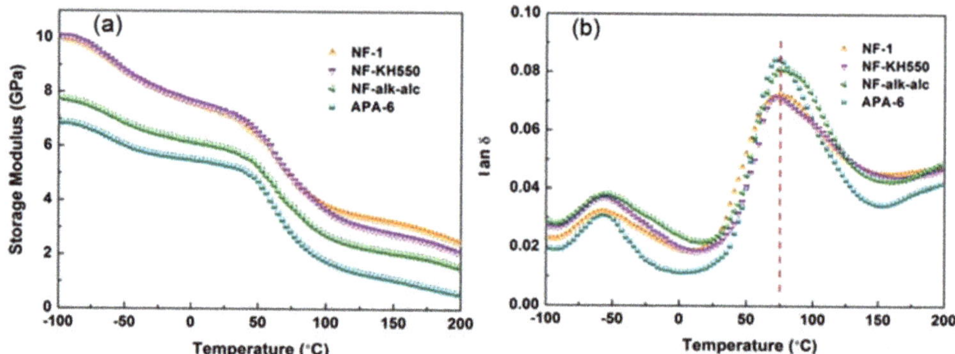

**Figure 12.** The dynamic modulus (**a**) and Tan δ (**b**) for different modified ramie fabrics reinforced APA-6 composites.

The interfacial properties for the ramie fiber reinforced APA-6 composites is better than the glass fiber reinforced APA-6 composites measured by SEM [35]. In this study, the fractured surfaces of the three composites were also characterized by SEM. All composites showed good fiber-to-matrix bonding expectedly, but there was no obvious difference. So the SEM images were not given in this article, instead of the interlaminar shear strength test. The interlaminar shear strength refers to the maximum load that the composites can bear per unit area parallel to the bonding surface. It mainly reflects the bonding properties of the composites under the applied load, which is very important for the safety and stability of the materials during applications. The interlaminar shear strength (ILSS) of these three composites prepared by the pretreated fiber, the coupling agent modified fiber and the alkali-modified fiber was found sequentially increased as shown in Figure 13. The NF-alk-alc composites had the worst mechanical properties but had the highest interlaminar shear strength. It indicated that the hydrogen bond on the surface of the cellulose was opened after the alkali treatment, and the content of free hydroxyl groups increased. Thus, it was very helpful to improve the interface interaction between the fiber and the matrix. The interlaminar shear strength of NF-KH550 composites was higher than that of NF-1 composites, which indicated that the interfacial interaction between the fiber and the matrix became better after coating a layer of coupling agent.

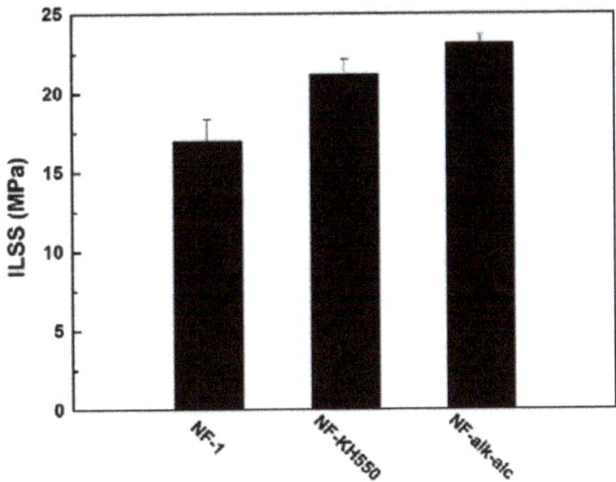


**Figure 13.** Interlaminar shear strength of different modified ramie fabrics reinforced APA-6 composites.

## 4. Conclusions

First, it was included that different modification methods had different effects on the structure of ramie fabrics. The uniformity of the structure was improved for the pretreatment and the coupling agent modified fabrics. The structure of the alkali-modified fabrics was greatly changed and the fabrics regenerated by ethanol tended to be uniform. The uniformity of the fabrics regenerated by isopropanol was found seriously degraded. Then, different modification methods also had different effects on the surface roughness of ramie fabrics. The strengths of the ramie fabrics modified by the pretreatment and silane coupling agent were improved compared with that of the untreated fabrics. The strength of the fabrics after the alkali treatments were severely reduced, but the elongation at break was greatly increased. The performance analysis of different modified ramie fabrics reinforced APA-6 composites was carried out, and the conversion, crystallinity and molecular weight of these three composites were found to be at a high level, which indicated that the polymerization of these three composites was well controlled. The coupling agent modified ramie fabrics composites and the pretreated ramie fabrics composites both had higher flexural modulus, tensile strength and dynamic mechanical properties than that of the alkali-modified ramie fiber composites. Alkali-modified composites had slightly lower mechanical properties, which however had the highest interlaminar shear strength and outperformed interface properties of the composites.

**Author Contributions:** Conceptualization, H.W. and Z.K.; Methodology, H.W. and Z.K.; Software, Z.K. and H.S.; Validation, Z.K.; Formal analysis, Z.K.; Investigation, H.S., E.Z. and Z.K.; Resources, H.W. and Z.K.; Data curation, H.W.; Writing—Original draft preparation, Z.K.; Writing—Review and editing, H.W., E.Z. and H.S.; Visualization, Z.K.; Supervision, H.W.; Project administration, H.W.; Funding acquisition, H.W. and Z.K.

**Funding:** This work was supported by the National Natural Science Foundation of China (51803104), the Special Fund Project to Guide the Development of Local Science and Technology by the Central Government and Shandong Taishan Scholar Program (Hui Wang).

**Conflicts of Interest:** The authors declare no conflict of interest.

## References

1. Faruk, O.; Bledzki, A.K.; Fink, H.P.; Sain, M. Biocomposites reinforced with natural fibers: 2000–2010. *Prog. Polym. Sci.* **2012**, *37*, 1552–1596. [CrossRef]

2. Luo, N.; Lv, Y.; Wang, D.; Zhang, J.; Wu, J.; He, J.; Zhang, J. Direct visualization of solution morphology of cellulose in ionic liquids by conventional TEM at room temperature. *Chem. Commun.* **2012**, *48*, 6283–6285. [CrossRef]

3. Panchal, P.; Ogunsona, E.; Mekonnen, T. Trends in advanced functional material applications of nanocellulose. *Processes* **2019**, *7*, 10. [CrossRef]

4. Zhou, M.; Yan, J.; Li, Y.; Geng, C.; He, C.; Wang, K.; Fu, Q. Interfacial strength and mechanical properties of biocomposites based on ramie fibers and poly(butylene succinate). *RSC Adv.* **2013**, *3*, 26418–26426. [CrossRef]

5. Rijswijk, K.V.; van Geenen, A.A.; Bersee, H.E.N. Textile fiber-reinforced anionic polyamide-6 composites. Part II: Investigation on interfacial bond formation by short beam shear test. *Compos. Part A Appl. Sci. Manuf.* **2009**, *40*, 1033–1043. [CrossRef]

6. Rijswijk, K.V.; Teuwen, J.J.E.; Bersee, H.E.N.; Beukers, A. Textile fiber-reinforced anionic polyamide-6 composites. Part I: The vacuum infusion process. *Compos. Part A Appl. Sci. Manuf.* **2009**, *40*, 1–10. [CrossRef]

7. Henning, F.; Karger, L.; Dorr, D.; Schirmaier, F.J.; Seuffert, J.; Bernath, A. Fast processing and continuous simulation of automotive structural composite components. *Compos. Sci. Technol.* **2019**, *171*, 261–279. [CrossRef]

8. Máirtín, P.Ó.; McDonnell, P.; Connor, M.; Eder, R.; Ó Brádaigh, C.M. Process investigation of a liquid PA-12/carbon fibre moulding system. *Compos. Part A Appl. Sci. Manuf.* **2001**, *32*, 915–923. [CrossRef]

9. Li, M.; Wang, S.K.; Gu, Y.Z.; Li, Y.X.; Potter, K.; Zhang, Z.G. Evaluation of through-thickness permeability and the capillary effect in vacuum assisted liquid molding process. *Compos. Sci. Technol.* **2012**, *72*, 873–878. [CrossRef]

10. Zhang, X.; Fan, X.; Li, H.; Yan, C. Facile preparation route for graphene oxide reinforced polyamide 6 composites via in situ anionic ring-opening polymerization. *J. Mater. Chem.* **2012**, *22*, 24081–24091. [CrossRef]

11. Xu, Z.; Gao, C. In situ polymerization approach to graphene-reinforced nylon-6 composites. *Macromolecules* **2010**, *43*, 6716–6723. [CrossRef]

12. Kashani Rahimi, S.; Otaigbe, J.U. The role of particle surface functionality and microstructure development in isothermal and non-isothermal crystallization behavior of polyamide 6/cellulose nanocrystals nanocomposites. *Polymer* **2016**, *107*, 316–331. [CrossRef]

13. Rahimi, S.K.; Otaigbe, J.U. The effects of the interface on microstructure and rheo-mechanical properties of polyamide 6/cellulose nanocrystal nanocomposites prepared by in-situ ring-opening polymerization and subsequent melt extrusion. *Polymer* **2017**, *127*, 269–285. [CrossRef]

14. Rijswijk, K.V.; Bersee, H.; Jager, W.; Picken, S. Optimisation of anionic polyamide-6 for vacuum infusion of thermoplastic composites: Choice of activator and initiator. *Compos. Part A Appl. Sci. Manuf.* **2006**, *37*, 949–956. [CrossRef]

15. Rijswijk, K.V.; Bersee, H.; Beukers, A.; Picken, S.; Geenen, A.V. Optimisation of anionic polyamide-6 for vacuum infusion of thermoplastic composites: Influence of polymerisation temperature on matrix properties. *Polym. Test.* **2006**, *25*, 392–404. [CrossRef]

16. Pillay, S.; Vaidya, U.K.; Janowski, G.M. Effects of moisture and UV exposure on liquid molded carbon fabric reinforced nylon 6 composite laminates. Compos. *Sci. Technol.* **2009**, *69*, 839–846.

17. Pillay, S.; Vaidya, U.K.; Janowski, G.M. Liquid molding of carbon fabric-reinforced nylon matrix composite laminates. *J. Thermoplast. Compos.* **2005**, *18*, 509–527. [CrossRef]

18. Yan, C.; Li, H.; Zhang, X.; Zhu, Y.; Fan, X.; Yu, L. Preparation and properties of continuous glass fiber reinforced anionic polyamide-6 thermoplastic composites. *Mater. Des.* **2013**, *46*, 688–695. [CrossRef]

19. Lee, G.-W.; Lee, N.-J.; Jang, J.; Lee, K.-J.; Nam, J.-D. Effects of surface modification on the resin-transfer moulding (RTM) of glass-fibre/unsaturated-polyester composites. *Compos. Sci. Technol.* **2002**, *62*, 9–16. [CrossRef]

20. O'sullivan, A.C. Cellulose: The structure slowly unravels. *Cellulose* **1997**, *4*, 173–207. [CrossRef]

21. Zin, E.; Scandola, M. Green composites: An overview. *Polym. Compos.* **2011**, *32*, 1905–1915. [CrossRef]

22. O'Donnell, A.; Dweib, M.A.; Wool, R.P. Natural fiber composites with plant oil-based resin. *Compos. Sci. Technol.* **2004**, *64*, 1135–1145. [CrossRef]

23. Bledzki, A.; Reihmane, S.; Gassan, J. Properties and modification methods for vegetable fibers for natural fiber composites. *J. Appl. Polym. Sci.* **1996**, *59*, 1329–1336. [CrossRef]

24. George, J.; Sreekala, M.; Thomas, S. A review on interface modification and characterization of natural fiber reinforced plastic composites. *Poly. Eng. Sci.* **2001**, *41*, 1471–1485. [CrossRef]

25. Gassan, J.; Bledzki, A. Possibilities to improve the properties of natural fiber reinforced plastics by fiber modification–jute polypropylene composites. *Appl. Compos. Mater.* **2000**, *7*, 373–385. [CrossRef]

26. John, M.J.; Anandjiwala, R.D. Recent developments in chemical modification and characterization of natural fiber-reinforced composites. *Polym. Compos.* **2008**, *29*, 187–207. [CrossRef]

27. Saheb, D.N.; Jog, J. Natural fiber polymer composites: A review. *Adv. Polym. Technol.* **1999**, *18*, 351–363. [CrossRef]

28. Li, X.; Tabil, L.G.; Panigrahi, S. Chemical treatments of natural fiber for use in natural fiber-reinforced composites: A review. *J. Polym. Environ.* **2007**, *15*, 25–33. [CrossRef]

29. Valadez-Gonzalez, A.; Cervantes-Uc, M.; Olayo, R.J.; Herrera-Franco, P. Effect of fiber surface treatment on the fiber–matrix bond strength of natural fiber reinforced composites. *Compos. Part B Eng.* **1999**, *30*, 309–320. [CrossRef]

30. Cartledge, H.C.; Baillie, C.A. Studies of microstructural and mechanical properties of nylon/glass composite part I the effect of thermal processing on crystallinity, transcrystallinity and crystal phases. *J. Mater. Sci.* **1999**, *34*, 5099–5111. [CrossRef]

31. James, E.M. *Polymer Data Handbook*, 2nd ed.; Oxford University Press: New York, NY, USA, 1999; pp. 180–181.

32. Knill, C.J.; Kennedy, J.F. Degradation of cellulose under alkaline conditions. *Carbohydr. Polym.* **2003**, *51*, 281–300. [CrossRef]

33. Mwaikambo, L.Y.; Ansell, M.P. Chemical modification of hemp, sisal, jute, and kapok fibers by alkalization. *J. Appl. Polym. Sci.* **2002**, *84*, 2222–2234. [CrossRef]

34. Gassan, J.; Bledzki, A.K. Alkali treatment of jute fibers: Relationship between structure and mechanical properties. *J. Appl. Polym. Sci.* **1999**, *71*, 623–629. [CrossRef]

35. Kan, Z.; Yang, M.; Yang, W.; Liu, Z.; Xie, B. Investigation on the reactive processing of textile-ramie fiber reinforced anionic polyamide-6 composites. *Compos. Sci. Technol.* **2015**, *110*, 188–195. [CrossRef]

MDPI

St. Alban-Anlage 66

4052 Basel

Switzerland

Tel. +41 61 683 77 34

Fax +41 61 302 89 18

www.mdpi.com

*Processes* Editorial Office

E-mail: processes@mdpi.com

www.mdpi.com/journal/processes

9 783039 287666